# Plasma Diagnostics
## Volume 2
*Surface Analysis and Interactions*

# Plasma-Materials Interactions

A Series Edited by

## Orlando Auciello

*Microelectronics Center of
North Carolina and
North Carolina State University
Research Triangle Park, North Carolina*

## Daniel L. Flamm

*AT&T Bell Laboratories
Murray Hill, New Jersey*

A list of titles in this series appears at the end of this volume.

# Plasma Diagnostics
## Volume 2
### *Surface Analysis and Interactions*

Edited by

## Orlando Auciello
*Microelectronics Center of*
*North Carolina and*
*North Carolina State University*
*Research Triangle Park, North Carolina*

## Daniel L. Flamm
*AT&T Bell Laboratories*
*Murray Hill, New Jersey*

ACADEMIC PRESS, INC.
Harcourt Brace Jovanovich, Publishers
Boston   San Diego   New York
Berkeley   London   Sydney
Tokyo   Toronto

**PHYSICS**

ACADEMIC PRESS, INC.
1250 Sixth Avenue, San Diego, CA 92101

*United Kingdom Edition published by*
ACADEMIC PRESS, INC. (LONDON) LTD.
24-28 Oval Road, London NW1 7DX

Library of Congress Cataloging-in-Publication Data
Plasma diagnostics.

(Plasma-materials interactions)
Includes bibliographies and index.
Contents: v. 1. Discharge parameters and chemistry
—v. 2. Surface analysis and interactions.
1. Plasma diagnostics.   I. Auciello, Orlando,
Date-      II. Flamm, Daniel L.   III. Series.
QC718.5.D5P54 1988      530.4'1      87-35161
ISBN 0-12-067635-4 (v. 1)
ISBN 0-12-067636-2 (v. 2)

Printed in the United States of America
89 90 91 92     9 8 7 6 5 4 3 2 1

# Contents

# Contributors

Numbers in parentheses refer to the pages on which the authors' contributions begin.

D. E. ASPNES (67), *Bell Communications Research, Inc., Red Bank, New Jersey 07701-7020*

B. K. BEIN (211), *Ruhr-Universitat Bochum, Institut für Experimentalphysik VI, D-4630 Bochum, PO Box 102148, Federal Republic of Germany*

R. P. H. CHANG (67), *Materials Science and Engineering Department, Northwestern University, Evanston, Illinois 60208*

WEI-KAN CHU (109), *Department of Physics and Astronomy, University of North Carolina, Chapel Hill, North Carolina 27514*

J. W. COBURN (1), *IBM Almaden Research Center, 650 Harry Road, San Jose, California 95120-6099*

B. L. DOYLE (109), *Sandia National Laboratories, Albuquerque, New Mexico 87185*

H. P. GILLIS (19), *m / s MSRL-70, Hughes Research Laboratories, Malibu, California 90265*

THOMAS M. MAYER (19), *Department of Chemistry, University of North Carolina, Chapel Hill, North Carolina 27514*

J. PELZL (211), *Ruhr-Universitat Bochum, Institut für Experimentalphysik VI, D-4630 Bochum, PO Box 102148, Federal Republic of Germany*

DONALD L. SMITH (19), *Xerox Palo Alto Research Center, 3333 Coyote Hill Road, Palo Alto, California 94306*

P. C. STANGEBY (157), *University of Toronto, Institute for Aerospace Studies, Downsview, Ontario M3H 5T6, Canada*

# Preface

The study of plasma–material interactions has evolved into an important and dynamic field of research. An understanding of the basic physical and chemical processes underlying these interactions is vital to the development of microelectronics, surface modification, fusion, space, and other key technologies of our age. Plasma processing is a critical technology for leading-edge microelectronics. For example, ultra large scale integrated circuits (ULSI) cannot be manufactured without plasma-assisted etching and plasma chemical vapor deposition. Similarly, the various plasma-surface phenomena—physical sputtering, chemical etching, particle trapping in solid walls etc.—must be understood and controlled to achieve self-sustained fusion reactions in future commercial power plants. Plasma interactions with surfaces of spaceships can produce harmful degradation, such as the undesirable etching of thermal blankets observed in the cargo bay of the Space Shuttle. These effects could jeopardize long-term missions in space. All of these problems are now being investigated by scientists and engineers around the world.

Unfortunately, scientific and technical information in these diverse fields is often published in journals aimed at a narrow specialized audience. One of the chief goals of this series on "Plasma–Materials Interactions", which we are now initiating, is to provide an interdisciplinary forum. We hope to disseminate knowledge of basic and applied physicochemical processes and plasma-processing art to the global community. The series is structured to make this information readily accessible to scientists, engineers, students and technical personnel in universities, industry, and national laboratories. We consider plasma–materials interactions to be one of the pivotal fields of research that will contribute to the technological revolution now under way. Therefore, we hope that this series will encourage the pursuit of new ideas and expand the horizons of science and technology in allied interdisciplinary fields.

Diagnostics and characterization techniques are prerequisites for understanding plasmas and solid surfaces exposed to plasmas. Unfortunately the necessary know-how is scattered throughout the literature, often in a form that is difficult to use. Consequently, we begin this series with an authoritative and up-to-date treatment of plasma and surface diagnostics written for an interdisciplinary audience. The authors are renowned specialists who explain how to set up, make, and interpret measurements and how to assess the validity of diagnostic data and detect complications. Finally, they present the theoretical background necessary to understand each technique with references to recent literature. Because the material is fairly comprehensive, the book is divided into two volumes.

Volume 1 contains seven chapters on the important diagnostic techniques for plasmas and details their use in particular applications. This part includes (1) optical diagnostics for low-pressure plasmas and plasma processing, (2) plasma diagnostics for electrical discharge light sources, (3) Langmuir probes, (4) mass spectroscopy of plasmas, (5) microwave diagnostics, (6) paramagnetic resonance diagnostics, and (7) diagnostics in thermal plasma processing.

Volume 2 covers diagnostics of surfaces exposed to plasmas and includes chapters on (1) quartz crystal microbalances for studies of plasma–surface interactions, (2) elemental analysis of treated surfaces by electron and ion spectroscopies, (3) spectroscopic ellipsometry in plasma processing, (4) ion beam analysis of plasma-exposed surfaces (Rutherford backscattering, elastic recoil detection, particle-induced X-ray emission and nuclear reaction analysis), (5) the interpretation of plasma probe data in fusion experiments, and (6) non-destructive photoacoustic and photothermal techniques for the analysis of plasma-exposed surfaces.

We hope that these, and subsequent books in this series, will be valuable to experts and newcomers alike. "Plasma–Materials Interactions" volumes on plasma etching technology and on plasma deposition and etching of polymers are now in press. We would welcome your suggestions for future volumes.


Orlando Auciello
Daniel L. Flamm

January, 1989

# 1    Quartz Crystal Microbalances for Studies of Plasma-Surface Interactions

J. W. Coburn

*IBM Almaden Research Center*
*650 Harry Road*
*San Jose, California*

## I.   Introduction

Since the pioneering work of Sauerbrey (1959), quartz crystal microbalances (QCMs) have been used extensively in science and technology. The low cost, compact size, and durability of these devices, combined with their high sensitivity for detecting small mass changes, makes QCMs very suitable for a wide variety of applications. Probably the most widepsread use of QCMs is in the area of thin film deposition, in which commercial units can be purchased to accurately control complex deposition processes. The high sensitivity of QCMs allows the measurement of submonolayer amounts of deposited or adsorbed material, and this capability has led to QCMs being used quite extensively in basic surface science studies as well as in analytical chemistry and outer space studies. These applications are discussed in some detail in a book edited by Lu and Czanderna (1984). Another significant application is in the area of particle-solid interactions such as sputtering yield measurements (EerNisse, 1974; Ullevig and Evans, 1980), electron-induced erosion (Schou et al., 1984), and ion assisted gas-surface reactions (Coburn and Winters, 1979). More recently, the operation of QCMs has been extended to include liquid environments (Kanazawa and Gordon, 1985), which allows important studies in electrochemical science (Melroy et al., 1986) and photoresist dissolution kinetics (Hinsberg et al., 1985) to be carried out.

The application of QCM methods to glow discharge studies have developed much more slowly than some of these other applications. One reason for this is the influence of the rather energy intensive environment within a glow discharge on the QCM operation. This is particularly true in rf glow discharges, where the electrical interference from the discharge and associated electrodes interferes with the QCM operation. However, the use of QCMs with a grounded surface of the quartz facing the discharge, careful shielding the QCM electrical connection, and the use of passive filters in the external electronics overcomes the rf interference problem in many instances. A second problem is the relatively large thermal power that is delivered to the QCM surface in some discharge environments. This problem is alleviated by using quartz crystals which are cut so as to minimize the effect of temperature changes on the QCM frequency (Lu, 1984) and by using commercially available water-cooled QCMs. One approach to improving QCM operation in glow discharges has been to introduce a magnetic field in the vicinity of the QCM to decrease the electron bombardment on the crystal surface. This approach has been used to monitor inert gas sputter deposition processes but may not be appropriate for reactive gas glow discharge processes in that the electron bombardment may play an important role in the surface modification process. However, it is clear that the mass change associated with a surface modification process is one of the most important pieces of information available and, consequently, one can anticipate an increasing use of QCMs in plasma process monitoring. The feasibility of this approach has been demonstrated in relatively low density reactive gas glow discharges (Coburn and Kay, 1979; Coburn, 1979; Nyaiesh and Baker, 1982).

The characteristics of a QCM which make it so attractive for in situ process monitoring are its small size, simplicity, low cost, durability, and high mass sensitivity. A typical commercially available QCM crystal and holder has a volume of less than 3 cm$^3$, and holders with volumes less than 1 cm$^3$, have been constructed. The Sauerbrey (1959) relationship between the QCM frequency and the mass is

$$\Delta f = \frac{f^2}{\rho N} \cdot \frac{\Delta m}{A},$$

where $\Delta f$ is the QCM frequency change (Hz) resulting from an areal mass change $\Delta m/A$ (gm-cm$^{-2}$), $f$ is the resonant frequency of the quartz crystal (Hz), $\rho$ is the density of quartz (2.65 gm-cm$^{-3}$), and $N$ is the frequency constant of quartz (1.67 × 10$^5$ Hz-cm). This expression assumes a uniform mass change over the surface of the crystal and is valid only for relatively small frequency changes ($\Delta f/f < 0.05$). This equation, when evaluated for 6 MHz crystals, gives a sensitivity factor of 12.3 nanograms-cm$^{-2}$ Hz$^{-1}$. With some care, frequency changes as small as 0.1 Hz can be measured, implying a mass sensitivity of the order of a nanogram-cm$^{-2}$. (1 monolayer of silicon, for example, is equivalent to 63.7 nanograms-cm$^{-2}$, 1 monolayer of tungsten is equivalent to 488 nanograms-cm$^{-2}$.)

## II. Experimental Considerations

### A. LOW-ENERGY ($\leq$ 30 eV) ION BOMBARDMENT

The installation of QCMs in plasma systems where only low-energy ions bombard the QCM surface is relatively straightforward. By far the most common example of this is the installation of the QCM in or on a grounded surface adjacent to a glow discharge with a low plasma potential. It is this kind of installation in which most applications of QCMs in glow discharges are expected. An example of such an arrangement is illustrated in Figure 1. Note that the QCM has been mounted coplanar with the anode surface so as to minimize any distortion of the glow discharge. Glow discharges are known to be sensitive to changes in the boundary wall geometry. Nonmagnetic vendor-supplied holders are quite suitable for this kind of mounting,

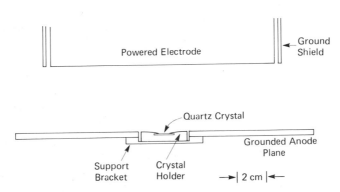

FIGURE 1.   QCM mounted in the ground plane of a planar system (Coburn, 1984).

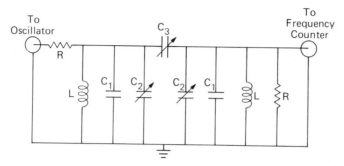

FIGURE 2. Filter used with a 6 MHz QCM in a 13.56 MHz rf glow discharge system. This filter was designed and constructed by D. E. Horne. $R = 1$ K$\Omega$, $L = 2.7$ $\mu$h, $C_1 = 82$ pf, $C_2 = 8 - 120$ pf, $C_3 = 6 - 60$ pf. The variable capacitors are adjusted so as to maximize the output signal from the QCM.

and water-cooling of the QCM can be easily incorporated. Care should be taken to shield the electrical lead to the QCM inside and outside the plasma system, and the use of a band pass filter tuned to the QCM operating frequency is recommended. The filter, an example of which is shown in Figure 2, is mounted between the vendor-supplied oscillator circuit and the frequency counter.

## B. HIGH-ENERGY ION BOMBARDMENT

The presence of energetic ion bombardment of a QCM in a glow discharge environment greatly complicates the experimental aspects of this method. The extent of this complication depends on 1) whether the QCM is operated at ground potential (with the high-energy ion bombardment being a result of a high plasma potential) or at a negative potential and 2) whether the process is a deposition process or an etching process. The first point has to do with the need to isolate the QCM electronics from ground. This is a complication but is not a major problem and can be accomplished in a straightforward manner. Also, the shielding of the electrical connection inside the plasma system is much more critical in that spurious discharges to the connection must not be allowed. The second point relates to a phenomenon well known in glow discharge sputtering, namely backscattering of sputtered material to the surface from which it came, following several collisions with atoms or molecules in the short mean free path discharge. This is a particularly serious problem in reactive ion etching, where the surface being etched is exposed to both reactive neutral species

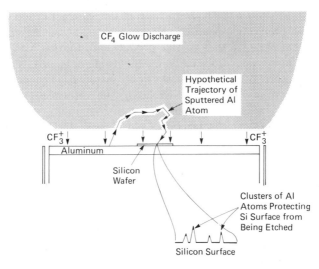

FIGURE 3. Schematic of a hypothetical trajectory of an involatile sputtered species and the resulting adverse effects in a plasma-assisted etching process (Coburn, 1984).

and energetic ions. In reactive ion etching, the etch product is volatile and etch rates are typically much larger than physical sputtering etch rates. The problem is illustrated in Figure 3, in which a silicon wafer is placed on a negatively biased aluminum electrode. Positive ions from the discharge ($CF_3^+$ in this example) will bombard both the silicon and the aluminum surfaces, and both surfaces will be fluorinated by the neutral fluorine atoms also arriving at the electrode surface. The silicon surface will be etched rapidly because of the volatility of silicon fluorides, whereas the aluminum surface will be etched very slowly by physical sputtering because aluminum fluorides are not volatile at temperatures reached in this reactive ion etching process. However, the species sputtered from the aluminum surface by the energetic ion bombardment will encounter many collisions with $CF_4$ molecules in the relatively short mean free path plasma environment, and some of these aluminum-containing species will be backscattered to the silicon surface, as illustrated in Figure 3. This transfer of material from one part of an electrode surface to another by backscattering from the discharge gas is sometimes referred to as *cross-talk* and is a serious problem in reactive ion etching technology (Vossen, 1976). When foreign atoms or molecules, which do not form volatile compounds with the discharge gas, arrive at a surface which does form volatile products, the foreign species tend to protect the underlying surface from the etching action, resulting in the formation of conical structures, as illustrated in Figure 3. This latter phenomenon is known as *micromasking*. The normal procedure in reactive ion etching is to ensure that, wherever possible, all surfaces exposed to energetic ion bombardment are such that volatile products are formed with constituents of the discharge gas. Thus carbon, other organic solids, silicon, silicon dioxide, etc., are often used as electrode materials. The problem in the QCM situation is that the QCM holder is not easily made of these materials, most often being constructed of aluminum, stainless steel, or other metals which can be machined more easily and which have greater mechanical strength. An example of the extent to which backscattering of sputtered involatile material can disrupt an etch rate measurement is shown in Figure 4. In this figure the etch rate of $SiO_2$ in a $CF_4$ glow discharge is

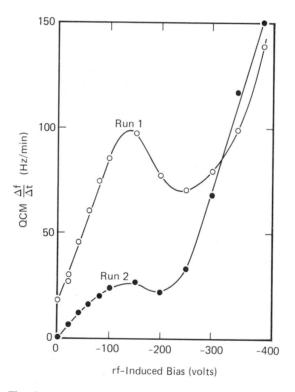

FIGURE 4. The plasma-assisted etch rate of a $SiO_2$ film (as given by the rate of change of the QCM frequency) as a function of the dc bias induced on the QCM by 13.56 MHz excitation. The QCM holder was gold-coated and, as the bias increases, gold is sputtered from the holder and is redeposited on the $SiO_2$ surface, as illustrated in Figure 3. The plasma parameters were a $CF_4$ pressure of 20 millitorr; 13.56 MHz rf power of about 100 W applied to the 100 cm² $SiO_2$ excitation electrode (separate and much larger than the power applied directly to the QCM); interelectrode spacing of 3.8 cm; a $CF_4$ flow rate of about 2 scc/min. See text for further discussion of these data (Coburn, 1984).

measured as a function of the rf-induced self-bias voltage on the QCM. The QCM holder is gold-plated metal and is biased at the same voltage as the $SiO_2$ thin film on the quartz crystal. Run 1 in Figure 4 was recorded on a fresh $SiO_2$ surface, whereas run 2 was recorded immediately after run 1. The data is interpreted as follows: Run 1—for ion energies less than 100 eV, the backscattering problem is not serious; for ion energies in the range between 100 and 250 eV, the backscattered gold sputtered from the holder causes a serious problem, obscuring the $SiO_2$ etch rate; and for ion energies greater than 250 eV, the backscattered gold appears to be sputtered away from the $SiO_2$ surface quite efficiently. Run 2—the much lower etch rate in the low bias voltage range indicates that the backscattered gold is still present on the surface from run 1 and the data is meaningless.

Whereas this backscattering of sputtered involatile material is a serious problem when using QCMs in a reactive ion etching environment, it should be emphasized that this is not a problem when QCMs are used in conjunction with ion beams, because the low gas pressure required for ion beam operation eliminates any significant backscattering of species sputtered from the QCM holder. Also, backscattering will not be a problem when QCMs are used to monitor plasma enhanced chemical vapor deposition processes in the presence of energetic ion bombardment. In this situation, the QCM holder is rapidly coated with the film of interest, and the material from which the QCM holder is constructed is of no concern.

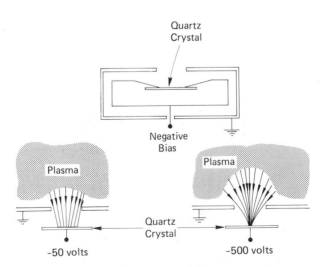

FIGURE 5.   Upper figure: Mounting a negatively biased QCM in a grounded shield. Lower figures: The focusing effect of the QCM bias voltage on the incident ion trajectories (Coburn, 1984).

It is clear that in order to operate a QCM in the presence of energetic ion bombardment in a reactive ion etching process, the QCM must be mounted in such a way that no surfaces made of materials that form involatile compounds are exposed to energetic ion bombardment.

One approach to this problem is to enclose the QCM in a grounded shield (or a shield kept near plasma potential) so that only the material of interest on the quartz crystal is subjected to energetic ion bombardment (see Figure 5—top figure). This approach has two major disadvantages: 1) The environment seen by the QCM in such an arrangement may differ substantially from the environment experienced by a much larger surface biased at the same potential (i.e., the QCM cannot be made part of a large equipotential surface, as in Figure 1). 2) There is a focusing of the ions incident on the QCM, as illustrated in the lower figures of Figure 5. This focusing action, combined with the large radial dependence of the QCM sensitivity (Pulker and Decosterd, 1984), causes the overall sensitivity of the QCM to be very dependent on the bias voltage. The consequences of this effect are shown in curve a of Figure 6, which is the dependence of the Si

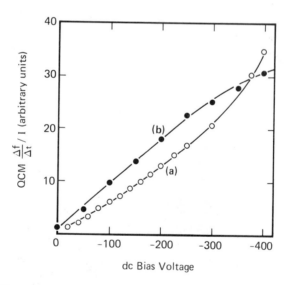

FIGURE 6.  Plasma-assisted etch rate of a Si film, as indicated by the rate of change of QCM frequency as a function of the dc bias voltage applied to the QCM. The etch rates are normalized with respect to the total ion current $I$ incident on the Si film. The plasma conditions were as tabulated in Figure 4. Curve a: Using a QCM mounting as shown in Figure 5. Curve b: Using the QCM mounting, as shown in Figure 5, but modified by the addition of a fine stainless steel grid over the aperture in the grounding shield. This mesh eliminates the focusing effect shown in Figure 5. See text for further discussion (Coburn, 1984).

etch rate on bias voltage in a $CF_4$ glow discharge. This data, unlike that shown in Figure 4, is completely reproducible. However, when the ion focusing is eliminated by placing a fine grid across the aperture in the ground shield, curve b in Figure 6 was obtained. Again, the data is very reproducible and may be a reliable measure of the dependence of Si etch rate on ion energy. However, when the Si-coated quartz crystal was removed from the system after this measurement, a grid pattern could be seen replicated on the Si surface. This observation casts some doubts on the reliability of curve b in Figure 6.

A second approach is to construct the QCM holder of a material which forms volatile products with the discharge gas being studied. This poses a rather difficult machining problem in some cases, and the simpler approach of depositing a relatively thick film of the material being etched on the existing QCM holder has been used more frequently.

## III. Applications of Quartz Crystal Microbalances in Reactive Gas Glow Discharges

### A. APPLICATIONS INVOLVING LOW-ENERGY ION BOMBARDMENT

The ease with which QCMs can be used in glow discharges under conditions where there are no high energy ions incident on the QCM (see Figure 1) suggests that the majority of the QCM applications in reactive gas glow discharges will be of this type. A major limitation of QCMs in plasma etching applications is that the material being etched must be deposited upon the quartz crystal by some deposition process prior to the installation of the QCM in the etching system. Often it is difficult to ensure that the thin film deposited on the QCM has the same etching characteristics as the bulk sample of interest. For example, the etching of single crystal silicon is unlikely to behave in the same way as the etching of an amorphous silicon thin film on the QCM. This problem obviously does not arise in monitoring deposition rates in reactive gas glow discharges.

Examples of the use of a QCM in measuring etch rates are shown in Figure 7. In this figure both the frequency [$\Delta f = f - f_0$] and the etch rate of an amorphous silicon film are plotted versus time for two plasma conditions, with the QCM at ground potential. The plasma potential in this system was approximately +20 volts with respect to ground, so only low-energy ions are incident on the amorphous Si film during these measurements. The uncertainty in the etch rate measurement is about $\pm 1$ Å/min. as judged from Figure 7a. The etch rate in Figure 7b is 15 to 20

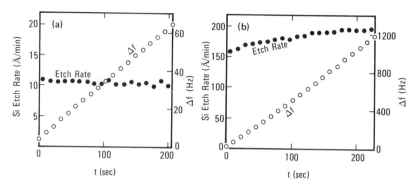

FIGURE 7.   QCM frequency versus time and the plasma-assisted etch rate of a thin Si film (previously deposited on the QCM) that can be deduced from the frequency data. These data were recorded using a QCM mounted as shown in Figure 1 and the system plasma potential was of the order of 20 V positive with respect to the grounded QCM. Plasma parameters: 13.56 MHz rf power of about 50 W applied to the 180 cm$^2$ SiO$_2$ upper electrode; interelectrode spacing of about 6 cm; curve a: CF$_4$ pressure was 15 millitorr at a flow rate of 9.4 scc/min; curve b: CF$_4$ pressure was 80 millitorr at a flow rate of 15 scc/min (Coburn, 1984).

times larger than that in Figure 7a and shows a definite increase with time. One must always be concerned about QCM frequency changes caused by heating of the QCM. In this case the QCM is water-cooled, and thermal effects are not believed to be significant. Real etch rate changes, as illustrated in Figure 7b, can be caused by desorption of impurity gas from surfaces in the plasma system. Inert gas plasmas of comparable power density can be used to assess the importance of thermal effects on the QCM frequency, provided care is taken to ensure none of the material sputtered from the target electrode reaches the QCM surface.

It is clear that this in situ etch rate or deposition rate measurement capability of the QCM would allow a large region of the parameter space associated with plasma reactors to be surveyed rather quickly. However, this method has not been used to any significant extent in plasma process development. It is suspected that the major obstacle to widespread use of QCMs in plasma-assisted etching studies is the aforementioned ion bombardment problem. Most plasma-assisted etching applications involve energetic ion bombardment, and it is clearly quite difficult to use QCMs under these conditions. It may be possible to correlate the etching behavior of a material on a QCM which is not subjected to ion bombardment with the etching of a heavily ion bombarded surface. However, then the QCM is nothing more than an indirect plasma diagnostic procedure, and it is not clear that more useful information is provided by the QCM than by other plasma diagnostic methods such as emission spectroscopy.

The two major concerns associated with using QCMs to measure etch rates are much less serious when QCMs are used to measure deposition rates. Clearly, the concern about the nature of the material involved is not applicable once deposition is underway. One might inquire as to differences in the initial nucleation of the deposited film on the surface of interest and on the QCM surface. Also, there is less interest in energetic ion bombardment during film deposition relative to the situation prevailing during etching, and even when ion bombardment is a major factor in deposition, there are no foreign species to be concerned about since all surfaces are coated with the deposited material including QCM holder. QCMs are widely used in deposition processes such as thermal evaporation and physical sputtering (Pulker and Decosterd, 1984) but have not as yet been widely used in reactive gas glow discharges such as those encountered in plasma-enhanced chemical vapor deposition. Two factors can be suggested for this: 1) The elevated substrate temperatures used in plasma-enhanced chemical vapor deposition complicate the QCM utilization somewhat and 2) The properties of the deposited films are of more interest than the deposition rate, and QCMs provide no significant information on film properties.

However, there is a situation which arises often in reactive ion etching processes which could make use of QCMs for process control in a relatively simple way. The deposition of halocarbon polymer plays an important role in several reactive ion etching processes. This phenomenon is often responsible for the etching directionality obtained in reactive ion etching and also is responsible for the large $SiO_2/Si$ etch rate ratios that can be obtained using fluorine-deficient fluorocarbon discharges. This latter result, first pointed out by Heinecke (1975), is now an important part of the reactive ion etching technology (Ephrath and Petrillo, 1982). However, the implementation of this selective etching of $SiO_2$ relative to Si requires careful control of the discharge chemistry, and in particular, the feed gas mixture. This etch process is carried out under energetic ion bombardment conditions, so a direct measurement of etch rates with QCMs is difficult. However, Figure 8 illustrates how a grounded QCM might be used to control this process. In Figure 8, the approximate etch rates of $SiO_2$ and Si are shown as a function of the percent $H_2$ added to the $CF_4$ etch gas. Note that these etch rates are for surfaces exposed to energetic ion bombardment. A surface at ground potential would revert from etching to polymerization conditions at rather low $H_2$ concentrations, and this is shown by the dashed line in Figure 8. This condition, where polymerization is observed on grounded surfaces and etching is observed on surfaces subjected to energetic ion bombardment, is very common in plasma-assisted etching processes. It may be feasible to use the polymerization rate on a grounded

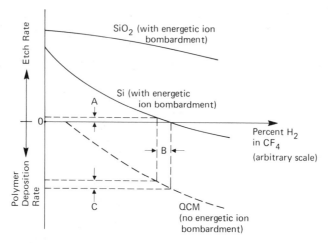

FIGURE 8.   Illustration of how the relatively easily measured rate of polymer deposition on a grounded (QCM) surface might be used to control a selective etching RIE process at a high $SiO_2/Si$ etch rate ratio (see text for further discussion). Coburn (1984).


surface as measured by a QCM to control the process as shown in Figure 8. The allowable Si etch rate is indicated by A, which requires the gas composition be kept in the range indicated by B. The control which is suggested is to keep the polymerization rate as monitored by a QCM in the range indicated by C, which in principle will keep the Si etch rate within the range indicated by A.


## B.   APPLICATIONS INVOLVING ENERGETIC ION BOMBARDMENT

Very little work has been reported in which QCMs are subjected to energetic ion bombardment in reactive gas glow discharges. The difficulties associated with this kind of application were discussed in Section II.B of this chapter. One approach to biasing QCMs in reactive gas plasmas is illustrated in Figure 9 (Us et al., 1986). In this design the entire unit including the flange can be biased with either rf or dc voltages. A commercially available ceramic vacuum break (Ceramaseal Model No. 808B3007.2) is used to electrically isolate the QCM mounting flanges from the grounded vacuum system. The microbalance and holder are also commercially available (Inficon Model No. IPN 750-040-G1), but the outer holder plate is modified. The electrical connection is carefully shielded, and

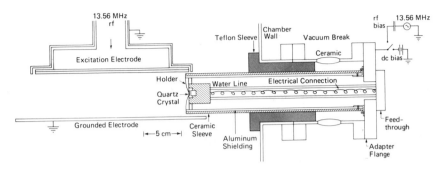

FIGURE 9. Schematic diagram of the mounting of one of two biasable QCMs. The second QCM has an identical orientation with respect to the glow discharge and is positioned 90° away from the QCM shown (i.e., directed into the plane of this figure) (Us et al., 1986).

spurious glow discharges inside the assembly are not possible because all internal components are at the bias potential. Spurious discharges external to the assembly are eliminated by the teflon sleeve. Ion bombardment of the outer aluminum shielding is prevented by the ceramic sleeve, and the entire front surface of the holder is coated with a film of the material to be studied in order to prevent backscattering of sputtered involatile material to the QCM surface. Carbon, which forms volatile products with most gases used in plasma-assisted etching, is used as an excitation electrode. A second identical biased QCM is mounted at 90° to the one shown in Figure 9. Normally, this second QCM is held at ground potential, and the etch rate of material on this grounded QCM is used to monitor the effect of the rf-biased QCM on the overall plasma characteristics. That is, at no time is the rf power applied to the biased QCM high enough to influence significantly the etch rate on the grounded QCM. Figure 9 illustrates the complexity of mounting QCMs in plasma systems when energetic ion bombardment is involved and should be contrasted with the grounded QCM mounting shown in Figure 1. However, the apparatus shown in Figure 9 is well suited for performing studies of the effect of energetic ion bombardment on plasma-surface interactions, and some examples of this work will be described below.

The ability to measure the etch rates of a material both with (biased QCM) and without (grounded QCM) ion bombardment provides an opportunity to simulate the etching directionality obtainable in plasma-assisted etching systems. The concept (Us et al., 1986) is illustrated in Figure 10, in which the etch rate of the grounded QCM is used to represent the lateral etch rate or undercut and the etch rate of the biased QCM is used to represent the vertical etch rate. The etch profile is characterized by

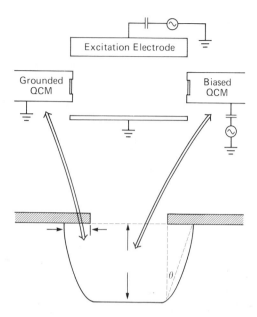

FIGURE 10.   Illustrative figure to describe the directionality simulation approach.

the angle $\theta$ = arctan [(lateral etch rate)/(vertical etch rate)]. In order that this simulation be realistic, the plasma potential of the system must be sufficiently low that the ion bombardment of the grounded QCM does not influence the etch rate. This technique has been applied to the etching of Si in $CF_4$ glow discharges, a system whose etching behavior is reasonably well established, and the results are shown in Figure 11. In this figure the sidewall angle $\theta$, as defined both in Figure 10 and by the expression above, is plotted as a function of three of the many parameters accessible in plasma process development. The three trends depicted in Figure 11 are well known to plasma process engineers and scientists, namely, the anisotropy of etching in the Si-F system can be improved by increasing the bias voltage, decreasing the pressure, or decreasing the fluorine/carbon ratio of the feed gas. This work is an example of how the influence of plasma parameters, feed gas mixtures, and substrate materials on etching directionality can be qualitatively determined relatively quickly with the biased QCM approach.

   This apparatus also affords the opportunity to look for energy thresholds in ion-assisted etching processes. In order to access the low ion energy regime in this work, it is preferable to use a dc glow discharge which has a plasma potential of about one or two volts with respect to ground. This

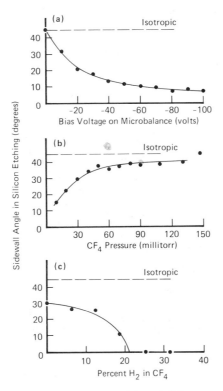

FIGURE 11.    Simulated sidewall angle obtained from QCM measurements of the etch rates of biased and grounded amorphous silicon films. 13.56 MHz rf power = 50 watts, electrode area = 180 cm². (a) $CF_4$ pressure = 20 millitorr; $CF_4$ flow rate = 9.4 sccm; carbon electrode. (b) bias voltage = −200 volts; $CF_4$ flow rate = 9.4 sccm; $SiO_2$ electrode. (c) total pressure = 20 millitorr; total flow rate = 9.4 sccm; bias voltage = −100 volts, $SiO_2$ electrode.

allows us to study the effect of ion bombardment in the energy range below 20 eV, which is the lower limit posed by the plasma potential in rf glow discharges. An example of this kind of measurement is shown in Figure 12 (Us et al., 1986) for the etching of Si in a $CF_4$ glow discharge. Note the onset of ion-assisted etching at about 20 eV ion energy. The decrease in the etch rate near zero bias voltage is believed to be caused by the rejection of electrons from the Si surface as the negative bias is applied and may be an indication of electron-assisted etching. The fact that the apparent threshold for ion-assisted etching (∼ 20 eV) coincides with the plasma potential of rf discharges in this system is coincidental. However, this allows us to have somewhat more confidence in the simulation of etching directionality discussed previously, because the ions incident on the grounded QCM will

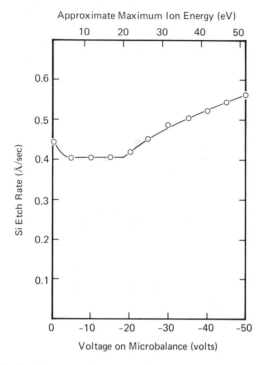

FIGURE 12.   The Si etch rate as a function of the DC bias voltage applied to the Si surface in a CF$_4$ glow discharge. CF$_4$ pressure = 50 millitorr; CF$_4$ flow rate = 9.4 sccm; excitation electrode voltage = −1500 V dc; plasma potential = +1–2 volts. (Us et al., 1986).

not cause ion-assisted chemistry. This is required if the etch rate of the grounded QCM is to be representative of the lateral etch rate or undercut.

A final example of the use of the biased QCM method is a study of the interplay between etching and polymerization in fluorocarbon glow discharges and the role of energetic ion bombardment. Figure 13 (Fracassi et al., 1987) is a plot of the etch rate or polymerization rate of a plasma perfluoropolymer thin film as a function of the rf induced self-bias voltage applied to the QCM for a range of feed gases and feed gas mixtures. As expected, for situations where there is little or no gas phase unsaturation (O$_2$, CF$_4$), ion-assisted etching of the plasma deposited fluoropolymer is seen. The polymer formation observed for pure CF$_4$ with no applied bias is a result of the fluorine consumed by the carbon excitation electrode used in this measurement. Note that the application of a very small bias voltage converts this deposition to etching. This result confirms the possibility of directional etching (via sidewall blocking) for very low incident ion en-

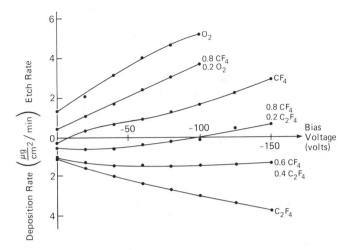

FIGURE 13. Etch rate/deposition rate of plasma perfluoropolymer thin films as a function of the 13.56 MHz rf-induced self-bias voltage on the sample (QCM) for six feed gases. Pressure = 10 millitorr; flow rate = 6 sccm; rf power = 30 watts to 182 $cm^2$ carbon excitation electrode; plasma potential ~ +20 volts (Fracassi et al., 1987).

ergies, a factor of importance in etching damage-sensitive materials. As the degree of unsaturation of the feed gas is increased by $C_2F_4$ additions, deposition begins to dominate over etching until, for pure $C_2F_4$ discharges, ion-enhanced polymer formation is observed (a factor of three increase in the deposition rate, with a −150 volt bias, as shown in Figure 13).

## IV. Summary and Acknowledgments

The methodology for operating QCMs in reactive gas glow discharge environments has been summarized and a few results have been presented. The significance of energetic ion bombardment in this technique has been emphasized, and it is suggested that the problems associated with the combination of QCMs and energetic ions may be the major reason for the sparse utilization of QCMs in reactive ion etching process development. The QCM approach would seem to be better suited for in situ monitoring of plasma deposition processes, as opposed to plasma etching processes, because during deposition energetic ions no longer interfere seriously with QCM operation, and in deposition applications there is no need to predeposit material on the QCM which is representative of the material of interest. It is anticipated that the low cost, compact size, durability, and high mass sensitivity of QCMs will result in greater use of these devices to monitor reactive gas plasma-surface interactions, particularly if the surface of interest is at ground potential.

The author is very appreciative of valuable collaborations with Natasha Us, Francesco Fraccassi, and Ernesto Occhiello and for the expert technical assistance of Robert Sadowski and Eric Isaacson. Much of the work discussed in this chapter was supported by the IBM General Technology Division in East Fishkill, New York.

## References

Coburn, J. W. (1979). *J. Appl. Phys.* **50**, 5210–13.

Coburn, J. W. (1984). In: *Applications of Piezoelectric Quartz Crystal Microbalances* (C. Lu and A. W. Czanderna, eds.). Elsevier, Amsterdam, pp. 221–49.

Coburn, J. W. and Kay, E. (1979). *IBM J. Res. Develop.* **23**, 33–41.

Coburn, J. W. and Winters, H. F. (1979). *J. Appl. Phys.* **50**, 3189–96.

EerNisse, E. P. (1974). *J. Vac. Sci. Technol.* **11**, 408–410.

Ephrath, L. M. and Petrillo, E. J. (1982). *J. Electrochem. Soc.* **129**, 2282–87.

Fracassi, F., Occhiello, E., and Coburn, J. W. (1987). *J. Appl. Phys.* **62**, 3980–81.

Heinecke, R. A. (1975). *Solid State Electron.* **18**, 1146–47.

Hinsberg, W. D., Willson, C. G., and Kanazawa, K. K. (1986). *J. Electrochem. Soc.* **133**, 1448–51.

Kanazawa, K. K. and Gordon, J. G. (1985). *Anal. Chem.* **57**, 1770–71.

Lu, C. (1984). In: *Applications of Piezoelectric Quartz Crystal Microbalances* (C. Lu and A. W. Czanderna, eds.). Elsevier, Amsterdam, pp. 19–61.

Lu, C. and Czanderna, A. W. (1984). *Applications of Piezoelectric Quartz Crystal Microbalances*. Elsevier, Amsterdam.

Melroy, O., Kanazawa, K., Gordon, J. G., and Buttry, D. (1986). *Langmuir* **2**, 697–700.

Nyaiesch, A. H. and Baker, M. A. (1982). *Vacuum* **32**, 305–308.

Pulker, H. K. and Decosterd, J. P. (1984). In: *Applications of Piezoelectric Quartz Crystal Microbalances* (C. Lu and A. W. Czanderna, eds.). Elsevier, Amsterdam, pp. 63–123.

Sauerbrey, G. (1959). *Z. Phys.* **155**, 206–222.

Schou, J., Sorensen, H. and Borgesen, P. (1984). *Nucl. Inst. Meth.* **B5**, 44–57.

Ullevig, D. M. and Evans, J. F. (1980). *Anal. Chem.* **52**, 1467–73.

Us, N. C., Sadowski, R. W., and Coburn, J. W. (1986). *Plasma Chem. Plasma Process.* **6**, 1–10.

Vossen, J. L. (1976). *J. Appl. Phys.* **47**, 544–546.

# 2 Elemental Analysis of Treated Surfaces

Donald L. Smith
*Xerox Palo Alto Research Center*
*3333 Coyote Hill Road*
*Palo Alto, California*

H. P. Gillis*
*Hughes Research Laboratories*
*Malibu, California*

Thomas M. Mayer
*Department of Chemistry*
*Venable Hall 045A*
*University of North Carolina*
*Chapel Hill, North Carolina*

## I. Plasmas and Surface Analysis

A gaseous plasma is an energetic mixture of charged particles and molecular fragments whose interaction with containment surfaces can produce

---

* Present address: Chemistry Department, Georgia Institute of Technology, Atlanta, GA 30332.

effects ranging from perturbation of the top monomolecular layer to structural damage extending hundreds of layers deep. Some technological processes, like plasma etching, are based on promoting such effects; while others, like atomic fusion, are plagued by them. In either case, elemental analysis of plasma-treated surfaces is central to the study of these interactions.

In this section, a summary of the nature of plasma and its effects on surfaces is presented, followed by a discussion of various surface analytical techniques and of the distinguishing features of the four to be reviewed in detail in this chapter. Reviews on electron and ion spectroscopies are presented in Sections II and III, respectively. All four techniques provide analyses of the chemical elements in the near-surface region of the sample, that is, the top ten monolayers or so. Brief tutorials on the techniques are followed by selected examples from the literature and suggestions for further work. The tutorials hopefully are sufficient to prepare the uninitiated reader for the applications examples and possibly are sufficient to allow productive interaction with analytical service laboratories offering these techniques. However, those planning to do their own surface analysis will need further study, and those using service laboratories will find that the value of the data so obtained is proportional to their own understanding of the techniques, since surface analysis is a subtle business fraught with both ambiguities in interpretation and with perturbations of the sample by the process of analysis.

The examples of applications from the literature were each selected for particular reasons and are not intended, as a whole, to represent a comprehensive review of the field. Some examples describe a revealing or innovative approach, and some illustrate a characteristic attribute or shortcoming of the analytical technique. The idea is to give the reader a spectrum of examples of what is possible in order to guide analytical choices and to stimulate further development of the techniques.

## A. PLASMAS

Plasmas receive energy from an applied electric field and dissipate it in various ways at the surfaces of and within the containment vessel. Energy is received by free electrons in the plasma. They are accelerated by the field until they achieve sufficient energy so that, upon collision with a gas molecule, they can cause electron ejection or bond-breaking, resulting in the creation of a positive ion or a molecular fragment (a free radical), respectively. Since the electrons are much more mobile than the ions, they escape to surrounding surfaces more readily, and thus the plasma develops a positive charge. Positive ions then become accelerated out of the plasma towards these surfaces and bombard them with an energy which increases with decreasing gas pressure and typically ranges from 1–1000 eV. At pressures below 10 torr or so, low ion-gas collision rates result in a plasma gas temperature much lower then the ion and electron "temperatures" (glow discharge plasma); whereas, at higher pressures, gas temperature can reach several thousand °C by energy transfer from charged particles (thermal plasma). In the former, much of the energy is dissipated in ion bombardment; in the latter, as heat. Thermal plasmas are discussed in detail in the chapter by Fauchais et al. Plasmas used specifically for lamps are discussed in the chapter by Waymouth. A more detailed discussion of charged particle bombardment of surfaces by plasmas is given by Thornton (1983).

Energy also arrives at plasma containment surfaces in the forms of highly reactive free radicals and of ultraviolet and x-ray photons. The photons are produced upon relaxation of plasma molecules electronically excited by electron collison. One important effect of these photons on materials is radiation damage to the gate dielectric layers of field-effect transistors. However, since this is a bulk rather than a surface effect, and since it is not amenable to elemental analysis anyway, it will not be discussed here.

The effects of ion bombardment on surfaces are manifold and include, roughly in order of increasing threshold energy, desorption of adsorbed gases, activation of chemical reactions among adsorbed gases or with the surface material, removal of surface atoms by momentum transfer (sputtering), intermixing of atoms and crystallographic damage within the top ten monolayers or so, and implantation of ions to hundreds of layers deep with accompanying crystallographic disordering. These effects of ions also must be kept in mind when using ions for analysis, either as the probe beam (Section III) or for sputter-removal of successive layers of material in depth-profile analysis (Section II). The kinetics of sputtering are described in Section III.A. Sputtering is important commercially as a means of volatilizing refractory material for film deposition, and surface analysis is useful there to measure the shift in surface composition of compound materials due to the different sputtering yields of its elements (Section III.C.3).

The neutral free radicals diffusing out of the plasma produce extensive chemical effects at surfaces due to the high reactivity of their unsatisfied (dangling) bonds. They can adsorb to form a chemically-bonded (chemisorbed) surface layer and, in some cases, can continue to build up to form thin films. This film-forming process is known as plasma-enhanced chemical vapor deposition and is used for the growth of amorphous Si and other semiconductors and also polymer films. However, since it represents the growth of a new material rather than the effect of a plasma on an existing material, it will not be addressed in this chapter.

Another chemical effect of considerable practical importance is plasma etching, in which plasma constituents react with solid materials to form gases, which evaporate and are pumped away. For example, Cl atoms, formed in the plasma by electron impact-dissociation of $Cl_2$ gas, can react with Si to form $SiCl_4$ gas. Alternatively, $Cl_2$ adsorbed on Si can be activated into a state of reactivity with Si by ion bombardment. Ion-activated etching proceeds only in the direction of ion arrival, so it is anisotropic. Such etching is crucial to the control of dimensions in the delineation of fine-line integrated circuitry, so plasma etching has become a key process in that industry. The term *reactive-ion etching* is often applied to plasma conditions in which chemically-active ions such as $Cl^+$ or ion bombardment-activated chemistry are believed to dominate the process, as opposed to neutral free-radical chemistry. However, it is becoming clear that all three are at work to some degree under most circumstances, so we will apply the generic term *plasma etching* throughout this chapter.

In the plasma etching of integrated circuits, attention must be paid to the details of the etching reaction, to the residues left behind on the surface, and to the damage produced in the layers remaining. Because of the power of elemental surface analysis to study such effects and because of the importance of plasma etching in the semiconductor industry, most of the work reported in the literature on analysis of plasma-treated surfaces relates to plasma etching, and this work comprises much of the discussion of Sections II and III. Nevertheless, the attributes and pitfalls which are mentioned for the various analytical techniques generally will apply to the investigation of other effects of plasmas on surfaces as well.

B. SURFACE ANALYSIS

Most analytical techniques involve directing a probe beam at the sample and detecting either changes in the reflected probe beam or particles emitted from the sample as a result of excitation by the probe beam. Techniques that are surface-sensitive have shallow escape depths for the detected beam or particles. The four surface analysis techniques that are the subject of this chapter are further characterized by providing data that directly identify the various chemical elements in the sampled volume.

Table 1 compares these four techniques. X-ray photoelectron spectroscopy (XPS) and Auger electron spectroscopy (AES) both detect electrons emitted with energies characteristic of their element of origin, under stimulation by an x-ray or an electron beam, respectively. They have similar

**Table 1.** Comparison of Surface Elemental Analysis Techniques

| Technique | Probe Beam | Particles Detected | Sampled Depth, Monolayers | Sensitivity, Atomic Fraction | Lateral Resolution, $\mu$m |
|---|---|---|---|---|---|
| Electron Spectroscopy | | | | | |
| Auger electron spectroscopy (AES) | 1–10 KeV electrons | Auger electrons | 2–30 | $3 \times 10^{-3}$ | 0.02 |
| X-ray photoelectron spectroscopy (XPS) | soft (1–2 KeV) x-rays | photo-electrons | 2–30 | $3 \times 10^{-3}$ | 150 |
| Ion Spectroscopy | | | | | |
| Secondary ion mass spectroscopy (SIMS) | 1–20 KeV ions | sputtered ions | 1–3 | $10^{-3}$–$10^{-9}$ | 0.1 |
| Ion scattering spectroscopy (ISS) | < 10 KeV $He^+$, $Ne^+$ | backscattered ions | 1 | $10^{-2}$ | 0.1 |

sampling depths and atomic fractional sensitivities. AES has an advantage in lateral spatial resolution, because an electron beam is relatively easy to focus; but, on the other hand, an electron beam is much more perturbing to the surface than is an x-ray beam. XPS gives additional information about the chemical bonding state of elements. There are other related analytical techniques which will not be discussed here. Scanning electron microscopy (SEM) detects the secondary electron emission stimulated by a probe electron beam, giving a topographical image but not an elemental analysis (see chapter by Joy). X-ray fluorescence emitted during SEM can be analyzed by energy dispersion (EDX) or wavelength dispersion (WDX) to give an elemental analysis, but the probe beam penetrates several microns and the x-ray escape depth is even longer, so this is not strictly surface analysis.

The other two techniques to be discussed here use ion beams as probes and detect either the energy loss of the backscattered probe ions, in ion scattering spectroscopy (ISS), or the masses of ions ejected from the surface material by the probe beam, in secondary ion mass spectroscopy (SIMS). Both have higher surface sensitivity than the electron spectroscopies, and SIMS has the highest atomic fractional sensitivity of any surface analysis technique. The techniques are destructive in that material is continuously sputtered away during analysis, making the analysis of very thin layers difficult; but, on the other hand, a profile of composition vs. depth is automatically obtained. Unfortunately, the depth resolution of this profile is not as good as the sampling depth would imply, because of ion-induced mixing and roughening effects that will be discussed later. If the plasmas whose surface effects one wants to study happen to have ion bombardment energies comparable to the 1–10 KeV needed for ion spectroscopy, then, rather than the beam being destructive, it becomes a simulation of the plasma itself. A technique related to ISS, known as Rutherford backscattering spectroscopy (RBS), uses MeV-range He ions, which penetrate and escape from much deeper into the sample (see the chapter by Doyle). Singer (1986) gives a survey of surface analysis techniques, with emphasis on applications in the semiconductor industry.

It is clear that each surface analysis technique has its own particular advantages and shortcomings, and these will become more apparent in the discussion of Sections II and III. Consequently, the most powerful approach and some of the best work to date involves using more than one technique. If the sample is simultaneously analyzed by multiple techniques, it is insured also that all data came from the sample while it was in the same condition. Some further developments along these lines are underway, and we look forward to others.

There are also various analytical techniques that use optical probes and detected signals; these are reviewed in the chapter by Donnelly, and ellipsometry specifically is reviewed in the chapter by Chang. The optical techniques are unique in being able to penetrate the plasma itself so that surfaces can be analyzed during plasma treatment, though not easily. This is not possible with electron or ion spectroscopy because of short collision lengths for the probe and signal particles, compared to the plasma cross section.

The incompatibility of elemental surface analysis with the plasma environment leaves three options, having various advantages and shortcomings. Examples of all three will be given in Sections II and III. The option which is easiest and least expensive, but most compromising of the data, is remote analysis, in which the sample is removed from the plasma equipment and inserted into the spectrometer. The intervening period of exposure to atmosphere will almost always result in adsorption of water and carbonaceous gases on the surface and will also alter reactive surface species before they can be examined. This method is most successful for stable layers that are also thick enough so that the few monolayers containing atmospheric contamination can be sputtered off at the start of analysis.

Atmospheric exposure can be avoided by going to the trouble of attaching the spectrometer to the plasma equipment and transferring the sample from one to the other environment under a vacuum. The quality of the vacuum and stability of the surface species will determine how much of the data is preserved. This vacuum transfer mode has been used to demonstrate explicitly by comparison the deleterious effects of atmospheric exposure in the remote mode.

The third option is the in situ technique, in which plasma treatment and analysis occur in the same chamber. This is less complicated than vacuum transfer and also lessens the amount of time between treatment and analysis, but it does compromise the plasma environment. In this mode, people use model beams of ions, reactive molecules, and radicals, which emulate the plasma whose effects are under study but which keep the pressure and thus the concentration of reactive species much lower for spectrometer compatibility. Independent control over the various plasma constituent particles arriving at the sample is also obtainable in this mode, making it a valuable tool for fundamental studies. SIMS is particularly amenable to this technique, since reactive species can be included in the ion probe beam itself, in which case treatment and analysis are carried out simultaneously. This mode is advisable for looking at shallow plasma effects by SIMS, because in any post-treatment mode the layer is likely to be sputtered away before an accurate analysis can be obtained.

Most in situ electron spectroscopy work has been done just *after* beam treatment, though it has been demonstrated that XPS can be done during sample exposure to beams from $Cl_2$ and $CCl_4$ plasmas (see Section II.C.2.c). The problem with analysis after rather than during treatment is that short-lived surface species, such as volatile etching products or reaction intermediates, are not going to be observable. To the authors' knowledge, no such transient species have yet been observed in model plasma beam work. Even if analysis is performed during treatment, there remains the following problem in the study of plasma-surface reaction mechanisms. Reaction rates scale linearly with pressure only for first-order reactions, that is, those in which the rate is proportional to the concentration of *one* reactant. So unless the reaction kinetics are understood (which they usually aren't, because that is what is being studied), extrapolation to the effects under real plasma conditions is tenuous. Nevertheless, the beam approach, especially using complementary spectroscopies simultaneously, remains the most powerful one for studying plasma-surface interaction.

## II. Electron Spectroscopy

In both electron spectroscopic techniques to be discussed here, AES and XPS, one measures the kinetic energy of electrons emitted from the near-surface region of the sample under the influence of a stimulating probe beam. From this energy spectrum, the elemental composition can be determined. In AES, the beam is KeV-range electrons, and in XPS, it is x-ray photons. Both techniques are of great utility in studies of the plasma etching process, since they can identify process residues, detect changes in surface stoichiometry induced by the process, identify deposits formed when etched surfaces are exposed to air or subsequent processing media, and provide mechanistic insight into the surface chemical reactions in the process.

XPS is to be preferred over AES in many such applications, since the electron beam of AES can damage the sample by inducing decomposition or desorption at the surface. Electron-stimulated desorption (Madey, 1986) is particularly effective for surface halogen species. On the other hand, AES gives far better spatial resolution, since the probe beam can be focused to a spot diameter as small as 200 angstroms and can be moved to different sites on the sample. In fact, a compositional "map" of a sample can be obtained by scanning Auger microscopy (SAM), which is routinely done in conjunction with scanning electron microscopy (SEM) in modern instruments. In XPS, the spot analyzed is usually a couple of millimeters in diameter. Newer commercial instruments include "small spot XPS" options in which the diameter of the analyzed spot is reduced to 100–200 $\mu$m by focusing of the x-ray beam or by restricting the acceptance angle of the analyzer; but considerable signal strength is lost in so doing. The complementary aspects of AES and XPS are exploited by having both in modern spectrometers; this requires a single energy analyzer, an electron gun, and an x-ray source.

Both AES and XPS have sensitivity limits around 0.3 atomic percent of the sampled volume. Sampling depth runs from 5 to 100 angstroms for both and is determined by the mean free path which the emitted electrons can travel before they experience inelastic collisions that alter the characteristic energy with which they were emitted (Seah and Dench, 1979). Quantitative analyses can be made, but caution must be exercised in the case of overlayers having a stoichiometry different than the substrate (Seah, 1983).

Application of these techniques will require further study beyond the discussion in this chapter. Recent sources include a multi-authored volume edited by Briggs and Seah (1983) and a textbook by Feldman and Mayer (1986). There are two older books which contain much useful information and give excellent documentation of the pioneering work by Siegbahn and his associates (Carlson, 1976; Czanderna, 1975). Specific topics are reviewed thoroughly in the multi-volume series, *Electron Spectroscopy: Theory, Techniques, and Applications* (C. R. Brundle and A. D. Baker, eds.; Academic Press). A wealth of practical information is in the two handbooks published by Perkin-Elmer (Davis et al., 1978; Wagner et al., 1979).

## A. Origin and Appearance of Spectra

The physical basis of XPS and AES is sketched in Figure 1, where we show the binding energy levels $E_b$ of electrons in an atom of a solid, using Si as the example. We will first explain the notation for electrons. $E_b$ levels are designated by the principal quantum number $n$ $(1, 2, \ldots)$ of the electron shell and by the orbital momentum quantum number $l$ (s, p, d, or f) of the subshell. Because of physical coupling of the orbital and spin angular momenta, in subshells where $l$ is not s (zero), the electron can take two new energy values, one above and one below the original "unperturbed" value. The one above (for spin quantum number $j = l + \frac{1}{2}$) corresponds to having the orbital and spin angular momenta roughly parallel, whereas the lower energy (for $j = l - \frac{1}{2}$) has them roughly antiparallel. The resulting two energy levels are distinguished by adding the $j$-value as a subscript, e.g., $2p_{3/2}$, $2p_{1/2}$. This "spectroscopic" notation is used in XPS. Binding energy levels in Figure 1 are also given in "x-ray notation," defined by representing $n$ by one of the capital letters K, L, M, N, O, as $n$ takes values 1, 2, 3, 4, 5. Subscripts 1, 2, etc. are added in order of decreasing binding energy as determined by $l$ and $j$, but $l$ and $j$ are not identified explicitly. This notation is used in Auger spectroscopy. The elementary theory of atomic structure, from which these state notations are drawn, is described by Eisberg (1961). The notations are described in more detail by Feldman and Mayer (1986) and by Briggs and Riviere (1983).

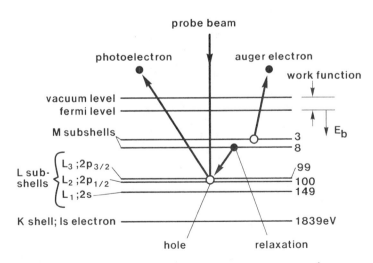

FIGURE 1. Electron spectroscopy processes for Si, showing binding energy ($E_b$) levels and ejection of a 2p photoelectron and of an LMM Auger electron.

For surface analysis, a probe beam of x-rays or electrons is directed at the sample, causing the ejection of an electron into the vacuum. This electron is shown coming from the 2p level in Figure 1. In XPS, the probe is an x-ray beam containing a strong line, usually the "K$\alpha$" line of Mg (1253.6 eV) or Al (1486.6 eV). The resulting ejected photoelectron has a kinetic energy equal to the x-ray photon energy less $E_b$, neglecting work function and other corrections amounting to a few eV, as discussed by Anthony (1983). In Figure 3, the 2p photoelectrons are used for Si analysis. (The electron directly ejected by an *electron* probe beam cannot be distinguished from the background, because the probe electrons lose energy by inelastic scattering as they penetrate the solid, and therefore the ejected electrons are spread over a wide range of kinetic energies.)

The processes involved in AES are also shown in Figure 1. The 1–10 KeV electron probe beam, by ejecting an electron from the L shell, has left a hole. This ionized atom is in a very excited state, and it can relax by an internal transition in which an electron from a higher shell (M) drops into L. This makes available energy equal to the difference between the two levels, which either can generate a photon (x-ray fluorescence) or can eject a

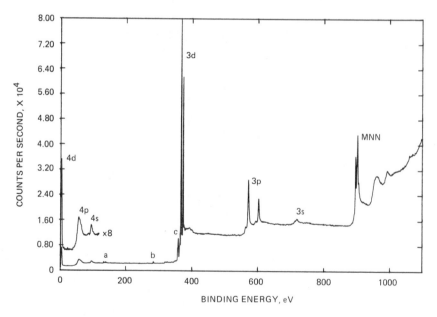

FIGURE 2.   XPS of Ag excited by MgK$\alpha$ and recorded with constant analyzer energy of 20 eV. (a) is "ghost" of Ag (3d) doublet produced by Al (K$\alpha$); (b) is C (1s) from slight surface contamination; (c) is Ag (3d) "satellite." (Courtesy of Hughes Research Laboratories.)

so-called Auger electron (named for the discoverer of the process) from the same or a higher level. The Auger electron escapes with kinetic energy characteristic of that transition, here one of the LMM transitions, for that particular element. Many such Auger transitions, all starting from the same initial shell, can be induced in a group of subshells. Consequently, spectra appear as groups of closely spaced transitions called *characteristic Auger series*, identified by the shell levels without subscripts. The Auger electron energy depends only on atomic energy levels. Therefore, it provides a unique signature for each element. The Auger transition is independent of the energy of the probe beam, so either an x-ray or an electron probe beam can be used for excitation. An electron beam is usually used for Auger spectroscopy, because it can be focused easily. Auger electron emission is more probable for light elements and x-ray fluorescence for heavy ones.

Silver is universally used in the testing and calibration of electron spectrometers, so in Figure 2 we show its spectrum excited by Mg (K$\alpha$). One sees the core levels 3s, 3p, and 3d, and the valence levels 4s, 4p, 4d, as well as the MNN Auger series of peaks from the various subshells of M and N. Note that the 3p and 3d levels are split into doublets by the spin-orbit

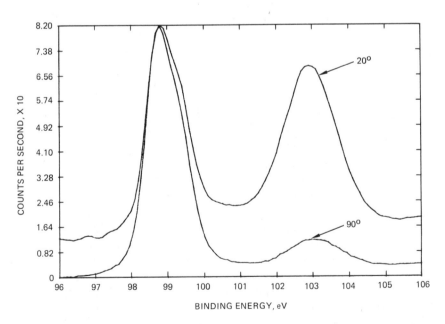

FIGURE 3. Variation of Si (2p) signal with take-off angle. Sample is Si with thin overlayer of SiO$_2$. Take-off angles specified relative to sample surface plane. (Courtesy of Hughes Research Laboratories.)

interaction. The ratio of intensities of the peaks in these doublets is useful in helping to identify elements in a complicated spectrum.

There are two ways to make use of the energy spectra of the emitted electrons. First, one can identify the elements present on the surface. Second, one can determine their state of chemical bonding. Element identification is accomplished by comparison with tabulated reference spectra, in which the kinetic energy values of the Auger transitions (Davis et al., 1978) or the x-ray photoelectrons (Wagner et al., 1979) are listed for each element.

Chemical bonding analysis using XPS is illustrated in Figure 3, which is a high-resolution scan of the Si(2p) peak for a sample of Si with a thin overlayer of $SiO_2$. In $SiO_2$, this peak appears at $E_b = 103$ eV, whereas in clean elemental silicon it appears at 99 eV. This shift occurs because bonding to O displaces valence electrons from Si and thus increases the $E_b$ of its remaining electrons. Auger peaks are much less useful for chemical shift analysis, because they are much broader. The absolute determination of the chemical bonding state requires very accurate measurement of the bonding energies of the various peaks, which are then compared with tabulated values of chemical shifts. This requires precise calibration of the system, a strong peak which can be scanned over a narrow energy range with high-energy resolution and signal level, and correction for static charge on insulating samples. Special attention must be given to charge compensation techniques (Wagner et al. 1979; Swift, Shuttleworth, and Seah 1983).


B.  INSTRUMENTATION

The essential components of an AES/XPS spectrometer are an electron gun, an x-ray source, and an energy analyzer, as shown in Figure 4. An ion sputtering gun is usually included for sample cleaning and depth-profile analyses. Additional components also may be included to enable SIMS, ISS, ultraviolet photoelectron spectroscopy (UPS), or low-energy electron diffraction (LEED). Commercial spectrometers universally operate at ultrahigh vacuum (UHV) in order to ensure cleanliness of the sample for the duration of the measurement. Almost all commercial spectrometers now are equipped with load-locks and sample insertion systems to allow rapid throughput without loss of UHV. These same capabilities allow spectrometers to be attached to plasma processing systems so that samples can be analyzed without exposure to air. Design, operation, and limitations of the various components are thoroughly reviewed by Riviere (1983). We will discuss beam sources, electron energy analyzers, signal treatment, and depth profiling separately below.

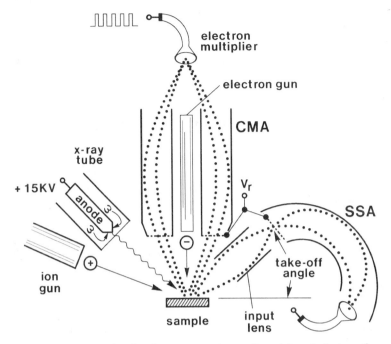

FIGURE 4. Instrumentation for electron spectroscopy. Dotted lines indicate pathways of emitted electrons being energy-analyzed. CMA = cylindrical mirror analyzer. SSA = spherical sector analyzer. $V_r$ = electron retarding potential.

Commercial XPS spectrometers universally have an x-ray source which can produce either Mg-K$\alpha$ or Al-K$\alpha$ radiation by electron bombardment of a Cu anode coated on one side with Al and on the other with Mg, as shown in Figure 4. The advantage of using two anodes is that the kinetic energy of photoelectrons shifts with photon energy, while that of Auger electrons does not; this aids in interpretation of a complicated spectrum. The disadvantage is that, through aging or damage, the source may emit small amounts of one radiation when the other has been selected; this produces small "ghost" lines displaced 233 eV from a parent photoelectron line. Other contaminating radiation can come from O-K$\alpha$, if the source is accidentally exposed to air and oxidizes the Al or Mg films, or from Cu-L$\alpha$ if the Al or Mg films become eroded away.

Recently, XPS spectrometers have been attached to synchrotron radiation sources, which can provide very intense and continuously energy-tunable monochromatic x-ray radiation. When the probe beam energy is tuned to just above $E_b$, the photoelectrons escape with very low kinetic energy and thus have an escape depth of only a few angstroms. This low kinetic energy makes the technique very surface sensitive.

The electron spectrometer usually is either a cylindrical mirror analyzer (CMA) or a spherical sector analyzer (SSA). Both are shown in Figure 4, though it would be awkward to use them together. In either case, pass energy of the analyzer is fixed, and the spectrum is scanned by changing a retarding voltage on grids or lenses which guide the emitted electrons from sample to analyzer. The way in which the retardation is accomplished determines whether energy resolution (peak width) will be constant throughout the spectrum and greatly influences quantitative analysis. Historically, the CMA has been preferred for AES because it has very high signal transmission at moderate energy resolution and therefore enables rapid acquisition of spectra. The SSA, with input lens, has been preferred for XPS because it has adequate throughput at a resolution high enough for chemical shift measurements. Thus, the practical guideline for choosing an analyzer is to know in advance the resolution required for the measurement to be done and then select the analyzer giving the highest signal level at that resolution. In fact, modern commercial instruments based on either the CMA or the SSA are adequate for applied work in both AES and XPS.

After energy analysis, the electrons are individually amplified to measurable pulse heights using a channel-type electron multiplier and are counted by a ratemeter or a digital computer. In XPS, the results are displayed as a spectrum of $N(E)$, the counts per second (cps) at kinetic energy $E$, versus $E$, the kinetic energy in electron volts (eV). In AES, the derivative $dN/dE$ is displayed instead (as in Figure 7) in order to bring out the Auger electron peaks, which are small compared to the very large continuum background of secondary electrons generated by inelastic scattering of the probe electron beam.

Depth profiling to arbitrary depths can be done by analyzing the surface as it is progressively sputter-eroded by an inert gas ion beam. However, ion sputtering can change the surface composition of the sample. It can also cause ion mixing and/or surface roughening with consequent reduction in depth resolution (see Section III.A). Therefore, shallow ( < 50 angstrom) depth profiling is preferably accomplished by variation of the "take-off" angle between the surface plane and the centerline of the analyzer, as shown in Figure 4. As the angle is reduced, electrons emitted from atoms below the surface encounter a longer path length before escaping, so their signal is attenuated. Hence, the depth of an element in the sample can be inferred by recording the change in strength of one of its lines as take-off angle is varied. This nondestructive technique is illustrated in Figure 3, where we see that the peak characteristic of the thin $SiO_2$ overlayer becomes more visible as the take-off angle is reduced. Take-off angle work is better done with a SSA than a CMA because of its much narrower cone of signal acceptance and is best done with a rotating SSA (rather than a rotating sample) so that the angle of the probe beam to the sample can remain fixed. One might expect that sweeping kinetic energy with a synchrotron source would also be useful for depth profiling, but because photoelectron emission probability also varies with photon energy, deconvolution becomes intractable.

## C. Applications in Plasma Etching

### 1. Remote Analysis

All of the samples discussed here were processed in commercial plasma etchers and were exposed to atmosphere on the way to analysis in standard electron spectrometers. This is the easiest way to make use of AES and XPS and is especially attractive when a central service-type lab is available with skilled operators to acquire and interpret spectra. However, the information which can be obtained is limited by the exposure to atmosphere. In general, one can identify the effects of side products, impurities, sputtered wall materials, and damaged layers. Sometimes alterations of surface stoichiometry in compound materials can be detected. Detailed information on reaction pathways is not obtained.

**a. Si and SiO$_2$.** Since the early days of plasma etching, C has been suspected to play an important role in the selectivity between Si and SiO$_2$ in etching with halocarbon gases (Morgan, 1985). It has been proposed that C is deposited on the Si and passivates it against further etching, whereas C on SiO$_2$ reacts with the O in the lattice to form volatile species and therefore exposes the surface to further etching.

Oshima (1979) tested these ideas using AES and XPS to identify surface contamination on Si and SiO$_2$ after etching in C$_2$F$_6$. Auger peaks for Si, C, O, and F were detected on both Si and SiO$_2$ after etching. The C and F contaminants on SiO$_2$ were only 5%–20% as large as on Si, which is consistent with the starting hypothesis. The bonding of C—F on Si was examined with XPS by measuring the chemical shift for C(1s) (see Figure 5). After etching, this peak showed three shoulders shifted up in binding energy by 2.9, 5.5, and 7.5 eV. These were attributed to CF, CF$_2$, and CF$_3$, respectively, which suggests that a CF$_x$ layer was formed during etching. Note that the CF$_x$ peaks are much weaker for the samples receiving greater ion bombardment from the plasma during etching. The as-etched surfaces

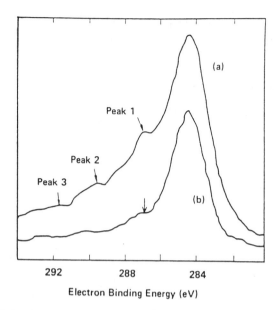

FIGURE 5. C (1s) XPS spectra from Si after etching. Sample (b) received greater ion bombardment than did (a). (Reproduced with permission from Oshima (1979).)

also showed very high sheet resistivity, presumably due to the fluorocarbon layer and to surface damage. Resistivity was restored to its original value and the surface contamination was removed by annealing for 30 minutes at 600°C in $H_2 + N_2$.

The Si(2s) peak on Si after etching also showed a second peak, displayed 3 eV up, which was attributed to Si—C bonding. However, it grew and shifted up another 1 eV upon annealing, and this was attributed instead to oxidation of the Si. An electron spectrometer with higher energy resolution than the 2 eV employed here might have been able to unravel this situation by resolving separate peaks due to Si—C, Si—O, and Si—$O_2$ bonding.

Oehrlein et al. (1985a) studied surface damage and contamination in the etching of Si with $CF_4 + 40\%$ $H_2$ using XPS, He ion channeling, H profiling, and Raman scattering techniques. They showed the contamination to consist mainly of a C—F layer, similar to Oshima's results. They found a heavily disordered layer reaching 30–50 angstroms below the surface, followed by a lesser degree of damage down to 250 angstroms. XPS was used to identify the surface contamination from O(1s), C(1s), F(1s), and O(KLL) features and to show that the thickness of the contamination layer was actually reduced in over-etching, that is, during a continuation of plasma exposure after the Si film had been etched through. In a similar study, XPS was used to identify a C—F—Cl contaminating layer produced on the underlying Si substrate during etching of $SiO_2$ with $CClF_3 + H_2$ and to show that subsequent cleaning with $O_2$ plasma did not suffice to remove the layer (Oehrlein et al. 1985b).

Two very interesting reports have been made on the utility of AES and XPS for examination of plasma etching system artifacts. Karulkar and Tran (1985) showed that oxygen stripping in a system previously used for $CF_4$ etching produced a fluorinated contamination layer which could not be removed by HF acid or a mixture of $H_2SO_4$ and $H_2O_2$. Tuppen et al. (1984) used AES and XPS to examine "persistent" and "nonpersistent" films formed during etching of $SiO_2$ on Si by $CHF_3$. Nonpersistent films could be removed by $O_2$ plasma and were found to be fluorocarbon polymers. Persistent films could not be removed by $O_2$ plasma and contained oxides and fluorides of Si, along with metal atoms sputtered from the electrodes of the etching reactor. Proper choices of electrode material and geometry were shown to reduce the production of persistent films.

**b. GaAs.** Plasma etching is a complicated process involving several physical and chemical effects whose interplay is not yet completely understood. This is especially apparent in applications to compound materials, where the question of different reactivities of the constituent atoms comes into play. It is useful to compare results with those from wet etching of the same materials, where the chemistry is generally better understood and the complications due to ion bombardment are not present. Accordingly, we start the discussion of GaAs by summarizing results from an XPS study of wet-etched samples.

Bertrand (1981) used XPS to analyze the surfaces of GaAs immediately after etching in HCl and in Br/methanol and again after the etched samples had a prolonged exposure to air. The objectives were to detect any changes in surface stoichiometry induced by etching, to detect any surface contamination from the etching, to identify the oxide layers formed by air exposure, and to correlate all of these effects with the height of Schottky barriers produced by deposition of contact metals upon similarly prepared surfaces. For comparison and reference, spectra were taken as well for clean Ga and As and for a variety of oxides of Ga and As. The analysis relied upon measurements of chemical shifts for As (3d) and Ga ($3p_{3/2}$). The principal results were as follows. The Br etch produced a stoichiometric surface, whereas the HCl etch produced a surface rich in As. After oxidation in air, both etched surfaces reverted to the Ga-rich composition characteristic of a native oxide overlayer. These findings correlated with Schottky barriers formed by deposition of Pb upon similarly prepared samples. After the Br etch, the barrier was 0.65 eV; after the HCl etch it was 0.4 eV. Subsequent air exposure caused both samples to give the barrier value 0.65 eV. In no case was contamination found from the etching solutions. All etched samples showed an observable oxygen signal, but no shifts in the Ga or As which would indicate oxidation. This was attributed to a small amount of O chemisorbed during loading of the samples into the spectrometer. These results provide an interesting connection to two areas of compound semiconductor technology where AES, XPS, and ultraviolet photoelectron spectroscopy have had wide utility: composition of native and deposited oxide layers and the defect model of Fermi level pinning (Wagner and Wilmsen, 1982 and 1985).

Plasma etching has become very attractive in GaAs technology because of its demonstrated ability in Si technology to etch fine-line features with good anisotropy. The main problems have been to obtain flat surfaces and anisotropic profiles without radiation damage or contamination. Most GaAs processes have been developed by analogy with Si etching and thus involve the effect of additive gases on plasmas of a halocarbon gas, usually $CCl_2F_2$ (DuPont's Freon 12).

In one study, the etch rate of GaAs depended strongly on the percent dilution of $CCl_2F_2$ with Ar, total flow rate and pressure being held constant (Chaplart et al., 1983). The rate was maximized for 60% Ar. AES results for two etched samples (one in pure $CCl_2F_2$, the other in 50% Ar) showed that the F/Cl ratio fell from 20/3 to 5/3 with Ar dilution. This correlated with the dependence of rate upon dilution and was interpreted in terms of reduced dwell time of etchant gases in the reactor as dilution was increased. XPS showed that the Ga $(3p_{3/2})$ peak had chemical shifts characteristic of Ga—F as well as Ga—As bonds. This is consistent with the fact that $GaF_3$ is nonvolatile, so that its removal is suspected to be the rate-limiting step in the process.

Some plasma processes for GaAs have used $Cl_2$ or $CCl_4$ to avoid the problem of nonvolatile fluorides. Oxygen is usually added to the C-containing gases to prevent polymer formation. In one study, addition of $O_2$ to $CCl_4$ caused black films, indicating roughened surfaces; this was attributed to an excess of highly reactive Cl radicals freed by the $O_2$ (Semura, Saitoh, and Asakawa, 1984). Again, by analogy with Si processing, the authors reasoned that addition of $H_2$ to $CCl_4$ would tie up some of the Cl atoms and bring the reaction under control. They found that for $H_2/CCl_4$ ratios below 0.3, isotropic profiles were produced, whereas good anisotropy and smooth surfaces were obtained for ratios above 0.6, as long as total pressure and power were kept within certain narrow ranges. A sample etched under these "good" conditions and examined by AES showed C(KLL), O(KLL), Cl(KLL), Ga(LMM), and As(LMM) characteristic series. The surface concentration of Cl was much less than that of C and O. After ion-sputtering to a depth of 10 angstroms, the Cl was eliminated, but the C and O persisted to 50 angstroms. These results were interpreted as showing that the sample was not contaminated by the etching species, but had adsorbed C and O during exposure to atmosphere between etching and analysis. No explanation was offered for the difference in depth distribution for Cl and for C and O, but mixing into the bulk by ion-sputtering during analysis might have caused this difference. Analysis of samples etched in $CCl_4 + O_2$ by AES (and especially SAM) might have revealed whether surface roughness was caused by oxidation of GaAs.

It is clearly important to determine the surface composition of GaAs after plasma etching. Since device fabrication is done in air, it is important to determine as well the composition of the oxide which grows upon exposure to air after etching and to compare it to native oxide. Comparison of plasma etching results with the findings of Bertrand (1981) for wet-etched samples will provide some insight into the effect upon the surface of ion bombardment during plasma etching. Such a study with XPS has been reported by Yabumota and Oshima (1985) for GaAs etched in $CCl_2F_2$. Their results illustrate well the power of quantitative XPS combined with chemical shift measurement to determine changes in surface composition and the use of take-off angle variation to accomplish nondestructive depth profiling. Taking measurements normal to the surface (take-off angle 90°) so that some subsurface information was obtained, they showed by comparing the relative intensities of O (1s) to Ga (3d) and As (3d) that the oxide grown in air after plasma etching had more oxygen than native oxide. This is very different from Bertrand's results and suggests that the ion-damaged surface after plasma etching is easily oxidized. Moreover, by making similar comparisons of relative intensities of peaks as a function of ion bias voltage, the authors concluded that As was easily depleted from the surface during etching with ion energy greater than 400 eV. Similarly, they concluded from intensity ratios at 90° that the surface oxide consisted of a mixture of oxides, one containing As and the other not, whose thicknesses depended on ion energy. They confirmed this conclusion by comparing spectra at several take-off angles from 90° to 25°. Over this range, surface sensitivity is highest for 25°, where the escape depth of As (3d) photoelectrons is calculated to be about 5 angstroms. The results (Figure 6) show shoulders on the high binding energy side of both the Ga and the As peaks, indicating oxidized states. The intensities vs. angle indicate a layer of $Ga_2O_3$ over a mixed layer of $Ga_2O_3$ and $As_2O_3$ over the GaAs substrate.

Many GaAs device applications require plasma-etching patterns into W over GaAs with good anisotropy and selectivity and minimal damage to the substrate. The usual W etching gas in Si technology is $CF_4 + O_2$, which has

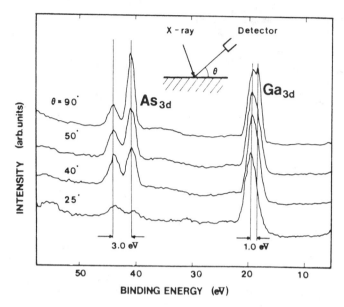

FIGURE 6. Ga (3d) and As (3d) spectra as take-off angle $\theta$ was varied from 90° to 25°. Sample bias voltage was 600 V. (Reproduced with permission from Yabumoto and Oshima (1985).)

the disadvantage here of producing surface oxides on the GaAs. Susa (1985) found that $CF_4 + N_2$ gave satisfactory results and used AES depth profiling to demonstrate that the process left minimal residues and caused no significant changes in substrate stoichiometry.

## 2. Analysis in Model Beam Etching Systems

In order to simulate the complex chemistry of plasma etching in more controlled circumstances, two beam approaches have been introduced. In reactive ion beam etching (RIBE), the ion source is operated with a reactive gas, e.g., $CF_4$, $Cl_2$; both the extracted ion beam and the efflux of reactive neutral species directed upon the substrate (Morgan 1985). Coburn and Winters (1979) introduced the method of ion beam-assisted etching (IBAE), in which a chemically inert ion beam (usually Ar) and a reactive molecular beam are applied simultaneously and independently to the substrate. IBAE has provided considerable insight into reaction mechanisms (Winters, 1985; Winters and Coburn, 1985; Barish et al., 1985); moreover, it shows some promise for becoming a practical process. Geis and coworkers have scaled up IBAE to larger samples, and Gillis and Gignac have examined problems of selectivity between Si and $SiO_2$ in IBAE (Geis et al., 1981; Demeo et al. 1985; Gillis and Gignac, 1986). Because these methods operate at substantially lower pressures than does plasma etching, AES and XPS can be conducted in the same chamber with considerable gain of information. We offer selected examples below.

**a. Si and $SiO_2$.** Coburn, Winters, and Chuang (1977) used AES and XPS to study the results of bombarding a Si surface with $CF_3^+$ produced by operating a standard ion-sputtering gun with $CF_4$. Three topics were considered: composition of the surface layer after bombardment, nature of the chemical bonding between the surface layer and the Si substrate, and the effect of ion energy on the composition and bonding. The implications for plasma etching are summarized in Coburn et al. (1977); details of the electron spectra are presented in Chuang et al. (1978). With the ion beam energy at 500 eV, the surface was quickly covered with C, to the extent that the Si Auger peak at 92 eV was not observable. At a beam energy of 3000 eV, a "steady-state" coverage was produced which showed Auger peaks for both Si and C. This was attributed to sputtering of the residual C layer. The etch rate, measured by a quartz crystal microbalance, was essentially independent of ion dose at the higher beam energies, but fell off with increasing dose at the lower beam energies. This suggested that the accumulated C layer impedes the etch rate. The authors examined the surface chemical bonding using both AES line-shape analysis on Si ($L_3M_{2,3}M_{2,3}$) and XPS chemical shifts for C (1s), Si (2p) and F (1s). The results showed that at low C coverage, all the F is bonded to Si. At full coverage, the F is bonded mainly to C. Therefore, the C layer serves as a blocking agent to protect Si from etching. Partial oxidation of the clean Si surface (by in situ bombardment with $O_2^+$) caused substantial reduction in the C accumulation until all the O was used up and the situation reverted to that described above. The authors speculated that on a film of $SiO_2$, oxygen in the lattice would prevent accumulation of C, allowing $SiO_2$ to be etched readily by F. These early experiments form the basis of the empirical procedure for improving the selectivity of etching $SiO_2$ relative to Si by increasing the C accumulation, already discussed in Section 1.a above.

Similar experiments were carried out on thermally grown $SiO_2$ on Si by Thomson and Helms (1985). For $CF_3^+$ beams at energy greater than 750 eV, they saw little C accumulation on the $SiO_2$, which they attributed to a rapid reaction of C and O. However, this may in fact be due to increased physical sputtering yield with ion energy, especially since the ion beam was incident at 49°, which is optimum for sputtering. Thompson and Helms did not report comparisons of $Ar^+$ and $CF_3^+$, as did the previous authors, to separate physical sputtering effects from etching. By examination of the Si (LVV) Auger peak, Thompson and Helms concluded that during the etching, the $SiO_2$ surface layer has Si atoms of reduced coordination, which they attributed to Si—F bonds. In the absence of complementary XPS data, this is a difficult interpretation. The deficiency in O here may be an example of the chemical reduction of oxide films by impact of energetic ions (Czanderna and Pitts, 1986).

Fundamental understanding of plasma etching mechanisms requires separation and analysis of the three basic kinetic steps (adsorption of reactant, formation of product, and removal of product), any one of which may be rate-limiting as conditions change. Chuang (1980) initiated studies of chemisorption of F on Si by using $XeF_2$, which undergoes dissociative chemisorption on surfaces and liberates Xe. Using a commercial electron spectrometer with relatively low resolution (1.5 eV), he concluded that the surface was covered by "$SiF_2$-like" species. McFeely and coworkers have taken this work further by using synchrotron radiation (for greater surface sensitivity, as explained in Section A), well-cleaned and characterized surfaces, and very careful deconvolution of experimental spectra (McFeely et al., 1984 and 1986; Morar et al., 1984; Shinn et al., 1984). They have shown that $SiF_x$ ($x = 1, 2, 3, 4$) species appear on the surface and exhibit chemical shifts of approximately 1, 2, 3, and 4 eV, respectively. The relative quantities of these species, but not the degree of chemical shift, are affected by the surface restructuring of the Si lattice. Ion bombardment of the $XeF_2$-dosed surfaces has been shown to influence primarily the amount of $SiF_3$ on the surface while stimulating the desorption of $SiF_4$ (McFeely and Yarmoff, 1986). These very powerful experimental techniques are expected to provide considerable insight into etching mechanisms under carefully controlled conditions.

Stinespring and Freedman (1986) have developed a UHV-compatible microwave discharge source for producing beams of F atoms. With it they have compared the chemisorption of $F_2$ and F on Si (111) using XPS. They find that $F_2$ undergoes dissociative chemisorption to produce $SiF_2$-like species, whereas F atoms are taken up to a much greater extent and lead to $SiF_4$-like species. These results suggest that diffusion of the relatively small F atoms into the Si lattice, predicted by the calculations of Seel and Bagus (1983), is important in etching environments rich in F atoms.

**b. III–V Compounds.** Barker, Mayer, and Burton (1982) have studied the etching of GaAs and InP using a Kaufman-type (broad-beam) ion source operated with $Cl_2$. Mass spectrometry showed the composition of the ion

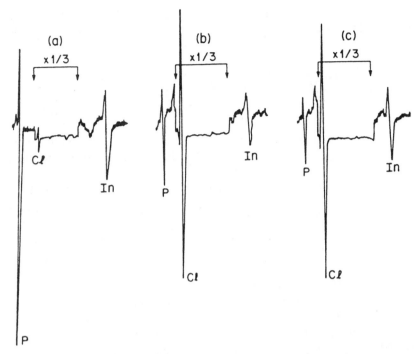

FIGURE 7. Auger spectra for InP (a) after cleaning with $Ar^+$; (b) after exposure to $5 \times 10^{-5}$ torr $Cl_2$; (c) after etching with 500 eV $Cl^+/Cl_2^+$ beam. (Reproduced with permission from Barker et al. (1982b).)

beam to be about 75% $Cl^+$ and 25% $Cl_2^+$. Auger spectra were recorded in situ before and after etching using a differentially-pumped spherical sector analyzer. Results are reproduced in Figure 7 for an InP sample after (a) cleaning by $Ar^+$ bombardment, (b) exposure to $5 \times 10^{-5}$ torr $Cl_2$ (no ion beam), and (c) etching with the $Cl^+/Cl_2^+$ ion beam at 500 eV for 5 minutes. The adsorption of a large amount of $Cl_2$ on the cleaned surface reduced the Auger signals from both In and P, with the greater reduction for P. This is due to electron scattering and perhaps to P depletion as well. After etching, the amount of Cl on the surface was the same as before, but the In/P ratio had grown from 0.56 to 1.19 through increase of In and decrease of P signals. This demonstrated that a surface layer enriched in In and depleted in P had been formed during etching, presumably because of the much lower volatility of $InCl_3$ than $PCl_3$. Similar experiments on GaAs gave no evidence of Ga enrichment after etching. Rate measurements as etching conditions were changed showed that if sufficient reactants are available, removal of the group III chlorides by either thermal or ion-stimulated desorption is the rate-limiting step.

c. Al. Smith and Saviano (1982) described a system in which a small plasma discharge was sampled through an orifice and directed at a target in UHV. This "plasma beam" contained both the ions and the neutral reactive species from a plasma operated at relatively high pressure (0.4 torr). Mass spectrometers permitted analysis of both beam species and volatile etching products leaving the sample during etching. In situ XPS enabled determination of surface composition of the sample. This experiment differs from other reported beam work in that XPS was done during etching, as well as before and after etching, and demonstrates that XPS can indeed be successfully employed in the inhospitable environment of $10^{-5}$ torr of $Cl_2$ that was present during beam operation. The windowed x-ray tube (Vacuum Generators Corp.) of this XPS was differentially pumped to prevent cathode contamination and consequent loss of the Al-K$\alpha$ emission lines; but the electron spectrometer could not be so pumped, so its channel electron multiplier remained the most vulnerable component of the system. In later unpublished work, Smith found that multiplier gain was more sensitive to $CCl_4$ and dropped noticeably during a run, although not so much as to preclude pulse counting.

Among several phenomena explored in this system, one observation, namely that oxidized Al etches only with ion bombardment while clean Al etches without ion bombardment, is of particular interest in the present discussion. Results reproduced in Figure 8 show simultaneous observation of the O (1s) XPS signal and the etching product $AlCl_3$ (ionized to $AlCl_2^+$ in the mass spectrometer) during the bombardment of an oxidized Al sample by 500 eV ions from a $Cl_2$ plasma. These results show that the etch rate increases as the native surface oxide layer is sputtered from the sample. The gradual nature of this transition is due to the uneven distribution of ions over the target. The long time scale (hours) required to clear the oxide compared to the seconds typical in commercial etchers is a consequence of the factor of $10^4$ pressure drop from plasma to target that is required to operate the mass spectrometer and the XPS. This is the price paid for real-time analytical access to the surface. Such access also offers the potential for XPS observation of transient surface species such as intermediaries in plasma-surface reactions.

Park, Rathbun, and Rhodin (1985) studied the effect of $Ar^+$ bombardment on the etching of Al by molecular beams of $Cl_2$, $Br_2$, $CCl_4$, and $CBr_4$ in UHV. Clean Al films etched spontaneously in both $Cl_2$ and $Br_2$. The Al (2p) XPS signal showed small, broad peaks due to chlorinated Al species at

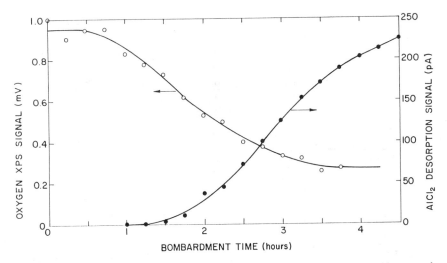

FIGURE 8. Variation in the surface O and etching product signals during oxide removal phase of Al etching in a $Cl_2$ plasma beam with ion energy 400–500 eV. (Reproduced with permission from Smith and Saviano (1982).)

higher binding energies than for the clean metal. The peaks were very similar with and without $Ar^+$ bombardment; intensities and binding energies were constant over the range of exposure and etch rates. Significant reductions in etch rate occurred when $Ar^+$ and $CCl_4$ (or $CBr_4$) were applied simultaneously. Only one C (1s) peak was seen, at 281.8 eV, showing that no partially-dissociated halocarbon species remain on the surface and that a carbide-like layer is probably formed. At saturation coverage with $CCl_4$, ion bombardment caused increase in C (1s) but a significant decrease in Cl (2p) with no shifts in binding energies. The halogenated Al species are easily removed by ion bombardment, but the carbide-like species are not.

The results of both of the above studies suggest that C or O impurities in chlorocarbon plasmas can cause anisotropic etching of Al by depositing upon sidewalls where, not being removed by ion bombardment, they block the spontaneous etching by halogen species.

### 3. Analysis by Vacuum Transfer

**a. Si and $SiO_2$.** Vossen et al. (1983) described a research system in which a small-scale rf plasma etching system (operated in ranges of parameters typical of production systems) is attached via transfer mechanisms to a UHV chamber equipped with commercial components for AES and XPS. This system enables wafers to be analyzed before and after realistic plasma processing without exposure to air. Results were presented for a film of $SiO_2/Si$ etched in $CF_4$. By examination of C (1s), Si (2s), Si (2p), O (1s), and F (1s) peaks, the authors concluded that the process had removed all the oxide and produced a F-rich surface layer in which F was bound to Si. In an attempt to remove this residue, the authors subsequently etched the sample in H plasma and again analyzed the surface. Three changes appeared in the spectrum: 1) O (1s) increased dramatically; 2) Si (2p) gained an additional small peak at 104 eV, which is 5 eV higher than that for elemental Si; and 3) F (1s) was shifted to higher binding energy. The authors attributed these changes to a F-O species attached to the Si and formed by etching residue materials freed in the chamber by the H plasma. This surface species was found to persist after exposure to air. Thus, the vacuum transfer technique allows determination of O contamination arising from etching by avoiding air exposure before analysis.

Suzuki et al. (1984) have developed an alternative to rf plasma etching which relies upon an electron cyclotron resonance (ECR) plasma operating at 2.45 GHz. In this case there is no cathode acceleration applied to the ions; they experience only the "plasma sheath" field which builds up between plasma and sample. For pressures around $10^{-3}$ torr, the ions arrive with an energy of around 20 eV, high enough to activate vertical etching and produce good anisotropy in Si, but low enough so that there is minimal damage and contamination. The selectivity is very high for Si etching relative to $SiO_2$ under these conditions in $SF_6$. When it is desired to etch $SiO_2$, a bias voltage is applied to the sample to increase the energy of incident ions (Suzuki et al., 1985). Ninomiya et al. (1984) have described a system in which an ECR plasma reactor is attached with sample transfer mechanism to a UHV system for AES and XPS. Results were presented for Si and $SiO_2$ etched in $SF_6$. The objectives were to identify etching residues and to see how these changed upon exposure to air. On the freshly etched Si surface, peaks were found for Si, S, F, C, and O. C and O were attributed to background gas in the reactor. No evidence of $SiO_2$ was seen in Si (2p). After standing in UHV for 24 hours, the surface composition showed no significant changes. After exposure to air for 20 minutes, the S (2p) peak decreased, perhaps due to formation of volatile oxides of S. The Si (2p) and O (1s) peaks shift to the values characteristic of $SiO_2$. After 18 days in the air, these changes were somewhat more pronounced, but no new features were found. These results clearly demonstrate that a brief exposure to room air alters the composition of a surface sufficiently that the identification of etching residues and after-effects is severely compromised. Results for $SiO_2$

were similar, except that C and S did not appear on the freshly etched surface, presumably because they formed volatile compounds with the oxygen in the film.

**b. GaAs and AlGaAs.** Microwave ion sources based on ECR plasmas have been used for RIBE in Japan. They are reported to be superior to Kaufman sources in this application by having greater efficiency, more uniform plasma, and no hot filaments (Asakawa and Sugata, 1985). One such system operates at UHV and is attached via sample transfer mechanisms to an AES system (Asakawa and Sugata, 1986). Results have been reported for RIBE of GaAs and AlGaAs in $Cl_2$, AES being used to monitor Cl residue after etching and to check the effectiveness of various in situ cleaning procedures for removal of O and C. It was shown that RIBE at room temperature left a substantial Cl residue, which could be removed by heating at 400°C for 10 minutes. On the other hand, RIBE at 200°C suppressed the Cl contamination. The authors used a separate ECR plasma source to create a beam of H or Cl radicals used in a novel way to clean wafer surfaces. They found by AES that heating to 400°C and exposing to

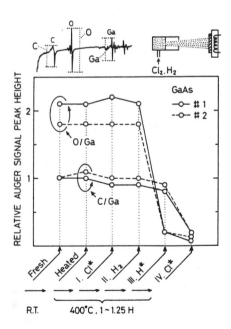

FIGURE 9.   Change of relative O and C AES peak heights on a GaAs wafer put through the designated cleaning steps. (Reproduced with permission from Asakawa and Sugata (1986).)

$H_2$ and to Cl radicals did not remove surface O and C. However, exposure to H radicals removed the O, after which exposure to Cl radicals removed the C to give a clean surface. The steps in this sequence for two wafers are reproduced here in Figure 9. At each step the wafer was held at 400°C. It is unlikely that this situation could have been understood without in situ surface analysis.


D.  FUTURE PROSPECTS

One can confidently expect to see continued application of AES and XPS to plasma etched surfaces in each of the three modes described here. However, two broad directions are expected to dominate. First, one can expect to see many more systems with electron spectrometers attached to realistic etching systems. The greatest need is for analysis of damage and contamination, and it has been clearly shown that air exposure compromises the results. AES will be less useful than XPS because the primary electron beam stimulates the desorption of many of the species of interest in etching chemistry. Shallow depth profiling will be done by take-off angle variation instead of ion sputtering because it is nondestructive. Thus, the further development of accurate models for quantitative analysis and for depth distributions, such as those introduced by Yabumoto and Oshima (1985) and by Ninomiya et al. (1984) is required. Second, the use of synchrotron radiation in XPS is expected to provide considerable mechanistic detail, through its unique power to maximize surface sensitivity. Further studies in the direction started by McFeely et al. (1984) will reveal the detailed dependence of initial halogenation on structural features. Coupled with thermal and ion-stimulated desorption studies, these methods will show the complete evolution of the reacted surface to volatile products. It will be especially exciting to see these methods extended from Si to dielectrics and compound semiconductors.


## III. Ion Spectroscopy

The two ion spectroscopic techniques to be discussed here, SIMS and ISS, are complementary analysis techniques that, in many ways, are ideally suited to characterization of surfaces treated in plasma environments. Both use energetic ions as a probe of surface composition and structure, and both give information primarily from the first few atomic layers of material (Buck, 1975; Feldman, 1986). The interactions of plasmas with surfaces, in a variety of environments from plasma fusion reactors to semiconductor material processing, is dominated by energetic particle bombardment of the surface. Surface erosion by sputtering or ion-assisted chemical etching, and surface compositional and structural modification, are primarily due to ion bombardment in the 0.1–10 keV regime.

For dynamic measurements of ion-surface interactions, SIMS and ISS are ideally suited since the analytical signals (secondary ion ejection and primary ion scattering) are generated as a natural consequence of the phenomenon to be studied. In this sense, the techniques are nonperturbing, and all that is required is to collect the signal that already exists. In applications to plasma processing, however, the plasma environment is not compatible with SIMS or ISS analysis. Thus, most dynamic measurements of ion–surface interactions have been performed in high or ultrahigh vacuum simulations of plasma process conditions. SIMS and ISS have also been applied in the more conventional remote mode to surface analysis and compositional depth profiling of materials exposed to plasma environments. In these applications, analysis is done separately from treatment, and one must consider the perturbing effect of the analysis beam. We will briefly discuss the fundamental nature of the two processes, followed by a number of examples of applications to plasma processing of materials.

## A.  SECONDARY ION MASS SPECTROSCOPY

Secondary ions are ejected from a surface during energetic particle impact as a result of a number of elastic and inelastic scattering processes. During an ion impact event, shown schematically in Figure 10, momentum transfer occurs both by elastic scattering of the primary ion with atoms of the solid and by recoiling atoms with other atoms; this imparts kinetic energy to a large number of atoms in the near-surface regions of the solid in a process known as the collision cascade. A small number of atoms, primarily in the first or second atomic layers, can acquire sufficient momentum perpendicular to the surface, and sufficient energy to overcome their binding forces, to be ejected from the surface. The yield of these sputtered atoms depends primarily on the mass and energy of the primary ion, the mass of the target, and the surface binding energy of the target atoms. For medium mass particles and energy in the keV range, sputter yields are generally on the order of 0.1–10 atoms/ion (Behrisch, 1981). In addition to atoms, a number of polyatomic clusters or molecular particles are also ejected from the surface.

A small fraction of these sputtered particles are ionized, either positively or negatively, by inelastic processes occurring during the collision cascade. Since ions can be collected and mass analyzed with great efficiency and

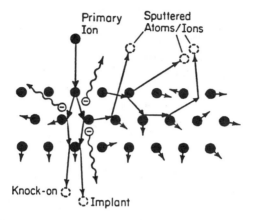

FIGURE 10. Schematic diagram of the "collision cascade," including sputtered atom and secondary ion ejection resulting from primary ion impact.

since they originate from a depth of usually < 5 angstroms, their measurement forms the basis of an extremely sensitive surface compositional probe with detection sensitivity to most elements of 1 ppb–1 ppm. Continuous measurement of secondary ion intensities as the surface is eroded by sputtering constitutes the most sensitive compositional depth profiling technique available.

Instrumentation required for SIMS is shown schematically in Figure 11. A beam of energetic ions is directed onto the sample in high or ultrahigh vacuum. A separate reactive gas jet may also be directed at the sample. Secondary ions sputtered from the sample are collected, energy-selected by an electrostatic filter and mass analyzed by either a quadrupole or magnetic

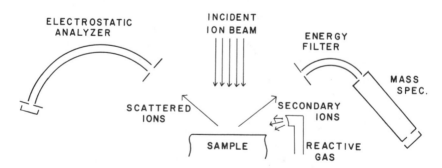

FIGURE 11. Elements of SIMS and ISS instrumentation, including ion source, reactive gas inlet, sample, energy filter, mass spectrometer, and electrostatic energy analyzer.

mass analyzer. See Figure 4 and accompanying text for additional discussion of electrostatic filters.

The yield of secondary ions (ions per incident primary ion) $S_A^+$, for element $A$ in a sample is given by

$$S_A^+ = \gamma_A C_A S, \tag{1}$$

where $S$ is the total sputter atom yield, $C_A$ is the atomic concentration of $A$, and $\gamma_A$ is the ratio of ionized particles of $A$ (positive or negative) to the total number of $A$ particles that are sputtered. The ratio of ionized to neutral sputtered particles, $\gamma_A$, is usually very dependent on the chemical nature of the surface, making quantitative interpretation of SIMS measurements quite difficult. It is beyond the scope of this article to review the various models of charge exchange at the surface that cause this effect, but the reader is referred to recent reviews (Oechsner and Zroubek, 1983; Yu, 1986). It is usually observed that positive ion yields are enhanced in the presence of electronegative gases such as $O_2$ or $Cl_2$, or by sputtering with electronegative ions such as $O_2^+$, while negative ion yields are enhanced by sputtering with low ionization potential ions, such as $Cs^+$. Accurate quantitative measurements of material composition require sputtering in a steady-state (unchanging) surface environment, and comparison of ion yields from a sample to a calibrated standard.

Other major quantitative problems with SIMS arise from ion mixing caused by disruption of the surface by ion impact, and from development of significant surface topography after high fluence ion bombardment. Ion mixing results in a general compositional scrambling of material over the range of the ion penetration into the solid. In the 1–10 keV regime, ion projected ranges are typically 10–100 angstroms. Compositional depth profiles are thus broadened, giving typical depth resolution of approximately 25 angstroms, even though the sputtered species originate from only the first few angstroms of material (Williams and Baker, 1981; Magee and Honig, 1982). Surface topography is generated from local variations in sputter yield due to composition and ion angle of incidence variations on a heterogeneous sample (Carter et al., 1983). Uneven surface erosion leads to substantial distortion of depth profiles (Auciello, 1984).

Both the variation in ion yield and ion mixing effects make quantitative analysis of surfaces and interfaces a particularly vexing problem. Steady-state surface conditions are not achieved until a dose of $> 10^{15}$ ions/cm$^2$ has accumulated, at which point ion mixing is extensive and the surface has been significantly eroded. Thus, quantitative depth profiling of the first 25 angstroms or so of a surface by SIMS is difficult.

By using a scanned focused ion beam or two-dimensionally imaging secondary ion optics, a two-dimensional SIMS analysis of a surface can be performed with submicron lateral resolution (Castaing and Zlodzion, 1962; Rudenauer et al., 1982). Combined with depth profiling, a multi-element three-dimensional composition profile of complex structures can be obtained (Furman and Morrison, 1980; Bryan et al., 1985).

### B. ION SCATTERING SPECTROSCOPY

Ion scattering spectroscopy observes the fraction of primary ions that are elastically scattered from the outermost surface layer of the solid. From conservation of energy and momentum, it is easy to show that the mass of the target atom can be obtained by measuring the energy of the scattered ion. The kinematics of the scattering process is shown in Figure 12, and the scattered ion energy is given by

$$\frac{E_1}{E_0} = \left[ \frac{\cos \theta + \left( A^2 - \sin^2 \theta \right)^{1/2}}{A + 1} \right]^2 \qquad (2)$$

for $A = m_2/m_1 > 1$ and $\theta > 90°$. Light ions, such as $He^+$ and $Ne^+$ are typically used as probes, since the probe ions must have lower mass than the target atoms.

Instrumentation for ISS is also shown schematically in Figure 11. Energy analysis of scattered ions is usually accomplished with a spherical sector analyzer located at a fixed angle with respect to the ion beam. Cylindrical mirror analyzers with a coaxial ion beam have also been used (Sparrow, 1976). Use of a pulsed ion beam allows time-of-flight analysis of scattered ion velocities (Chen et al., 1977).

The kinematics of ion scattering in ISS are identical to the more well known technique of Rutherford backscattering spectroscopy (RBS) (see chapter by Doyle and Chu; also Chu et al., 1978). In contrast to RBS, which usually employs MeV $He^+$ as probe particles, ISS typically employs

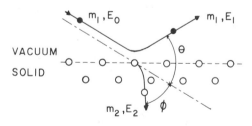

FIGURE 12.    Kinematics of elastic scattering from surfaces. m = particle mass; E = energy.

low energy ( < 10 keV) ions. This results in a much shorter range of probe ions into the solid such that scattering is confined to the first few atomic layers.

The chief advantage of ISS lies in its high degree of surface selectivity. This is a result of the high probability that an ion penetrating the first atomic layer will be neutralized via inelastic scattering processes and thus not observed. Since the energy analyzers typically employed will only detect ions, particles that have been neutralized are not observed. The observed signal then is due primarily to binary elastic collisions with the topmost atomic layer. The intensity of the ions scattered from $A$ atoms on the surface is given by

$$I_A = I_0 N_A \sigma_A (1 - P_n) \tag{3}$$

where $I_0$ is the primary beam intensity, $N_A$ is surface density of $A$ atoms, $\sigma_A$ is the scattering cross section of the primary ion with $A$ and $P_n$ is the neutralization probability.

Although $\sigma_A$ can be calculated with good accuracy for $E_0 > 1$ keV, $P_n$ is not generally known, so that absolute measurements require standards. In contrast to SIMS, however, $P_n$ does not appear to be a strong function of surface composition, making relative composition measurements much more reliable (Taglauer et al., 1979). Because $P_n$ is almost unity for rare gas ions —.999 for He (Taglauer, 1980; Overbury et al., 1981)—the sensitivity of the technique is much less than that of SIMS. Typical minimum detectable surface concentrations are about 0.01 monolayer. Greater sensitivity can be achieved using ions with lower $P_n$, such as $Li^+$ or other alkalis (Taglauer, 1980; Overbury et al., 1981), but observation of multiple scattering events become dominant, complicating the analysis.

Although ISS is a very surface-specific measurement, it suffers, as does SIMS, from compositional mixing and surface roughening effects due to ion bombardment. Consequently, depth profiling by sputter erosion has depth resolution comparable to SIMS. Ion mixing effects can be seen quite readily in sputtering of adsorbed gases (Taglauer et al., 1979; Barish et al., 1985). Nevertheless, ISS using $He^+$ ions is less destructive than SIMS using $Ar^+$ because of the reduced momentum transfer from the probe beam. This means both a lower sputter-erosion rate and a higher dose threshold before mixing effects become apparent. Alternatively, ion dose can be reduced by using alkali ions, as discussed above.

Geometrical shadowing and blocking of ion trajectories can also be exploited for detailed measurements of surface geometry (Aono et al., 1981; Niehus and Comsa, 1984), defect structures (Aono et al., 1983), and atomic vibrational amplitudes (Souda et al., 1983) on crystalline surfaces. Observation of light atoms on surfaces, particularly H, is difficult due to small cross sections and unfavorable kinematics (see Eq. 2). However, H and other light atoms have been successfully measured by forward recoil scattering (Chen et al., 1977).

## C. Applications in Plasma-Surface Interactions

### 1. In situ SIMS

SIMS has been applied to in situ investigations of ion assisted etching processes in a number of instances. In this application, SIMS can provide information on three aspects of the surface etching process. 1) SIMS primarily samples the population of etch products that are ejected via collisional sputtering-type mechanisms. Thus, it can give insight into the type of chemical species formed and ejected from surfaces during a collision cascade event. 2) With proper calibration, SIMS can be used as a quantitative measure of surface concentration of a particular species. 3) Changes in the surface chemical or electronic structure are reflected in the variation in secondary ion yields.

Examples of the first and third aspects above are shown in Figures 13 and 14, which show secondary ions emitted from Si and $SiO_2$ surfaces during etching with $Cl_2$ and fluorocarbon gases (Barish et al., 1985; Miyamura et al., 1982). Figure 13 shows secondary ion intensities observed while sputtering Si with 1 keV $Ne^+$ upon addition of varying $Cl_2$ pressure. Exposure of the surface to $Cl_2$ leads to formation and ejection of a variety of $SiCl_x^+$ ions, as well as an increase in the yield of $Si^+$. The intensity of all Si-containing ions becomes constant at $Cl_2$ pressure above $2 \times 10^{-7}$ torr, at which point the surface is nearly covered with a monolayer of adsorbed Cl. In the submonolayer regime, the intensities of all ions increase both due to changing ionization/neutralization probability (as for $Si^+$) and due to formation of new surface species.

Independent measurements of Cl surface coverage for these experiments (by ISS) and neutral product ejection under similar conditions (Kolfschoten et al., 1984) allow qualitative conclusions to be made about formation and removal of etch products and production of secondary ions. The strong enhancement of $Si^+$ yield upon Cl adsorption can be understood in terms of the bond-breaking model of chemical enhancement of secondary ion yields (Yu, 1986; Oechsner and Zroubek, 1983). In this view, the ionization probability of a Si-containing ion should be related to the number of Cl atoms bonded to it on the surface. Then the $Si^+$ intensity will be the sum of contributions from different bonding configurations involving Si ($SiCl_x$,

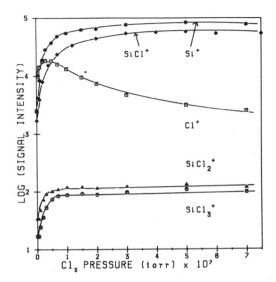

FIGURE 13. Positive secondary ions observed from sputtering Si by 1 KeV $Ne^+$ in the presence of $Cl_2$ gas. (Reproduced with permission from Barish et al. (1985).)

$x = 0$ to 4); that is,

$$I_{Si} = \sum_{x=0}^{4} f_x Y_x P_x^+, \qquad (4)$$

where $f_x$ is the fraction of surface Si bonded to $x$ chlorine atoms, $Y_x$ is the partial sputtering yield of Si from a $SiCl_x$ surface species, and $P_x^+$ is the ionization probability of Si emitted from $SiCl_x$. Similar intensity expressions for $SiCl_x^+$ ions can be written.

The coverage dependences of $Si^+$ and $SiCl^+$ yields are similar to each other and also large for coverages between 0.05 and 1 monolayer. This suggests that the ionization probability is similar for both ions, that they originate from similar bonding environments, and that changes in their intensities reflect changes in the relative abundance ($f_x$) of surface species responsible for their production.

The $SiCl_2^+$ and $SiCl_3^+$ intensities exhibit similar coverage dependence but are markedly weaker than $Si^+$ and $SiCl^+$. Again, this suggests that they originate from similar bonding configurations and hence have similar ionization probability. Their intensities, therefore, reflect changes in the relative abundance of higher $SiCl_x$ bonding configurations.

The relative intensities of $SiCl^+$ vs. $SiCl_2^+$ and $SiCl_3^+$ are in accord with observations of sputtered neutral species, where SiCl is the dominant

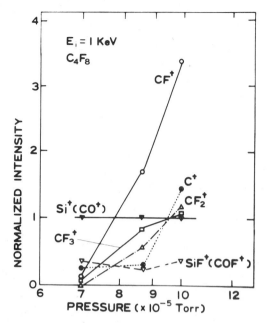

FIGURE 14.   Secondary ions emitted from a Si surface being etched by ions from a $C_4F_8$ discharge. (Reproduced with permission from Miyamura et al. (1982).)

product at submonolayer coverage (Kolfschoten et al., 1984). Together, these measurements suggest that the dominant Cl surface bonding environment involves only one Cl atom bound per Si atom.

$SiCl^+$ is observed to be the major $SiCl_x$ ion at much higher $Cl_2$ exposures as well, even though the more highly chlorinated neutral species are observed in increasing abundance. This observation emphasizes the fact that secondary ions reflect only the population of species that are collisionally ejected from the surface with relatively high energy. For this population, it is unlikely that large numbers of polyatomic ions will be observed in any instance. And of course, SIMS gives no direct information concerning the nature of species emitted by thermal or recombination processes.

Secondary ions emitted from a $SiO_2$ surface during exposure to an ion beam and neutral flux from a $C_4F_8$ plasma are shown in Figure 14 (Miyamura et al., 1982). In addition to species originating from the target ($Si^+$, $SiF^+$), a number of C-containing ions are observed, due to adsorption and accumulation of fluorocarbon species on the surface. It is difficult to identify the chemical species responsible for these ions, but it is well established that exposure of Si and $SiO_2$ to $CF_x$ ion and neutral fluxes results in the deposition of a thin fluorocarbon film which eventually passivates the Si surface (Coburn et al., 1977).

A number of attempts have been made to quantify secondary ion emission for measurements of surface coverage and sputter yields in ion-assisted etching processes. Si etching by $F_2$ was investigated (Knabbe et al., 1982), and $SiF^+$ yield was calibrated as a function of total surface $F$ concentration by independent measurements of coverage with a quartz crystal microbalance. An empirical calibration was found whereby $I_{SiF} = (N_F)^a$, with the exponent $a$ about equal to 2 over most of the submono-layer regime, shown in Figure 15. Using this calibration, the sputter-removal cross section for F from the surface was measured from the decay of the $SiF^+$ signal after irradiating a dosed surface with energetic $Ar^+$. The initial slope of the $SiF^+$ decay curve is related to the removal cross section

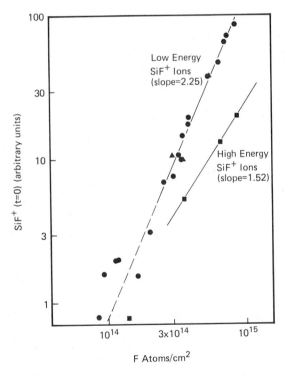

FIGURE 15. $SiF^+$ secondary ion intensity as a function of surface F atom coverage on Si; sputtered by 1 KeV $Ar^+$. (Reproduced with permission from Knabbe et al. (1982).)

by

$$\sigma_F = \frac{1}{aI_0} \cdot \frac{d \ln(I_{\text{SiF}})}{dt}, \tag{5}$$

where $I_0$ is the $\text{Ar}^+$ current density. Measurements of $\sigma_F$ over a range of submonolayer concentrations showed an increase of a factor of 5 from 0.05 to 1 monolayer, perhaps suggesting a marked change in $\text{Si}-\text{F}$ bonding environment over this range.

Similar measurements were made on chlorinated Si surfaces (Barish et al., 1985); the $\text{SiCl}^+$ signal was calibrated by independent measurements of surface coverage by ISS (see Section C.3). Cross section measurements for Cl removal as a function of coverage and ion angle of incidence were made in a fashion similar to Knabbe et al. (1982). Barish et al. (1985) also observed that $\text{SiCl}^+$ yield was sensitive to prior history of surface treatment. Cl previously implanted or ion-mixed into subsurface layers noticeably enhances ion yields during subsequent measurements, complicating the measurements.

In situ SIMS measurements of this type are a convenient and sensitive way to observe surface reactions and composition changes. Quantitative measurements of sputter cross sections of adsorbed gases are possible, provided careful calibration of coverage is performed by a complementary technique. SIMS has advantages over AES in that the surface is not altered by the probe electron beam, a substantial problem for halogens because of electron-stimulated desorption. Secondary ions are produced as a natural consequence of the ion-surface interaction in a plasma or ion beam environment. Interpretation of secondary ion yields in terms of detailed surface chemical structure remains the major problem. Without complementary independent measurements, it is thus far impossible to separate the effects of surface structure, sputter yields, and ionization probability.

## 2. Remote SIMS

SIMS also has been applied to the analysis of surfaces after exposure to plasma environments in the more conventional remote mode using sputter depth profiling. The primary interest is measurement of the extent of surface and subsurface modification by the plasma. Since the depth of such surface modification is usually only 10–50 angstroms, depth profiling by SIMS typically yields only qualitative results, as outlined earlier. This is because yields of all ions change dramatically until the sputtering surface reaches a steady state and because ion mixing and roughening broaden

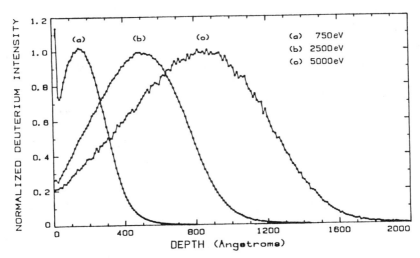

FIGURE 16. SIMS depth profiles of D implanted into Si at 750, 2500, and 5000 eV. (Reproduced with permission from Magee et al. (1980).)

depth profiles severely. Nevertheless, SIMS has been used frequently to measure compositional modifications, such as metal contamination in Si due to sputtering of apparatus parts (Gildenblat, 1983), Cl incorporation by ion mixing into Si targets (Mizutani et al., 1985), and F incorporation into Si after exposure to $CHF_3$ discharge (Wu et al., 1986), to mention just a few examples. In all of these cases, however, secondary ions characteristic of the impurity were observed even after sputter erosion to depths many times the probable original depth profile. Only qualitative assertions about the true depth profile and concentrations are possible.

The most advantageous use of SIMS in this application has been for the measurement of H and D implantation into metal targets, important in plasma-wall interactions in fusion reactors. In this instance, the range of $H^+$ of $D^+$ ions into the solid is sufficiently greater than the depth resolution of the SIMS profiling analysis, so that accurate profiles can be obtained. Ultraclean vacuum conditions in the SIMS instrument must be maintained for analysis of H, however. Figure 16 (Magee et al., 1980) shows typical depth profiles obtained for D implanted into Si at 750–5000 eV. Since the depth of implantation is so large (> 100 angstroms), effects of ion yield variations at the surface and profile broadening by ion mixing are relatively insignificant. Since there is a lack of alternative convenient methods for measuring H concentrations and profiles in solids, SIMS stands out as a capable, if not ideal, technique.

SIMS measurements of surface modification of fragile materials, such as polymers or other organic molecules, is particularly difficult due to extensive damage to the material caused by ion bombardment. Analysis using very low total dose ($< 10^{14}$ ions/cm$^2$) of probe ions—so called *static SIMS* (Benninghoven, 1970)—can be made such that the surface is not altered substantially during the analysis. This static SIMS technique has been applied to examining modification of polymethyl methacrylate (PMMA) surfaces by oxygen plasmas (Simko et al., 1985a). Again, quantitative measurements are difficult, but correlation of characteristic secondary ions with surface functional groups is possible, yielding a wealth of information on the chemical modification of polymer surfaces.

An application of the three-dimensional imaging capability of SIMS relevant to modification of polymers by radiation is shown in Figure 17 (Simko et al., 1985b). In this case, a film irradiated by 3 keV electrons

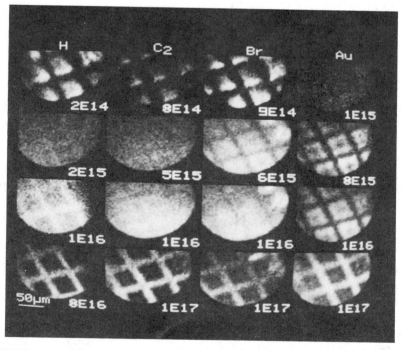

FIGURE 17.   Image depth profiles of electron-bombarded PMMA exposed to Br$_2$. Masses 1 (H$^-$), 24 (C$_2^-$), 79 (Br$^-$), and 197 (Au$^-$) are displayed at various sputter doses (ions/cm$^2$). The PMMA film was spun on a gold-coated substrate, irradiated through a 500 lines/inch mask, and exposed to Br$_2$, which reacts with the irradiated PMMA. (Reproduced with permission from Simko et al. (1985b).)

through a grid mask was exposed to gaseous $Br_2$, which is incorporated into the irradiated areas of PMMA. Spatially resolved Br images and image depth profiles were obtained by SIMS analysis using $Cs^+$ sputtering.

### 3. ISS

ISS has been employed most often in studies of adsorption and sputtering of gases on metals and semiconductors (Taglauer et al., 1974, 1979, 1980; Barish et al., 1984) and of surface layer composition changes in alloys and compounds caused by sputtering (Schwartzfager et al., 1981) or plasma etching (Ameen and Mayer, 1986a). Often it has been used to complement other analysis techniques or as a surface composition reference to study charge exchange or sputtered particle excitation (Taglauer et al., 1979). These measurements exploit the extreme surface selectivity of ISS.

In the combined SIMS–ISS study of Cl adsorption on Si (Barish et al., 1985), ISS was used as an in situ monitor of Cl surface coverage and as a calibration of the secondary ion neutralization probability for $SiCl_x^+$ ions. An example of measurement of steady-state Cl coverage on Si during sputtering by $Ne^+$ is shown in Figure 18. Over the range of 0–1 monolayer coverage, the scattered ion yield does not demonstrate obvious changes in

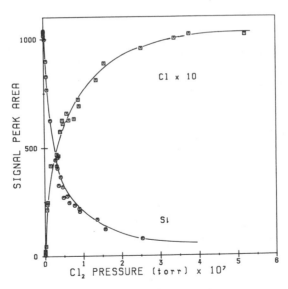

FIGURE 18.    ISS measurement of steady-state Cl coverage on Si during sputtering with $Ne^+$. (Reproduced with permission from Barish et al. (1985).)

neutralization probability, and the surface selectivity is evident from the disappearance of the Si signal due to overlying Cl.

These authors used transient measurements to study the sputter yield of Cl from Si as a function of coverage and of mass and angle of incidence of the primary ion. These studies have complemented other measurements using SIMS, AES (Barker et al., 1982b), sputtered neutral mass spectrometry (Kolfschoten et al., 1984), and Rutherford backscattering spectrometry (Mizutani et al., 1985) to yield a comprehensive understanding of the reaction and etching of Si surfaces with Cl.

ISS has proved indispensible in measurements of near-surface composition changes in alloys and compounds due to segregation, sputtering, and reactive ion etching. Segregation at interfaces is recognized to be a ubiquitous phenomenon, a result of the tendency of surfaces to minimize free energy. For a multicomponent material, segregation leads to an excess surface concentration of one element and rejection of the others into the bulk. The depth of this effect may be only a few atomic layers. Surface segregation, combined with unequal sputtering yields for elements in most multicomponent materials, gives rise to complex compositional profiles in the near-surface region of sputter-eroded materials.

This phenomenon has been overlooked, misinterpreted, and generally difficult to study by the electron spectroscopies (AES, XPS) because of the

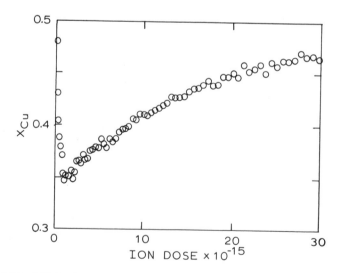

FIGURE 19.  ISS depth profile of an altered layer produced by sputtering a Cu–50% Ni alloy at 200°C. $X_{Cu}$ = atomic fraction Cu. (Reproduced with permission from Schwartzfager et al. (1981).)

relatively long escape depth of the Auger or photoelectron compared to the depth of alteration. The single-layer sensitivity of ISS has allowed quantitative study of surface segregation phenomena in sputtering situations (Brongersma et al., 1978).

Figure 19 shows a depth profile of a sputter-eroded Cu-Ni alloy surface obtained by ISS (Schwartzfager et al., 1981). The excess Cu concentration at the surface is the result of segregation. While excess Cu concentration exists at the surface, Cu is depleted from the subsurface layers as a result of diffusion from the bulk. While the actual segregation driving force (surface chemical potential) acts only over a few atomic layers, transport processes in the bulk (thermal or radiation-enhanced diffusion) result in an altered composition profile which may extend a few tens of angstroms deep.

Near-surface composition changes in III-V semiconductors and metal silicides exposed to plasma etching environments have also been examined by a variety of techniques. GaAs and InP etched in $Cl_2$ have been examined by AES (Barker et al., 1982a) and found to be depleted of the

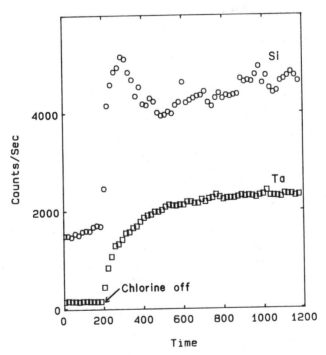

FIGURE 20. $Ne^+$ ISS depth profile of $TaSi_2$ immediately after etching by $Ne^+ + Cl_2$. (Reproduced with permission from Ameen and Mayer (1986b).)

group V element in the near-surface region. ISS measurements on GaAs, however (Ameen and Mayer, 1986a), have shown that As segregates to the surface as a result of Cl adsorption, yielding an excess surface concentration, with depletion in the subsurface layers.

Similar experiments on $TaSi_2$ are shown in Figure 20 (Ameen and Mayer, 1986b). In this case, the sample was sputtered by $Ne^+$ in the presence of $Cl_2$ and then depth-profiled after the $Cl_2$ was removed. Si is seen to have an initially high surface concentration due to segregation and a depletion in the subsurface down to approximately 25 angstroms. SIMS measurements on $TaSi_2$ and $TiSi_2$ under identical conditions yield rather bizzare results because ion yields of elemental and chloride ions are changing dramatically as the degree of surface chlorination changes. At the moment, these results are impossible to understand without more detailed ISS measurements of surface composition.


D. ASSESSMENT AND FUTURE PROSPECTS

The application of standard analysis techniques such as SIMS and ISS to plasma-surface interactions has been a fruitful exercise and will remain so, primarily because the field is as yet virtually unexplored. Already, though, the inherent limitations of the techniques are apparent, and the horizons are in sight. Fortunately, a number of extensions and improvements of these techniques are visible on the horizon, which will push our materials analysis capability well into the future. The most vexing problems in ion beam analysis techniques are the matrix effects, which give such large variation in ion yields with target composition in SIMS, and the ion mixing effects, which limit depth resolution and chemical and physical structure analysis in both SIMS and ISS depth profiling.

A thorough understanding of secondary ion yields does not seem to be forthcoming in the immediate future. General quantitative profiling of surfaces and interfaces by SIMS is thus likely to remain a problem. An attractive alternative is to measure the sputtered neutral flux of material from the surface. Although a more difficult measurement than collection of secondary ions, emission of neutral particles reflects changes only in sputter yields rather than in local chemical environment. Ionization and mass analysis of the sputtered neutral flux can be performed by electron impact in a low-pressure, high-density plasma (Oechsner, 1984) or by resonant (Kimock et al., 1984) or nonresonant (Becker and Gillen, 1984) multiphoton ionization. The capability of these post-ionization techniques is improving rapidly and will likely prove equal to or better than SIMS in absolute sensitivity and much better than SIMS in quantitative application. For example, measurements of sub-ppb quantities of Fe in Si by resonant multiphoton ionization of sputtered Fe atoms have recently been reported (Pellin et al., 1987).

The low sensitivity of ISS is due predominantly to ion surface neutralization, as described earlier. Emerging use of alkali probe ions will give much greater sensitivity and allow analysis with greatly reduced beam doses, thus minimizing the damaging effects of the probe beam on fragile surfaces.

Ion mixing and other near-surface effects associated with ion bombardment, such as surface segregation and enhanced diffusion, are not well understood as yet and are characterized for only a few cases (Kelly, 1985). These effects are the primary contributors to near-surface modification of materials by plasmas, and they are also the main hindrance to our ability to analyze very thin films by ion bombardment techniques. The usual approach to quantitative depth profiling is comparison to standards implanted with known depth distributions of impurities or standards consisting of thin films with known composition and sharp interfaces. These methods are not satisfactory, however, for most surface modification studies, where the total depth of the modified material may be only a few tens of angstroms. A generally acceptable method of quantitatively analyzing such structures is not yet available.

Nondestructive depth profiling techniques are attractive from the point of view of avoiding the ion mixing and roughening effects of sputter erosion. Rutherford backscattering (RBS) accomplishes this by using penetrating, high-energy probe ions ($> 1$ MeV He$^+$); however, depth resolution is rather poor ($> 50$ angstroms). At low energy ($< 5$ KeV), ISS is very surface-specific because of short penetration depth and inelastic neutralization processes. The middle ground between ISS and RBS is beginning to be explored, using medium energy ($< 200$ KeV) ion probes with a shorter range than RBS and electrostatic analyzers with sufficient energy resolution to obtain $< 20$ angstroms depth resolution (Smeenk et al., 1981; Oehrlein et al., 1985a). Alkali ion probes (Li$^+$) at $< 10$ KeV energy offer the possibility of nondestructive probing of depths $< 20$ angstroms if multiple scattering effects and inelastic energy losses in this energy regime can be quantified.

The two-dimensional imaging capability of SIMS will find increasing use in applications related to microstructure fabrication, where compositional variations of submicron lateral dimension are of interest. Focused ion beam probes are now available giving lateral resolution of $< 1000$ angstroms (Levi-Setti et al., 1986).

## References

Ameen, M. S. and Mayer, T. M. (1986a). *J. Appl. Phys.* **59**, 969.
Ameen, M. S. and Mayer, T. M. (1986b), unpublished.

Anthony, M. T. (1983). In: *Practical Surface Analysis* (D. Briggs and M. P. Seah, eds.). Wiley, New York, p. 429.

Aono, M., Oshima, C., Zaima, S., Otani, S., and Ishizawa, Y. (1981). *Jpn. J. Appl. Phys.* **20**, L829.

Aono, M., Hou, Y., Souda, R., Oshima, C., Otani, S., and Ishizawa, Y. (1983). *Phys. Rev. Lett.* **50**, 1293.

Asakawa, K. and Sugata, S. (1985). *J. Vac. Sci. Technol.* **B2**, 402.

Asakawa, K. and Sugata, S. (1986). *J. Vac. Sci. Technol.* **A4**, 677.

Auciello, O. (1984). Chapter 11, In: *Ion Bombardment Modification of Surfaces: Fundamentals and Applications* (O. Auciello and R. Kelly, eds.). Elsevier, Amsterdam.

Barish, E. L., Vitkavage, D. J., and Mayer, T. M. (1985). *J. Appl. Phys.* **57**, 1336.

Barker, R. A., Mayer, T. M., and Burton, Randolph H. (1982a). *Appl. Phys. Lett.* **40**, 583.

Barker, R. A., Mayer, T. M., and Pearson, W. C. (1982b). *J. Vac. Sci. Technol.* **B1**, 33.

Becker, C. H. and Gillen, K. T. (1984). *Anal. Chem.* **56**, 1671.

Behrisch, R. (1981). *Sputtering by Particle Bombardment*. Springer-Verlag, Berlin.

Benninghoven, A. (1970). Z. *Physik* **230**, 403.

Bertrand, P. A. (1981). *J. Vac. Sci. Technol.* **18**, 28.

Briggs, D. and Riviere, J. C. (1983). In: *Practical Surface Analysis* (D. Briggs and M. P. Seah, eds.). Wiley, New York, p. 87.

Briggs, D. and Seah, M. P. (1983). *Practical Surface Analysis*. Wiley, New York.

Brongersma, H. H., Sparnaay, M. J., and Buck, T. M. (1978). *Surf. Sci.* **71**, 657.

Bryan, S. R., Woodward, W. S., Griffis, D. P., and Linton, R. W. (1985). *J. Microsc.* **138**, 15.

Buck, T. M. (1975). In: *Methods of Surface Analysis* (A. W. Czanderna, ed.). Elsevier, Amsterdam, p. 75.

Carlson, T. A. (1976). *Photoelectron and Auger Spectroscopy*. Plenum, New York.

Carter, G., Navinsek, B., Whitton, J. L. (1983). In: *Sputtering by Particle Bombardment II* (R. Behrisch, ed.). Springer, Berlin, p. 231.

Castaing, R. and Slodzion, G. (1962). *J. Microsc.* **1**, 395.

Chaplart, J., Fay, B., and Linh, Nuyen T. (1983). *J. Vac. Sci. Technol.* **B4**, 1050.

Chen, Y. S., Miller, G. L., Robinson, D. A., Wheatley, G. H., and Buck, T. M. (1977). *Surf. Sci.* **62**, 133.

Chu, W. K., Mayer, J. W., and Nicolet, M. A. (1978). *Backscattering Spectrometry*. Academic Press, New York.

Chuang, T. J. (1980). *J. Appl. Phys.* **51**, 2614.

Chuang, T. J., Winters, H. F., and Coburn, J. W. (1978). *Appl. Surf. Sci.* **2**, 514.

Coburn, J. W., Winters, H. F., and Chuang, T. J. (1977). *J. Appl. Phys.* **48**, 3532.

Coburn, J. W. and Winters, H. (1979). *J. Appl. Phys.* **50**, 3189.

Czanderna, A. W. (1975). *Methods of Surface Analysis*. Elsevier, New York.

Czanderna, A. W. and Pitts, J. R. (1986). *Nucl. Instr. Methods* **B13**, 245–249.

Davis, L. E., MacDonald, N. C., Palmberg, P. W., Riach, G. E., and Weber, R. E. (1978). *Handbook of Auger Electron Spectroscopy*. Perkin-Elmer Corp., Eden Prairie, MN.

Demeo, N. L., Donnelly, J. P., O'Donnell, F. J., Geis, M. W., and O'Connor, K. J. (1985). *Nucl. Instr. Methods* **B7**, 814.

Eisberg, R. M. (1961). *Fundamentals of Modern Physics*. Wiley, New York.

Feldman, L. C. and Mayer, J. W. (1986). *Fundamentals of Surface and Thin Film Analysis*. North Holland, New York.

Furman, B. K. and Morrison, G. H. (1980). *Anal. Chem.* **52**, 2305.

Geis, M. W., Lincoln, G. A., Efremow, N., and Piacentini, W. J. (1981). *J. Vac. Sci. Technol.* **19**, 1390.

Gildenblat, G., Health, B. A., and Katz, W. (1983). *J. Appl. Phys.* **54**, 1855.

Gillis, H. P. and Gignac, W. J. (1986). *J. Vac. Sci. Technol.* **A4**, 696.

Karulkar, P. C. and Tran, N. C. (1985). *J.Vac. Sci. Technol.* **B3**, 889.

Kelly, R. (1984). *Surf. Interface Anal.* **7**, 1.

Kimock, F. M., Baxter, J. P., Pappas, D. L., Kobrin, P. H., and Winograd, N. (1984). *Anal. Chem.* **56**, 2782.

Knabbe, E. A., Coburn, J. W., and Kay, E. (1982). *Surf. Sci.* **123**, 427.

Kolfschoten, A. W., Haring, R. A., Haring, A., and DeVries, A. E. (1984). *J. Appl. Phys.* **55**, 3813.

Levi-Setti, R., Wang, Y. L., and Crow, G., (1986). In: *Secondary Ion Mass Spectrometry: SIMS V* (A. Benninghoven, R. J. Colton, D. S. Simons, and H. W. Werner, eds.). Springer, Berlin.

Madey, T. E. (1986). *J. Vac. Sci. and Technol.* **A4**, 257.

Magee, C. W., Cohen, S. A., Voss, D. E., and Brice, D. K. (1980). *Nucl. Instr. Meth.* **168**, 383.

Magee, C. W. and Honig, R. E. (1982). *Surf. Interface Anal.* **4**, 35.

McFeely, F. R., Morar, J. F., Shinn, N. D., Landgren, G., and Himpsel, F. J. (1984). *Phys. Rev.* **B30**, 764.

McFeely, F. R., Morar, J. F., and Himpsel, F. J. (1986). *Surf. Sci.* **165**, 277.

McFeely, F. R. and Yarmoff, J. A. (1986). *Proceedings of ICPS* **18**, Stockholm (to be published).

Miyamura, M., Tsukakoshi, O., and Komiya, S. (1982). *J. Vac. Sci. Technol.* **20**, 986.

Mizutani, T., Dale, C. J., Chu, W. K., and Mayer, T. M. (1985). *Nucl. Instr. Methods* **B7 / 8**, 825.

Morar, J. F., McFeely, F. R., Shinn, N. D., Landgren, G., and Himpsel, F. J. (1984). *Appl. Phys. Lett.* **45**, 174.

Morgan, R. A. (1985). *Plasma Etching in Semiconductor Fabrication.* Elsevier, New York.

Niehus, H. and Comsa, G. (1984). *Surf. Sci.* **140**, 18.

Ninomiya, K., Suzuki, K., Nishimatsu, S., Gotoh, Y., and Okada, O. (1984). *J. Vac. Sci. Technol.* **B2**, 645.

Oechsner, H. and Zroubek, Z. (1983). *Surf. Sci.* **127**, 10.

Oechsner, H. (1984). In: *Thin Film and Depth Profile Analysis* (H. Oechsner, ed.). Springer, Berlin.

Oehrlein, G. S., Tromp, R. M., Tsang, J. C., Lee, Y. H., and Petrillo, E. J. (1982a). *J. Electrochem. Soc.* **132**, 1441.

Oehrlein, G. S., Ransom, C. M., Chakravarti, S. N., and Lee, Y. H. (1985b). *Appl. Phys. Lett.* **46**, 686.

Oshima, M. (1979). *Surf. Sci.* **86**, 858.

Overbury, S. H., Heiland, W., Zehner, D. M., and Datz, S. (1981). *Surf. Sci.* **109**, 239.

Park, S., Rathbun, L. C., and Rhodin, T. N. (1985). *J. Vac. Sci. Technol.* **A3**, 791.

Pellin, M. J., Young, C. E., Calaway, W. F., Burnett, J. W., Jorgensen, B., Schweitzer, E. L., and Gruen, D. M. (1987). *Nucl. Instr. Methods* **B18**, 446.

Rudenauer, F. G., Pollinger, P., Studnicka, H., Gnaser, H., Steger, W., and Higatsburger, M. J. (1982). In: *Secondary Ion Mass Spectrometry: SIMS III* (A. Benninghoven, J. Giber, J. Laszlo, M. Riedel, and H. W. Werner, eds.). Springer, Berlin.

Schwartzfager, D. G., Ziemecki, S. B., and Kelly, M. J. (1981). *J. Vac. Sci. Technol.* **19**, 185.

Seah, M. P. (1983). In: *Practical Surface Analysis* (D. Briggs and M. P. Seah, eds.) Wiley, New York, p. 181.

Seah, M. P. and Dench, W. A. (1979). *Surf. Interface Anal.* **1**, 2.

Seel, M. and Bagus, P. S. (1983). *Phys. Rev.* **B28**, 2023.

Semura, S., Saitoh, H., and Asakawa, K. (1984). *J. Appl. Phys.* **55**, 3131.

Shinn, N. D., Morar, J. F., and McFeely, F. R. (1984). *J. Vac. Sci. Technol.* **A2**, 1593.

Simko, S. J., Griffis, D. P., Murray, R. W., and Linton, R. W. (1985a). *Anal. Chem.* **57**, 137.
Simko, S. J., Bryan, S. R., Griffis, D. P., Murray, R. W., and Linton, R. W. (1985b). *Anal. Chem.* **57**, 1198.
Singer, P. H. (1986). *Semiconductor International*, July and August issues.
Smeenk, R. G., Tromp, R. M., and Saris, F. W. (1981). *Surf. Sci.* **107**, 429.
Smith, D. L. and Saviano, P. G. (1982). *J. Vac. Sci. Technol.* **21**, 768.
Souda, R., Aono, M., Oshima, C., Otani, S., and Ishizawa, Y. (1983). *Surf. Sci.* **128**, L236.
Sparrow, G. R. (1976). *Industrial Res.* **18**, 81.
Stinespring, C. D. and Freedman, A. (1986). *Appl. Phys. Lett.* **48**, 718.
Susa,N. (1985). *J. Electrochem. Soc.* **132**, 2762.
Suzuki, K., Ninomiya, K., Nishimatsu, S. (1984). *Vacuum* **34**, 953.
Suzuki, K., Ninomiya, K., Nishimatsu, S., and Okudaira, S. (1985). *J. Vac. Sci. Technol.* **B3**, 1025.
Swift, P., Shuttleworth, D., and Seah, M. P. (1983). In: *Practical Surface Analysis* (D. Briggs and M. P. Seah, eds.). Wiley, New York, p. 473.
Taglauer, E., Beitat, U., Marin, G., and Heiland, W. (1974). *Appl. Phys. Lett.* **24**, 437.
Taglauer, E., Heiland, W., and MacDonald, R. J. (1979). *Surf. Sci.* **90**, 661.
Taglauer, E. (1980). *Nucl. Instr. Methods* **168**, 751.
Taglauer, E., Englert, W., Heiland, W., and Jackson, D. P. (1980). *Phys. Rev. Lett.* **45**, 740.
Thomson, D. R. and Helms, C. R. (1985). *Appl. Phys. Lett.* **46**, 1103.
Thornton, J. A. (1983). *Thin Solid Films* **107**, 3.
Tuppen, C. G., Heckingbottom, R., Gill, M., Heslop, C., and Davies, G. J. (1984). *Surf and Interface Anal.* **6**, 267.
Vossen, J. L., Thomas, J. H., III., Maa, J.-S., Mesker, O. R., and Fowler, G. D. (1983). *J. Vac. Sci. Technol.* **A1**, 1452.
Wager, J. F. and Wilmsen, C. W. (1982). *J. Appl. Phys.* **53**, 5789.
Wager, J. F. and Wilmsen, C. W. (1985). In: *Physics and Chemistry of III-V Compound Semiconductor Interfaces*. Plenum, New York.
Wagner, C. D., Riggs, W. M., Davis, L. E., Moulder, J. F., and Mullenberg, G. E. (1979). *Handbook of X-ray Photoelectron Spectroscopy*. Perkin-Elmer Corp., Eden Prairie, MN.
Williams, P. and Baker, J. E. (1981). *Nucl. Instr. Methods* **182 / 183**, 15.
Winters, H. F. (1985). *J. Vac. Sci. Technol.* **B3**, 9.
Winters, H. F. and Coburn, J. W. (1985). *J. Vac. Sci. Technol.* **B3**, 1376.
Wu, I.-W., Bruce, R. H., Mikkelsen, J. C. Jr., Street, R. A., Huang, T. Y., and Braun, D. (1986). Paper presented at MRS Spring Meeting, Palo Alto, CA.
Yabumoto, N. and Oshima, M. (1985). *J. Electrochem. Soc.* **132**, 2224.
Yu, M. L. (1986). *Nucl. Instr. Methods* **B** (to be published).

# 3 Spectroscopic Ellipsometry in Plasma Processing

### D. E. Aspnes
*Bell Communications Research, Inc.*
*Red Bank, New Jersey*

### R. P. H. Chang
*Materials Science and Engineering Department*
*Northwestern University*
*Evanston, Illinois*

PLASMA DIAGNOSTICS
Surface Analysis and Interactions

## I. Introduction

### A. PERSPECTIVE

In this chapter we discuss methods and applications of spectroscopic ellipsometry (SE) to the study of surfaces, interfaces, and materials of interest to plasma processing. Ellipsometry is the optical characterization technique in which the change in polarization state of light that occurs upon non-normal-incident reflection from a specular surface of a sample is measured and interpreted to determine physical properties of the sample. In SE, measurement and analysis are done as a function of the wavelength, $\lambda$, of light. General reviews and representative applications in materials science and semiconductor technology can be found in *Optical Characterization Techniques in Semiconductor Terminology* (1981), *Spectroscopic Characterization Techniques in Semiconductor Technology* (1983), *Proceedings of the International Conference on Ellipsometry* (1983), and Aspnes (1985).

SE measurements typically cover the two-octave quartz-optics or visible-deep ultraviolet wavelength range extending from about 800 to 200 nm (energy range from about 1.5 to 6.0 eV). In most presently operating spectroscopic ellipsometers, this range is determined by photomultiplier sensitivity and quartz transmission limitations in the near infrared and deep ultraviolet, respectively. Optical properties in this spectral range are due primarily to electronic transitions between bonding (valence) and antibonding (conduction) levels. Consequently, ellipsometric data contain information about chemical properties such as molecular configurations, electronic structure, and alloy compositions. In addition, the strong dependence of electronic wavefunctions on crystalline order makes ellipsometric spectra a good source of information about microstructural properties such as disorder, density, topography, and grain sizes (Aspnes, 1985). (Microstructure as used here refers to spatial inhomogeneities of the order of 1–100 nm, a size range now often referred to as nanostructure or mesostructure.)

Absorption coefficients of materials that are opaque in this spectral region tend to be very large, of the order of $10^6$ cm$^{-1}$, so transmission measurements are relatively useless and optical data must be obtained by reflecting light from specular surfaces. The correspondingly short optical penetration depths of $\sim 100$ Å imply a high surface sensitivity, which can be a nuisance when trying to accurately determine bulk properties but very useful for film thickness measurements, surface and interface analysis, and real-time monitoring of cleaning and deposition processes under reactive or high-pressure ambient conditions.

In addition to the above, SE has other characteristics that are well suited for plasma processing applications. It requires nothing more of a sample than that its surface be specular. It is nondestructive and hence can be used to provide information about unstable systems or about layers or interfaces buried beneath other layers. It can be used in any transparent ambient, including environments that are too dense or too reactive for other types of probes. It provides macroscopic or spatially averaged information, thereby automatically summarizing microstructurally complex systems in a simple, easily assimilated manner.

On the other hand, the relatively narrow accessible wavelength range generally prevents clear identification of atomic species and their chemical bonding. Lateral resolution, fundamentally limited by the wavelength of light, can be a problem. Also, optical measurements determine optical, not physical, properties: Attributes such as microstructure, film thicknesses, sample densities, and even dielectric functions themselves must be inferred from ellipsometric data by model calculations. This is undesirable for several reasons. A model never represents a sample to an arbitrary level of accuracy, and good physical insight may be needed to construct it. Further, the model must be evaluated numerically, so a computer, good software, and an accurate data base are essential. Finally, optical data contain information about every part of a sample to which the light penetrates, so the analysis of graded or otherwise complicated multilayered systems may be difficult or impossible.

Nevertheless, in many applications these difficulties are not essential, and much useful information about materials and processes can be obtained by SE. Although single-wavelength ellipsometry is an old technique, dating back to the last century, most of our present diagnostic capabilities are a direct result of the development of the fast, accurate, automatic spectroscopic ellipsometers of the last decade. These instruments have generated a wealth of accurate data on a wide range of materials. Not only do these data provide the accurate base needed for analysis, but they have also allowed basic modelling assumptions to be tested and experience to be gained. During the same period, new theoretical limits to the dielectric response of microscopically inhomogeneous systems have been developed (Bergman, 1980, 1982; Milton, 1980), allowing definitive microstructural information to be obtained, even for samples of which nothing is known about composition and microstructure.

## B. Performance

The capability of SE to determine various physical properties such as defects, surface roughness, crystallinity, film thickness, etc., that are of special interest in plasma processing has been established. In this section

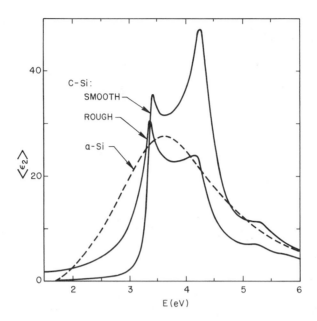

FIGURE 1. A comparison of the imaginary parts of the dielectric functions of crystalline (*c*-) and amorphous (*a*-) silicon, and the imaginary part of the pseudodielectric function of a microscopically roughened *c*-Si wafer (Aspnes, 1982).

we briefly summarize these capabilities and compare SE to other analytical techniques that provide similar information.

*Defects, disorder, and polycrystallinity*:  Degradation of visible-near UV optical spectra of crystalline semiconductors is detectable when concentrations of dopants or point defects exceed about $10^{18}$ cm$^{-3}$ and becomes quite obvious when such concentrations exceed about $10^{19}$ cm$^{-3}$ (about 0.01% of the total number of lattice sites) (Aspnes et al., 1982; Erman et al., 1983; Viña and Cardona, 1984). As shown in Figure 1, the dielectric properties of the totally disordered or amorphized material that results from prolonged exposure of crystalline material to a plasma are completely different from those of the corresponding material in single-crystal form (Aspnes, 1982). Note that this pronounced difference occurs even though crystalline (*c*-) and amorphous (*a*-) silicon have the same structure, that is, silicon atoms tetrahedrally coordinated with four other Si atoms. Optical spectra of partially disordered materials can usually be represented by appropriate physical (effective-medium) mixtures of these crystalline and amorphous phases and can be analyzed to give the relative volume fractions of these phases to within a few percent (Aspnes, 1982). Polycrystalline materials can be similarly represented, but a void fraction must be included because these materials are typically less dense than corresponding materials in single-crystal form (Bagley et al., 1981). In principle, grain-size information can also be obtained, but virtually no calibration data are currently available. However, it is known that the last remnants of crystalline behavior in the optical spectra of disordered GaAs vanish when the characteristic dimension of the grains falls below about 1.5 nm (Aspnes et al., 1982).

*Layer thicknesses:* Changes in surface layer thicknesses of less than 0.01 nm (less than the thicknesses to which physical interfaces can even be defined) are routinely detectable by ellipsometry (Azzam and Bashara, 1977). However, the accurately measured quantity is not the metric thickness, $d$, but rather the optical thickness $nd$, where $n$ is the refractive index. Although $nd$ can be determined to within 1% or 2%, an equivalently accurate determination of $d$ requires a correspondingly accurate determination of $n$ or vice versa. This is usually possible only if $d$ is greater than about 10 nm. Also, because layers exposed to plasmas may show continuously varying properties with $d$, the metric thickness of a plasma-generated surface layer may be a matter of definition.

*Microstructure:* As stated above, the properties determined by SE are macroscopic averages. Thus while SE can summarize complicated systems in a few simple parameters, cross-sectional transmission electron microscopy (XTEM) must be used if the microstructure must be accessed directly. However, as shown in Figure 2, under certain circumstances SE can provide more detailed macroscopic information than XTEM (Vedam et al., 1985). Rutherford backscattering provides areal density and micro-

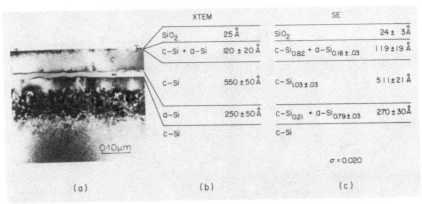

FIGURE 2. Comparison of XTEM and SE analyses of a $c$-Si wafer implanted with 80 keV Si ions to a fluence of $1 \times 10^{16}$ cm$^{-2}$ (Vedam et al., 1985).

crystalline orientational information, which is complementary to the volume density and grain-size information returned by SE.

*Surfaces and interfaces:* The 0.01 nm sensitivity of SE to changes in surface layer thicknesses suggests that it is a good technique for characterizing surfaces, especially during deposition and processing. A major advantage is that spectral differences allow microscopic roughness, surface contamination, and crystalline damage to be distinguished from each other. While electron spectroscopies have better surface sensitivities, this is of no advantage for analysis in reactive ambients or of buried interfaces. And while Raman and infrared spectroscopies have been applied to surface analysis, these techniques require layers about 5 nm thick for adequate signals.

## II. Fundamentals

In this section we describe the principles upon which SE is based and provide the equations needed for the reader to develop numerical calculations.

### A. The Dielectric Function

Optical propagation and the interaction of light waves with materials are described by Maxwell's Equations, which in cgs units have the form (Jackson, 1975)

$$\nabla \times E + \frac{1}{c}\frac{\partial B}{\partial t} = 0; \qquad \nabla \cdot B = 0; \tag{1}$$

$$\nabla \cdot D = 0; \qquad \nabla \times H = \frac{1}{c}\frac{\partial D}{\partial t}. \tag{2}$$

Material properties enter explicitly via the constitutive relations

$$D = \tilde{\epsilon}E; \qquad B = \tilde{\mu}H, \tag{3}$$

where $\tilde{\epsilon}$ is the dielectric function and $\tilde{\mu}$ is the magnetic permeability. At optical frequencies, $\tilde{\mu} = 1$, so the material is described entirely by $\tilde{\epsilon}$. The real and imaginary parts of $\tilde{\epsilon}$ are defined as $\tilde{\epsilon} = \epsilon_1 + i\epsilon_2$ and $\tilde{\epsilon} = \epsilon_1 - i\epsilon_2$ in the physics and engineering conventions, respectively. We use the physics convention here.

As written, these are *macroscopic* equations expressing time and space relationships between *observable* or *average* values of the corresponding microscopic quantities. The derivation of these macroscopic equations from microscopic theory shows that $\tilde{\epsilon}$ is related to the microscopic properties of a material according to (Jackson, 1975)

$$D = \tilde{\epsilon}E = E + 4\pi P, \tag{4}$$

where $P$ is the *dipole moment per unit volume*

$$P = V^{-1} \sum q_i \Delta x_i, \tag{5}$$

where $V$ is the volume, and $\Delta x_i$ is the displacement of the charge $q_i$ as a result of the applied field $E$.

Eq. (5) summarizes the essential physics of the dielectric response of materials. Note that $\tilde{\epsilon}$ is proportional to the ease by which the charges $q_i$ can be displaced. Thus metals, in which the electrons move freely, have large values of $\tilde{\epsilon}$, while values for oxides, where the charges are tightly bound, are small. Note also that the dipole contributions are additive so that the individual responses from different categories of charge (free electron, bound electron, ion, etc.) can be treated individually. However, the displacement $\Delta x_i$ depends on the *local* field at $q_i$, which is generally different from the average value, $E$, and generally must be calculated self-consistently. The sum over the volume represents the averaging process. Finally, the volume normalization factor shows that $\tilde{\epsilon}$ is also a function of density. Similar averaging over microscopic electric and displacement fields is used to derive the effective medium theories used in SE.

### B. OPTICAL PARAMETERS

Assuming that all fields have the form

$$E(r, t) = E_0 e^{ik \cdot r - i\omega t}, \tag{6}$$

nontrivial solutions of the wave Eqs. (1)–(3) require that

$$\frac{c^2 k^2}{\omega^2} = \tilde{\epsilon}. \tag{7}$$

This *dispersion equation* relates the square of the *propagation vector* $k$ to $\tilde{\epsilon}$. The *complex refractive index* is now defined as $\tilde{n} = ck/\omega = \tilde{\epsilon}^{1/2} = n + i\kappa$, where $\tilde{n}$ is the *ordinary index of refraction* and $\kappa$ is the *extinction coefficient*. $\kappa$ is occasionally defined as $\tilde{n} = n (1 + i\kappa)$, but fortunately this usage is rare. The optical response of a composite medium is described more naturally in terms of $\tilde{\epsilon}$, while propagation is described more naturally in terms of $\tilde{n}$. Since the intensity $I \sim |E|^2 \sim e^{-2 \operatorname{Im}(k)z}$, (assuming propagation in the positive $z$ direction), the absorption coefficient $\alpha$ is given by $\alpha = 4\pi\kappa/\lambda = \omega\epsilon_2/(nc)$ (Azzam and Bashara, 1977). Thus $\epsilon$ is the fundamental

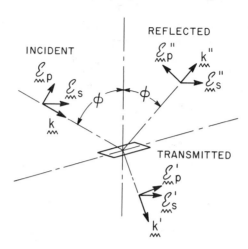

FIGURE 3. Electric field directional conventions used to calculate the complex reflectances at a plane-parallel interface where the angle of incidence is $\phi$ (Aspnes, 1976).

parameter that not only describes the dielectric response but also the optical properties of materials.

## C. POLARIZATION

If $z$ is taken to be the propagation direction, the local $x$ and $y$ coordinates are usually defined such that they are parallel ($p$) and perpendicular ($s$ = German *senkrecht*) to the plane of incidence, as shown in Figure 3 (Aspnes, 1976). In this local coordinate system, the plane wave can be written in terms of *complex field coefficients* $E_p$ and $E_s$

$$E(r, t) = \mathrm{Re}\left[\left(\hat{x}E_p + \hat{y}E_s\right)e^{ikz - i\omega t}\right]. \tag{8}$$

Consequently, a plane wave can carry at most four pieces of information, represented by the two amplitudes and two phases of $E_p$ and $E_s$. Reflection from a specular surface transforms the incident coefficients $E_p^i$ and $E_s^i$ into the outgoing coefficients $E_p^o$ and $E_s^o$ according to $E_p^o = r_p E_p^i$ and $E_s^o = r_s E_s^i$, where $r_p$ and $r_s$ are *complex reflectances*. At any given wavelength $\lambda$ and angle of incidence $\varphi$, the reflectance properties of a laterally isotropic, specular sample are completely described by $r_p$ and $r_s$, no matter how complicated its structural variation normal to the surface may be. Since $r_p$ and $r_s$ themselves only contain four pieces of information, the need for making many measurements under different experimental conditions (as, for example, different values of $\lambda$) is obvious.

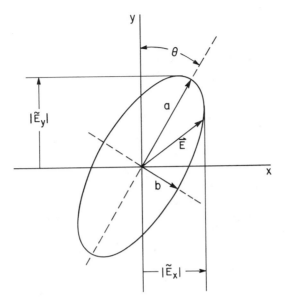

FIGURE 4. Locus of the electric field in the *xy* (*ps*) plane at fixed *z* for a plane wave in a general state of polarization (Aspnes, 1985).

## D. PHOTOMETRY AND ELLIPSOMETRY

To see how an ellipsometric measurement relates to $r_p$ and $r_s$, we consider the path traced out by $E(r, t)$ in the *ps* plane as shown in Figure 4. Since the *p* and *s* components of the field oscillate in simple harmonic motion about the origin, the path is elliptical. This *polarization ellipse* has a size and a shape. The size is clearly related to the *intensity* $I \sim (|E_p|^2 + |E_s|^2)$. Instruments that measure intensity (power) are *photometers*, the optical analog of wattmeters. A reflectometer is a common example.

The shape is an intensity-independent quantity that needs two parameters, such as the azimuth angle of the major axis and the major-minor axis ratio for its specification. One intensity-independent representation is the *polarization state* χ, defined as

$$\chi = \frac{E_p}{E_s}. \tag{9}$$

If $E_p$ and $E_s$ are in phase, the trajectory is a line and χ is real (linear polarization). If $E_p$ and $E_s$ have the same amplitude and differ in phase by 90°, $\chi = \pm i$ (circular polarization). Instruments that deal with polarization states are *ellipsometers*. They are the optical analogs of impedance bridges.

Specifically, a *reflectometer* is a photometer that is used to determine the ratio of outgoing to incident intensities. It therefore provides the perspectives $R_p = |r_p|^2$ and $R_s = |r_s|^2$ of a sample for *p*- and *s*-polarized light, respectively. Analogously, an *ellipsometer* is an optical instrument that is used to determine the ratio of outgoing to incident polarization states. It, therefore, provides the perspective

$$\frac{\chi^o}{\chi^i} = \left(\frac{E_p^o}{E_s^o}\right)\left(\frac{E_p^i}{E_s^i}\right) = \frac{r_p}{r_s} \tag{10}$$

$$= \rho = (\tan \psi)e^{i\Delta} \tag{11}$$

where $\rho$ is the *complex reflectance ratio*. $\rho$ is often represented in the literature by the angles $\psi$ and $\Delta$. The way $\rho$ is actually determined by the ellipsometer configuration most often used in SE is described in detail in Section III.


## E. Optical Modelling

A major drawback of optical analysis is that optical instruments measure optical properties, not the film thicknesses, densities, alloy compositions, relative volume fractions of constituents, grain sizes, or even dielectric functions, which the user really wants. These sample properties must be inferred from the optical data. To do this we need a *model*, that is, a set of mathematical expressions that give $\rho$ in terms of parameters describing the physical properties of the sample. For a multilayer system, the set of equations comprising the model can be written schematically as

$$\rho = \rho\left(\tilde{\epsilon}_s; \ldots \tilde{\epsilon}_j, d_j, \ldots; \epsilon_a; \varphi; \lambda \ldots\right) \tag{12}$$

where $\tilde{\epsilon}_s, \ldots \tilde{\epsilon}_j, \ldots$, and $\tilde{\epsilon}_a$ are the dielectric functions of the substrate, the different layers, and the ambient, respectively, the $d_j$ are the layer thicknesses, $\varphi$ is the angle of incidence, and $\lambda$ is the wavelength. More parameters can be added if necessary. Once the model has been selected, its parameters are then determined either by using the optical data to obtain their values directly or by using least-squares regression analysis to systematically establish the parameter values (and their uncertainties) that minimize the discrepancy between calculated and measured spectra. The determination of $n$ and $d$ for a $SiO_2$ film on Si by single-wavelength null ellipsometry is an example of the first approach, which is usually effective only for very simple physical systems. The construction and evaluation of models will be discussed after the necessary equations have been given.

## F. FRESNEL EXPRESSIONS

If the dielectric response of the left- and right-side media across a boundary are $\tilde{\epsilon}_b$ and $\tilde{\epsilon}_c$, respectively, and if $E_b^l$, $E_b^r$, $E_c^l$, and $E_c^r$ are the complex field coefficients of the leftward- and rightward-propagating waves in the two media, then the Fresnel expressions (Azzam and Bashara, 1977) for $s$-polarized light can be written

$$E_b^l = \frac{1}{2}\left(1 + \frac{\tilde{n}_{c\perp}}{\tilde{n}_{b\perp}}\right)E_c^l + \frac{1}{2}\left(1 - \frac{\tilde{n}_{c\perp}}{\tilde{n}_{b\perp}}\right)E_c^r; \qquad (13)$$

$$E_b^r = \frac{1}{2}\left(1 - \frac{\tilde{n}_{c\perp}}{\tilde{n}_{b\perp}}\right)E_c^l + \frac{1}{2}\left(1 + \frac{\tilde{n}_{c\perp}}{\tilde{n}_{b\perp}}\right)E_c^r. \qquad (14)$$

The corresponding expressions for $p$-polarized light are

$$H_b^l = \frac{1}{2}\left(1 + \frac{\tilde{\epsilon}_b \tilde{n}_{c\perp}}{\tilde{\epsilon}_c \tilde{n}_{b\perp}}\right)H_c^l + \frac{1}{2}\left(1 - \frac{\tilde{\epsilon}_b \tilde{n}_{c\perp}}{\tilde{\epsilon}_c \tilde{n}_{b\perp}}\right)H_c^r; \qquad (15)$$

$$H_b^r = \frac{1}{2}\left(1 - \frac{\tilde{\epsilon}_b \tilde{n}_{c\perp}}{\tilde{\epsilon}_c \tilde{n}_{b\perp}}\right)H_c^l + \frac{1}{2}\left(1 + \frac{\tilde{\epsilon}_b \tilde{n}_{c\perp}}{\tilde{\epsilon}_c \tilde{n}_{b\perp}}\right)H_c^r, \qquad (16)$$

where

$$\tilde{n}_{j\perp} = \left(\tilde{\epsilon}_j - \tilde{\epsilon}_a \sin^2\varphi\right)^{1/2}. \qquad (17)$$

Propagation from one medium to the next can be treated by multiplying the coefficients at one boundary by $e^{ik_j d}$ or $e^{-ik_j d}$ according to the direction of propagation to get the corresponding coefficients at the other boundary. Here, $d$ is the film thickness. The correct sign of the exponent is that for which the wave amplitude decreases (in case of finite absorption) in the propagation direction.

Since the boundary condition is that no back-reflected component can exist in the substrate, optical functions are most conveniently evaluated by starting at the substrate, applying the above boundary and propagation equations at each layer, then evaluating $r_p = H_p^{\text{out}}/H_p^{\text{in}}$; $r_s = E_s^{\text{out}}/E_s^{\text{in}}$ at the surface. Except for several simple cases discussed below, *this is a numerical problem* because these ratio equations are complicated and involve complex quantities.

## G. CRYSTAL OPTICS

Anisotropic (crystal-optics) systems are more difficult to treat because reflectance and propagation generally cannot be separated into independent, linearly polarized $s$ and $p$ modes. A succinct systematic formalism for dealing with crystal optics has been given by Yeh (1979, 1980). Everything is expressed in matrices readily adapted for computer programming.

## H. The Two-phase Model

The two-phase (substrate-ambient) system is one case where the equations discussed in Section IIF reduce to a usefully simple analytic form

$$r_s = \frac{n_{a\perp} - \tilde{n}_{s\perp}}{n_{a\perp} + \tilde{n}_{s\perp}}; \qquad r_p = \frac{\epsilon_s n_{a\perp} - \epsilon_a \tilde{n}_{s\perp}}{\epsilon_s n_{a\perp} + \epsilon_a \tilde{n}_{s\perp}}, \qquad (18)$$

where the subscripts s and a on the optical functions refer to substrate and ambient, respectively. From these expressions it follows that

$$\rho = \frac{r_p}{r_s} = \frac{\sin^2 \varphi - \tilde{n}_{s\perp} \cos \varphi}{\sin^2 \varphi + \tilde{n}_{s\perp} \cos \varphi}. \qquad (19)$$

This expression can be inverted to give $\tilde{\epsilon}_s$ directly in terms of $\rho$

$$\frac{\tilde{\epsilon}_s}{\epsilon_a} = \sin^2 \varphi + \sin^2 \varphi \tan^2 \varphi \left[ \frac{(1 - \rho)}{(1 + \rho)} \right]^2. \qquad (20)$$

No equivalent analytic expressions exist to give $\epsilon_s$ in terms of the reflected intensities, $R_p$ and $R_s$.

The two-phase model applies to any multilayer system where the optical absorption in the first layer is sufficient to completely suppress back reflections from deeper layers. By this criterion most metal films thicker than about 200 Å can be treated as bulk materials for visible-near UV optical analysis.

## I. The Three-phase Model, Thin-film Limit

When a film is present on a substrate, the optical equations are best solved numerically. But if the film thickness $d$ is much smaller than $\lambda$, useful analytic expressions can again be obtained

$$r_s \cong r_s^{\,o} \left[ 1 + \frac{4\pi i d n_a \cos \varphi}{\lambda} \frac{\tilde{\epsilon}_s - \tilde{\epsilon}_o}{\tilde{\epsilon}_s - \epsilon_a} \right]; \qquad (21)$$

$$r_p \cong r_p^{\,o} \left[ 1 + \frac{4\pi i d n_a \cos \varphi}{\lambda} \frac{\tilde{\epsilon}_s - \tilde{\epsilon}_o}{\tilde{\epsilon}_s - \epsilon_a} \frac{1 - (1/\tilde{\epsilon}_o + 1/\tilde{\epsilon}_s)\epsilon_a \sin^2 \varphi}{1 - (1/\epsilon_a + 1/\tilde{\epsilon}_s)\epsilon_a \sin^2 \varphi} \right], \qquad (22)$$

$$\rho \cong \rho^{\,o} \left[ 1 + \frac{4\pi i d n_a \cos \varphi}{\lambda} \frac{\tilde{\epsilon}_s (\tilde{\epsilon}_o - \epsilon_a)(\tilde{\epsilon}_s - \tilde{\epsilon}_o)}{\tilde{\epsilon}_o (\tilde{\epsilon}_s - \epsilon_a)(\tilde{\epsilon}_s \cot^2 \varphi - \epsilon_a)} \right]. \qquad (23)$$

The last equation can be put in *pseudodielectric function* form, where the pseudodielectric function $\langle \tilde{\epsilon} \rangle$ is that quantity calculated from $\rho$ in the two-phase model, even though overlayers are present. $\langle \tilde{\epsilon} \rangle$ is often used instead of $\rho$ because it represents the data in a form more nearly related to the quantity of interest in materials analysis. The original data $\rho$ can always be recalculated from $\langle \tilde{\epsilon} \rangle$. To first order in $d/\lambda$ (Aspnes, 1976)

$$\langle \tilde{\epsilon} \rangle = \tilde{\epsilon}_s + \frac{4\pi i d n_a}{\lambda} \frac{\tilde{\epsilon}_s (\tilde{\epsilon}_s - \tilde{\epsilon}_0)(\tilde{\epsilon}_0 - \epsilon_a)}{\tilde{\epsilon}_0 (\tilde{\epsilon}_s - \epsilon_a)} \left[ \frac{\epsilon_s}{\epsilon_a} - \sin^2 \varphi \right]^{1/2} ; \quad (24)$$

$$\cong \epsilon_s + \frac{4\pi i d n_a}{\lambda} \tilde{\epsilon}_s^{3/2}, \quad (25)$$

where Eq. (25) applies if $|\tilde{\epsilon}_s| \gg |\tilde{\epsilon}_0| \gg \epsilon_a$. In the last expression $\tilde{\epsilon}_0$ has dropped out completely, showing that if $|\tilde{\epsilon}_s|$ can be made large enough (as is the case for semiconductors in the near UV), then relative surface quality can be assessed without having to identify the overlayer material.

We can use the thin-film limit to estimate the effect of overlayers on the accuracy of optical analyses. Taking Si as an example, for which $\tilde{\epsilon}_s \cong 0 + i46$ near 4.3 eV (2800 Å), Eq. (25) predicts a decrease of $\langle \epsilon_2 \rangle$, the apparent value of $\epsilon_2$, by about 0.5 per Å of overlayer. This is consistent with the calculated absorption coefficient $\alpha \sim 2 \times 10^6$ cm$^{-3}$, which indicates an optical penetration depth of only 50 Å. Thus it is reasonable that a 1 Å film should have a 1% effect on the optical data under these conditions. Note that Eq. (25) shows that dielectric films on dielectric substrates affect $\Delta$ but not $\psi$, a rule long known from conventional null ellipsometry.

## III. Instrumentation—A Spectroscopic Ellipsometric System for Plasma Processing

As a typical example of ellipsometric instrumentation, we discuss in this section a spectroscopic ellipsometric system (Theetan et al., 1980) designed to monitor thin film growth kinematics and to nondestructively characterize surface, interface, and thin film properties in a plasma. A simplified schematic diagram is shown in Figure 5. It consists of a spectroscopic ellipsometer attached to a plasma reactor (Chang, 1977). The ellipsometer is a photometric type related to previously described (Hauge and Dill, 1973) Fourier-transform rotating-analyzer ellipsometers, but with the source and detector interchanged. The optical elements form the sequence lamp, rotating polarizer, sample, fixed analyzer, monochromator, and photomultiplier. Because the monochromator is placed between the reactor and the detector, the light generated by the reactor is essentially eliminated before reaching the detector, so stray light rejection and the signal-to-noise ratio are greatly

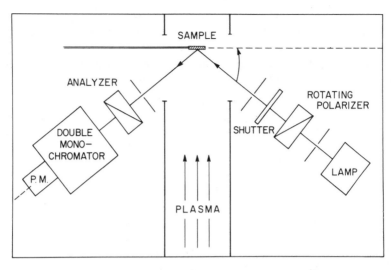

FIGURE 5. Schematic diagram of the automatic variable-incidence-angle spectroscopic ellipsometer used for in situ measurement of plasma-exposed samples (Theeten et al., 1980).

improved. The angle of incidence can be varied continuously, thereby allowing sensitivities to buried interfaces to be controlled. The acquired data are interpreted by optical modelling, as will be discussed in the next section.

The optical components are as follows. The light source is a feedback-stabilized high-brightness 500 W Xe arc lamp that provides an essentially line-free continuum over the accessible energy range of 1.5–5.6 eV. The rotating polarizer is a quartz Rochon prism that depolarizes the residual polarization of the lamp flux as well as generates the linearly polarized light with which the sample is illuminated. The shutter chops the incident beam to allow the background to be determined, thereby eliminating effects of drift in the electronic components and isolating the signal fraction of the light. Strain-free fused-quartz windows mounted on standard $2\frac{3}{4}$ inch conflat flanges provide optical access to the plasma reaction chamber. To minimize oxide nonuniformity effects, a 0.3 mm aperature restricts the illuminated area on the sample to approximately $0.3 \times 0.6$ mm$^2$. The fixed analyzer is a UV calcite Glan-Thompson prism, which limits the high-energy spectral range to 5.6 eV but which reduces overall system length. Wavelength dispersion is provided by a quarter-meter double grating mono-chromator operated at a typical resolution of 1 nm. A photomultiplier used as the detector determines the low-energy spectral limit of 1.5 eV.

The polarizer is mounted in a hollow-shaft DC motor that is feedback stabilized for rotation at speeds typically near 25 rps. An attached hollow-shaft optical encoder triggers a 14-bit analog-to-digital converter (ADC) that provides 1024 data conversions per revolution. The entire lamp, polarizer, and shutter assembly is mounted on Teflon-supported rails designed to slide on the optical table to provide angles of incidence from 60° to 76°. The angle of incidence is determined geometrically by laser alignment and further checked by ellipsometric measurements on a reference $SiO_2$-Si sample.

The detector output voltage is given by

$$I(t) = I_{av}\left[\sin^2 P \sin^2 A \tan^2 \psi + \cos^2 P \cos^2 A \tan^2 \psi\right.$$
$$\left. + \left(\tfrac{1}{2}\right)\tan \psi \cos \Delta \sin 2P \sin 2A\right], \qquad (26)$$

where $I_{av}$ is the average intensity. The azimuth angles $P$ and $A$ of the polarizer and analyzer, respectively, are measured from the plane of incidence. In operation, $A = A_0$ is fixed, and $P = \omega_0 t$, where $\omega_0$ is the mechanical angular frequency. Under these conditions Eq. (26) becomes

$$I(t) = a \cos 2P + b \sin 2P + c, \qquad (27)$$

where the coefficients $a$, $b$, and $c$ are determined by Fourier analyzing the measured intensity. It follows that

$$\tan \psi = \tan A_0 \sqrt{\frac{c + a}{c - a}} \qquad (28)$$

$$\cos \Delta = b/\sqrt{c^2 - a^2}. \qquad (29)$$

Data reduction and overall experimental control are provided by the computer system outlined in Figure 6. The system features two processors: a fast interface to control peripherals and accumulate on-line data and a central processing unit (CPU) to reduce and analyze the data and to interact with the control terminal and various input/output and data storage devices. The interface frees the CPU from operational details, and its 1024 circulating locations allow the output of the ADC to be accumulated in real time while the CPU is analyzing the previous set of data.

A typical time sequence of operations is given in Table 1. In steps 4 and 5, the plasma is turned on for about 1 second by an interface-controlled switch while the background data are transferred and the ellipsometric parameters $(\Delta, \psi)$ of the previous data set are computed, displayed, and stored. The actual plasma exposure time is accurately controlled by a clock. In steps 1, 2, and 3, the plasma is shut off while ellipsometric data are accumulated and transferred, the shutter is closed, and the background data accumulated. The ellipsometric and background data are accumulated for

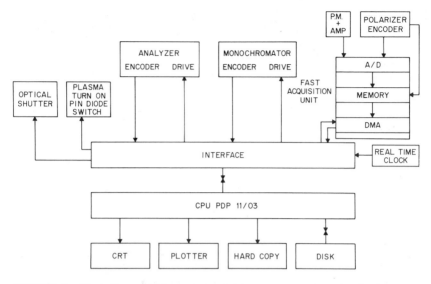

FIGURE 6.   Block diagram of the control and data processing system, showing interconnections among the peripherals, interface, and on-line computer (Theeten et al., 1980).

**Table 1.**   Typical time sequence of computer operations during a plasma experiment. BGD = background data.

| n<br>Time Sequence | CPU Job Status | Fast Acquisition<br>Unit | Plasma | Shutter |
|---|---|---|---|---|
| 1 | Reduces Previous Data | Accumulates Data | off | open |
| 2 | Transfer Data | | off | closed |
| 3 | Reduces Previous BGD | Accumulates BGD | off | closed |
| 4 | Transfer BGD | | on | open |
| 5 | Computes Previous $(\Delta, \psi)$<br>Displays and Stores;<br>Controls the Plasma<br>Exposure Time | No Accumulation | on | open |
| | Return to Step #1 | | | |

24 and 4 turns of the polarizer, about 1 and 0.04 seconds, respectively. By chopping the plasma exposure in this way, normally fast processes can be slowed down to allow the ellipsometer to follow fast kinetic phenomena. For slow reactions, chopping is not necessary, and measurements can be taken with uninterrupted plasma exposure. It is thus always possible to obtain spectroscopic ellipsometry data under quasiequilibrium conditions.

In thin film processing, the sample is placed in a vacuum chamber such that its surface is nearly normal to the plasma column. A plasma with the desired gas mixture is produced at a pressure of $10^{-3}$ torr by a pair of rf electrodes. Typical rf power is a few ten to a few hundred watts at 27 MHz. A magnetic field up to one kg confines the plasma along the axis of the vacuum chamber. The sample can be biased with respect to the plasma for special applications.

## IV. Optical Modelling

### A. CONSTRUCTION

Equation (12) represents the set of equations that allows the optical properties of a given sample to be calculated analytically in terms of a data base and parameters that are intrinsic to the sample, i.e., those sample parameters that are independent of experimental conditions. For surface and thin film analysis, parameters such as $\tilde{\epsilon}_s$ are expected to be part of a data base, possibly as a result of measurements performed on a bare substrate before overlying films were deposited. Any macrostructurally homogeneous material can be represented phenomenologically by a dielectric function, but if a material is microscopically inhomogeneous its dielectric function will itself be a function of the microstructural parameters of the material and the dielectric functions of its individual constituents. If microstructural analysis is the objective, this macroscopic dielectric function must be modelled further by effective medium theory, as discussed below.

As an example of how a model might be constructed, we consider a hypothetical film deposited on the oxidized surface of a semiconductor, as shown in Figure 7. The model must include at least four phases: substrate, oxide, layer, and ambient. These are represented by the set of Fresnel reflectance and propagation equations including as parameters the dielectric functions $\tilde{\epsilon}_s$, $\tilde{\epsilon}_{ox}$, $\tilde{\epsilon}_f$, and $\epsilon_a$, and the thicknesses $d_1$ and $d_2$. In principle $d_1$ and $d_2$ are unknown, but $\tilde{\epsilon}_s$, $\tilde{\epsilon}_{ox}$, and $\epsilon_a$ should be available in a data base. If our objective is to obtain $d_1$, $\tilde{\epsilon}_f$, and $d_2$, the four-phase model shown in Figure 7 is probably adequate, and the values of the parameters could be obtained by matching the calculated optical response to the data via

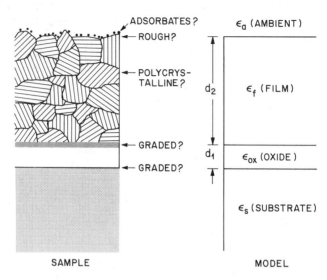

FIGURE 7.   Hypothetical sample consisting of a film deposited on the oxidized surface of a semiconductor. One possible model representation is shown at the right (Aspnes, 1985).

simplex or least-squares regression (Keeping, 1962) analysis. If finer detail is required, e.g., surface or interface information, the model must be extended to include these phases as well. There is no guarantee, however, that the data would be adequate to deal with a more complicated model.

Experience and some basic knowledge of the structure of a sample are the best guides to choosing correct models. We will consider the polycrystalline ($p$-) Si film whose $\langle \epsilon \rangle$ spectrum is shown in Figure 8 as an example (Bagley et al., 1981). We suppose that the dielectric responses of crystalline ($c$-) and amorphous ($a$-) silicon and the effect of microscopic roughness on $\langle \epsilon \rangle$ for the crystalline material are already in our data base. The data of Figure 1 show that $c$-Si is clearly identifiable by the so-called $E_1$ and $E_2$ peaks near 3.4 and 4.2 eV, respectively, amorphous Si is recognized by the single broad peak near 3.6 eV, and microscopic roughness is characterized by a depression of the $E_2$ peak relative to $E_1$. Thus the weak features at 3.4 and 4.2 eV in Figure 8 show that the $p$-Si film must contain $c$-Si, the broad background centered at 3.6 eV shows that the $p$-Si film must also contain $a$-Si, and the depression of $E_2$ relative to $E_1$ shows that the surface must be microscopically rough. In addition, the low overall amplitudes of $\langle \epsilon_1 \rangle$ and $\langle \epsilon_2 \rangle$ show that the material is considerably diluted by voids. Thus this microscopically inhomogeneous material would be modelled as an effective-medium composite of $c$-Si, $a$-Si, and voids, with an overlayer of

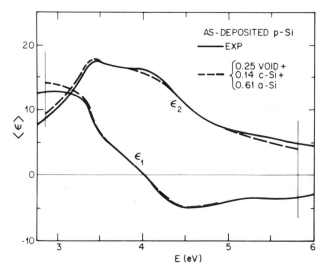

FIGURE 8. Solid lines: pseudodielectric function spectrum of a polycrystalline Si film deposited at 650°C by low-pressure chemical vapor deposition. Dashed lines: representation calculated with three-parameter effective medium model (Bagley et al., 1981).

increased void fraction to describe the relative reduction of the $E_2$ peak. A microscopically rough overlayer is in fact quite common on polycrystalline films, as the length scale of boundary irregularities is of the order of the grain sizes. In this example, the major features of the model have been identified from the optical spectrum itself. The dashed curves show the calculated model representation for the parameters indicated in the figure. The good fit for both real and imaginary parts of $\langle \tilde{\epsilon} \rangle$ over the entire spectral range using only three parameters shows that the model is reliable.

## B. SOLUTION

Once an appropriate model is established, it can be evaluated and sample properties obtained either by using the optical data as a template and applying least squares regression analysis to systematically determine the parameter values that give the best fit to the data (as done in the above example) or by using the optical data themselves as part of the data base and solving the model equations implicitly for one or more of the parameters of the model. The former approach is generally more useful for complicated samples, while the latter is commonly applied to simpler problems such as the determination of the dielectric function of a uniform, isotropic substrate or film from ellipsometric data or the determination of the thickness of an oxide film.

In least-squares regression analysis (Keeping, 1962), correlation matrices and parameter uncertainties are obtained as a byproduct of the fitting procedure and provide a built-in check of the capability of the data to determine model parameters. More model parameters always mean a better fit, but if the added parameters are highly correlated with others already in the model, or if the data are relatively independent of the added parameters, the uncertainty becomes very large.

A more stringent, but more difficult, test of the validity of a model is to use the optical data to determine one or more of the model parameters as a function of the experimental variables. For example, film thickness can be determined on a wavelength-by-wavelength basis in a three-phase model, but if $d$ turns out to be wavelength-dependent, there is clearly something wrong with either the data or the model.

Sometimes this approach can be used to determine parameters on the basis of expected physical behavior. An important case is a film on a substrate that has distinctive optical features (Arwin and Aspnes, 1984). If we solve the three-phase model implicitly for the dielectric function of the film using an educated guess for the thickness of the film, we can write the result to first order as

$$\langle \tilde{\epsilon}_o \rangle = \tilde{\epsilon}_o + (\langle d \rangle - d_o) \frac{\partial f}{\partial d} \bigg/ \frac{\partial f}{\partial \tilde{\epsilon}_o}, \qquad (30)$$

where $\langle d \rangle$ and $d_o$ are the guessed and correct thicknesses, respectively, and $\langle \epsilon_o \rangle$ and $\epsilon_o$ the approximate and correct overlayer dielectric function. The only part of Eq. (30) that contains $\epsilon_s$ (and its characteristic optical structure) is the coefficient multiplying $(\langle d \rangle - d_o)$. Thus if $\langle d \rangle$ had been chosen correctly, the substrate-related structure would have vanished and $\langle \tilde{\epsilon}_o \rangle$ would have equaled $\tilde{\epsilon}_o$. Thus both $\tilde{\epsilon}_o$ and $d_o$ can be determined without either being known *a priori*. The same remarks apply to interference oscillations in thick semitransparent films (Aspnes et al., 1984).

The above examples imply (correctly) that the most important single variable under direct experimental control is $\lambda$. The wavelength is important not only because it can be varied without affecting the sample but also because the wavelength dependence of different materials is unique and distinctive and can lead to useful identifications. Moreover, absorption can be used to advantage. The penetration depth of light in the deposited polycrystalline Si example given above did not exceed a few hundred Å, so the oxidized $c$-Si substrate underlying the $p$-Si film could be ignored. Of course, other wavelength ranges could have been used to give information about buried layers.

Other parameters that might be varied include $\varphi$ or $\epsilon_a$, but the former does not give much analytic leverage for bulk materials or thin films because Snell's Law shows that propagation in a sample is always refracted toward the surface normal, thereby reducing the effect of $\varphi$; moreover, a change of ambient can induce undesired modifications if the sample is porous. Comparing results for films of different thicknesses is not recommended, as there are many examples of material systems where film microstructure evolves with increasing film thickness.

## C. MICROSCOPICALLY INHOMOGENEOUS MATERIALS: EFFECTIVE MEDIUM MODELS

On an atomic scale all materials are inhomogeneous. But macroscopically homogeneous materials can also consist of separate regions that possess their own dielectric identities yet have dimensions that are small compared to the wavelength of light. The dielectric function of these materials can be described in terms of the dielectric functions of its constituents and its microstructure by effective medium theory. Here, the *effective medium* is the equivalent macroscopically homogeneous material that has the same dielectric properties over the accessible wavelength range as the microscopically inhomogeneous material. The microstructural parameters (density, grain size, and grain shape, composition, connectedness of the various phases, etc.) of the actual sample appear in the mathematical expression of $\tilde{\epsilon}$ as averages over the corresponding microscopic parameters and thus represent a summary of the detailed microstructural properties of the sample.

The derivation of $\tilde{\epsilon}$ for these materials is done by analogy with Eq. (5): given the microstructure, one solves for the microscopic electric and displacement fields $e(r)$ and $d(r)$, volume-averages them to obtain the corresponding macroscopic observables $D$ and $E$, then evaluates $\tilde{\epsilon}$ according to the constitutive relation $D = \epsilon E$. For example, if the sample is laminated with separate regions of dielectric functions $\tilde{\epsilon}_a$ and $\tilde{\epsilon}_b$, and if $E$ is applied parallel to the laminations, then $e(r) = E$ trivially. Since normal $D$ is conserved, we have $d(r)$ equal to either $\tilde{\epsilon}_a E$ or $\tilde{\epsilon}_b E$, depending on whether $r$ is located within $a$ or $b$, respectively. Upon averaging we obtain $D = f_a \tilde{\epsilon}_a + f_b \tilde{\epsilon}_b$, where $f_a$ and $f_b$ are the relative volume fractions of $a$ and $b$ in the composite (note $f_a + f_b = 1$), from which

$$\tilde{\epsilon} = f_a \tilde{\epsilon}_a + f_b \tilde{\epsilon}_b. \tag{31}$$

The electrical analog of this effective medium model is two capacitors in parallel. If $E$ is applied perpendicular to the laminations, $D = d(r)$ trivially, and an average over $e(r)$ leads to

$$\frac{1}{\tilde{\epsilon}} = \frac{f_a}{\tilde{\epsilon}_a} + \frac{f_b}{\tilde{\epsilon}_b}. \tag{32}$$

The electrical analog is two capacitors in series.

Note that the sample *microstructure* is the same in both cases, yet the effective medium expressions are completely different. The difference is due to the depolarization or screening charge that develops on the boundaries between regions in microscopically inhomogeneous samples, which is a function of the relative direction of the local electric field and the normal vector of the local boundary. As the more polarizable species supplies the screening charge and thereby tends to shield itself more effectively from the applied field, the less polarizable species (e.g., voids) tend to contribute more heavily to the dielectric function of a composite than what would be expected from simple volume-fraction arguments alone.

The examples discussed above represent the two extreme cases where screening charge is nonexistent and maximal, respectively. Since any real microstructure must fall between these extremes, Eqs. (31) and (32) give absolute bounds (the Wiener limits (1912)) to the dielectric response no matter what its microstructure. More stringent unit theorems have been derived by Hashin and Shtrikman (1962) when the volume fractions of the two constituents are known (as, for example, determined by chemical analysis) and by Bergman (1980, 1982) and Milton (1980) when, in addition, the sample is known to be isotropic in two or three dimensions. All have the remarkable property that the respective bounds are circular arcs in the complex $\epsilon$ plane and can be constructed with nothing more than a compass or straightedge.

As the real purpose of an effective medium theory is to take into account the effect of boundary charge, it is clear that *any* inhomogeneous medium can only be described to arbitrary accuracy by its own effective medium expression tailored to its own specific distribution of inhomogeneous regions. Since this is usually neither practical nor possible, one compromises by *assigning* a solvable microstructure to a sample, then determining to what extent it can be used to describe the data. For a two-phase cermet or aggregate microstructure, these "standard" effective medium expressions have the form derived from the electrostatic solution of an isolated sphere in an infinite medium

$$\frac{\tilde{\epsilon} - \tilde{\epsilon}_h}{\tilde{\epsilon} + 2\tilde{\epsilon}_h} = f_a \frac{\tilde{\epsilon}_a - \tilde{\epsilon}_h}{\tilde{\epsilon}_a + 2\tilde{\epsilon}_h} + f_b \frac{\tilde{\epsilon}_b - \tilde{\epsilon}_h}{\tilde{\epsilon}_b + 2\tilde{\epsilon}_h}, \tag{33}$$

where $\tilde{\epsilon}_h$ is a "host" dielectric function that has the value $\tilde{\epsilon}_a$ or $\tilde{\epsilon}_b$ if the composite consists of isolated inclusions of $b$ in $a$ or of $a$ in $b$, respectively (cermet microstructure) or $\tilde{\epsilon}$ if the phases are randomly mixed (aggregate microstructure). The corresponding expressions describe the effective medium models of Garnett (1906) and Bruggeman (1935), respectively. The self-consistent nature of the Bruggeman expression appears to better describe real materials, where the microstructure tends to be random; the composite dielectric function used to model the polycrystalline Si sample of Figure 8 was calculated with the Bruggeman expression. More terms can be added to describe mixtures of more than two phases.

In the more general case of nonspherical inclusions, the factors of 2 in Eq. (33) can be replaced by a screening parameter $K$. The values $K = 0$ and $\infty$ correspond to laminar samples with the field oriented perpendicular and parallel to the laminations, respectively. The value $K = 1$ corresponds to cylindrical inclusions (2-D symmetry).

As an application, we determine the effect of microscopic voids on $\tilde{\epsilon}$. Taking $\tilde{\epsilon}_a = 1 + i0$ and assuming that $|\tilde{\epsilon}_b| \gg 1$, $f_a \ll 1$, we have from Eq. (33)

$$\tilde{\epsilon} = \tilde{\epsilon}_b \left( 1 - \tfrac{3}{2} f_b \right), \tag{34}$$

which shows that voids reduce $\tilde{\epsilon}_b$ everywhere at a uniform relative rate of $\tfrac{3}{2}$ times the relative void fraction (following the rule that the less polarizable species dominate).

Microscopic roughness represents an important application of effective medium theory to plasma processing. Microscopically rough surfaces almost invariably result from plasma etching. These microscopically rough overlayers can generally be represented by the Bruggeman model as a 40%–60% mixture of voids and substrate.

## V.  Representative Examples

In this section we give four representative examples of applications of scanning ellipsometry in plasma processing. These examples show how information on surface, interface, and bulk film properties can be obtained under different experimental conditions. Other relevant examples of surface cleaning and real-time monitoring of film deposition are also discussed.

## A.  PLASMA OXIDATION OF GaAs

Plasma oxidation is a low-temperature (below 500°C) technique of growing native oxide films on various substrate materials. Here, we show how the oxide growth kinetics of a GaAs sample can be monitored in situ, in real time, with the spectroscopic ellipsometric system described in Section III. A detailed knowledge of oxidation kinetics is essential to understand how the oxides are formed. In situ real-time measurements can also be applied to study surface modification and other processing techniques discussed in this section. Owing to instrumentation limitations, kinetic studies are usually performed at fixed wavelength and angle of incidence, with $\lambda$ and $\varphi$ being chosen for optimum sensitivity to a specific property of the sample. The temporal evolution of the data performs the equivalent role of spectral dependence in the analysis of static samples.

The sequence of events for studying the plasma oxidation of GaAs is as follows: Data for the surface in its initial state are acquired. The oxidation plasma is turned on, the data are transmitted from the interface to the CPU, and the CPU analyzes the data. The oxidation plasma is then turned off and more surface data are acquired. The sequence is then repeated until the experiment is completed. The duration of the oxidation pulse is adjusted to meet the needs of the experimenter and the performance capability of the computer.

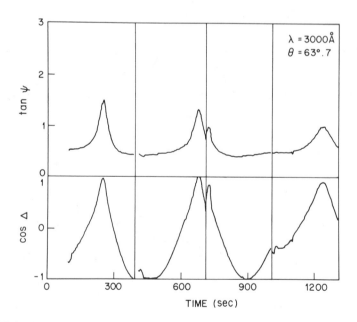

FIGURE 9.   Typical results of in situ, real time ellipsometric measurements of the plasma oxidation of GaAs. Data were acquired at $\lambda$ = 3000 Å and $\phi$ = 63.7° as a function of plasma exposure time. Measurements were interrupted for 15 minutes each at 400, 720, and 1040 second cumulative exposure times for spectral and angular analyses (Theeten et al., 1980).

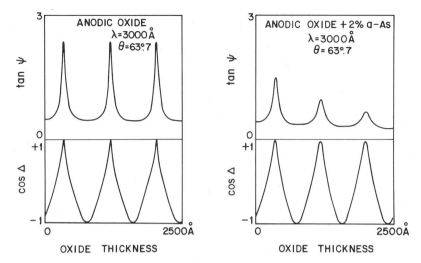

FIGURE 10.   Simulated kinetic data for (a, left) a transparent, anodic native oxide and (b, right) a physical mixture of 98% anodic oxide and 2% amorphous arsenic, to be compared to the data shown in Figure 9 (Theeten et al., 1980).

A typical series of measurements during the plasma oxidation of GaAs are shown in Figure 9 (Theeten et al., 1980). Oxidation was interrupted at 400, 720, and 1040 seconds cumulative exposure for a sufficient time to allow spectroscopic measurements to be obtained at various angles of incidence. The $\cos \Delta$ curve oscillates between $-1$ and $+1$ as the thickness of the oxide increases. For the 3000 Å, 63.7° conditions of Figure 9, one period corresponds to an oxide thickness of approximately 800 Å. The rate of growth is thus directly obtainable from the $\cos \Delta$ curve. Here, it is about 100 Å/min. The data show directly that the rate of growth decreases as the oxide thickness increases. About 570 seconds are required to complete the second oscillation, compared to about 440 seconds for the first one. This is due to the fact that the bias voltage of the sample was kept constant. This oxide growth behavior is in agreement with previously published kinetic curves (Chang et al., 1979).

The $\tan \psi$ curve exhibits a peak associated with each $\cos \Delta$ oscillation. Note that the heights of these peaks are not constant but decrease as the oxide thickness increases. As illustrated schematically by the model calculation in Figure 10, this decrease shows that the oxide is optically absorbing at 3000 Å. If the plasma oxide were transparent, like the anodic oxide (Aspnes et al., 1977) then both $\tan \psi$ and $\cos \Delta$ would be cyclic, with a repeat thickness of about 800 Å, as shown in Figure 10a. In particular, the height of the tan $\psi$ peaks would be constant. However, if one now assumes that the plasma-grown oxide is a Bruggeman effective-medium mixture of 98% anodic oxide and 2% unoxidized As, then the tan $\psi$ peak heights decrease, as shown in Figure 10b.

The decrease occurs because the *a*-As fraction is optically absorbing (Greaves and Davis, 1976), so the constructive interference that gives rise to the tan $\psi$ maxima becomes weaker as the oxide thickness increases. A comparison between Figures 9 and 10 indicates that 2% *a*-As incorporated in an otherwise transparent oxide gives a reasonable representation of the kinetic data. The dielectric function of the plasma-grown oxide is analyzed more extensively below. The example shows that the kinetic tan $\psi$ curve can be used as an indication of oxide composition.

Inspection of Figure 9 also shows that (cos $\Delta$) does not exactly reach $+1$ in its third oscillation. This effect, much less prominent than the decrease in

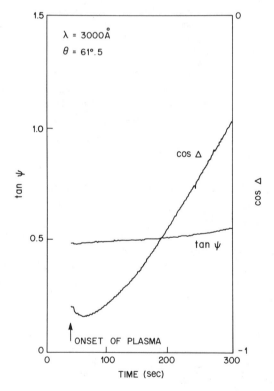

FIGURE 11. Kinetic data showing typical transient behavior at the resumption of plasma oxidation (Theeten et al., 1980).

the tan $\psi$ peaks, is due to a thickness nonuniformity of the oxide layer within the area probed by the beam. It is seen most clearly in the very narrow positive extreme of cos $\Delta$ and is also analyzed more completely below. Thus, kinetic data also provide an estimate of the uniformity of the deposit.

Another interesting feature of the data of Figure 9 is the anomalous behavior of tan $\psi$ and cos $\Delta$ when plasma oxidations are resumed after the spectral-measurement interruptions. Generally speaking, when plasma oxidation is restarted after a sufficiently long shutdown, the oxide thickness first decreases. More detailed data concerning this phenomenon are shown in Figure 11. At the resumption of plasma oxidation, (cos $\Delta$) decreases. We speculate that the oxide layer must first charge up to a critical voltage to enable either $O^-$ ions to migrate toward the oxide-substrate interface or $Ga^+$ and $As^+$ ions to migrate from the substrate toward the oxide-vacuum interface before oxidation actually begins. Before this voltage is reached, erosion of the oxide by the ongoing bombardment from plasma ions is the dominant mechanism. A detailed study of this effect will be reported elsewhere (Chang, in press).

In this series of experiments (Theeten et al., 1980), oxidation was performed under different conditions. Not surprisingly, different optical spectra were obtained depending on the rate of growth. For example, the spectroscopic tan $\psi$ and cos $\Delta$ data of Figure 12 were obtained from samples of comparable 1600 Å thickness but grown at very different rates (Theeten et al., 1980). The first sample was oxidized in about 30 seconds, while the second, in a lower density plasma, was oxidized in 50 minutes.

A three-phase model analysis of the data of Figure 12 yields the dielectric function $\epsilon_o$ shown in Figure 13. The film thicknesses were estimated by adjusting them to minimize the imaginary part of $\epsilon_o$. Results for the slow-grown oxide, given in Figure 13a, show much more prominent interference-related artifacts (oscillations) than those for the fast-grown oxide, shown in Figure 13b. As these artifacts are sample-, not oxide-, dependent, a three-phase model description of the slow-grown sample is less adequate than one for the fast-grown sample. In both cases, we note the striking feature that the Im($\epsilon_o$) curve is greater than zero, even at low energies. Since the anodic oxide film is totally transparent below 4.8 eV (Aspnes et al., 1977), the long absorption tails in Figure 13 are clear indications of unoxidized material in the plasma-grown oxides.

To obtain a quantitative estimate of the amount of absorbing material, we use the anodic GaAs oxide as a reference and calculate the effect of including small fraction of $a$-As using the Bruggeman effective medium model. The result for 98% anodic oxide and 2% $a$-As is also plotted in Figure 13. The agreement between the $\epsilon_o$ data and the calculated composite

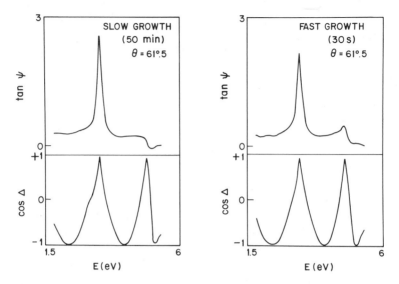

FIGURE 12.  (tan $\psi$) and (cos $\Delta$) data for two GaAs plasma oxides grown at 100 times different rates of growth (Theeten et al., 1980).

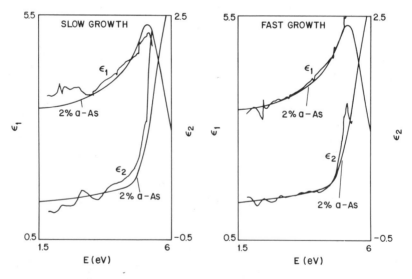

FIGURE 13.  Results of a three-phase-model analyses of the data of Figure 12. Calculated $\epsilon$ spectra for a physical mixture of 98% anodic oxide and 2% amorphous arsenic are also shown (Theeten et al., 1980).

dielectric function is satisfactory to 4.8 eV. Above 4.8 eV, deviations occur, indicating that the plasma-grown oxide and the anodic oxide are different materials. To choose one as a reference for the other is not justified at higher energies. However, below 4.5 eV the description of the plasma-grown oxide as a physical mixture of anodic oxide and $a$-As is a good approximation and can be used to estimate the amount of absorbing material.

The effect shown in Figure 13 is not unique to $a$-As. Optical absorption could also be simulated by physical mixtures of the anodic oxide with $a$-GaAs or $a$-Ga, although the amount of absorbing material would differ slightly in composition. The presence of $a$-Ga is unlikely on chemical grounds, but the existence of $a$-GaAs cannot be ruled out. Nevertheless, the assumption that $a$-As is the absorbing phase appears more reasonable from the greater stability of $Ga_2O_3$ compared to $As_2O_3$ (Wilmsen, 1976). Furthermore, the presence of $a$-As in similar oxides has been detected with Raman scattering (Rhinewine et al.).

The results of a complete five-parameter, four-phase-model analysis (Theetan et al., 1980), using as adjustable parameters an interface thickness $d_i$, the oxide thickness $d_o$, the fraction $f_{ai}$ of $a$-As in the interface layer, the

**Table 2.** Least squares analysis of the data of Figure 12 for a fast-grown plasma oxide. To illustrate the relative influence of the various parameters, only the thicknesses are varied for this table while compositions (percentage of $a$-As) of the layers are incremented. The minimum deviation, as well as the best thicknesses, for each set of compositions are given.

| Interface Layer \ Top Layer | Pure Anodic Oxide | 1% $a$As | 2% $a$As | 3% $a$As |
|---|---|---|---|---|
| No Film | 7.19 1589 Å | 3.52 1597 Å | 0.78 1585 Å | 1.69 1566 Å |
| 25% $a$-As | 5.80 1590 Å/350 Å | 0.95 1504 Å/118 Å | 0.67 1558 Å/32 Å | 1.69 1566 Å/0 Å |
| 50% $a$-As | 2.80 1570 Å/400 Å | 1.23 1539 Å/112 Å | 0.73 1566 Å/29 Å | 1.69 1566 Å/0 Å |

corresponding fraction $f_{ao}$ in the oxide, and the angle of incidence will now be discussed. We use the angle of incidence as an adjustable parameter because its actual value might differ slightly from its nominal alignment setting. The four phases are substrate, interface, oxide, and ambient.

Table 2 gives the least-squares deviations between the data of Figure 12b and model simulations when only the thicknesses are allowed to vary and the compositions are changed by finite increments. From Table 2, it is seen

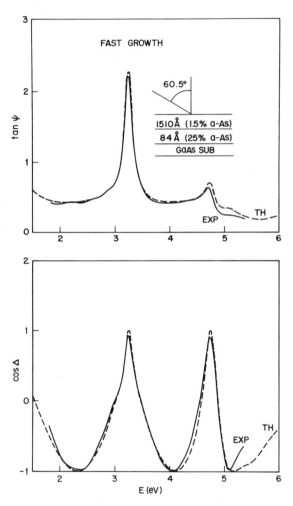

FIGURE 14.   Best-fit simulation of the data of Figure 12b, as discussed in the text (Theeten et al., 1980).

that the agreement is significantly improved if *a*-As is assumed to be present in the top layer and if a transition layer is also assumed to be present. The fit is very sensitive to the composition of the top layer and, to a lesser extent, to the composition of the interface. The best fit in Table 2 is obtained for a 1558 Å top layer of 98% anodic oxide and 2% *a*-As, a 32 Å interface of 75% anodic oxide and 25% *a*-As, and an angle of incidence of 61°. If all five parameters are allowed to vary independently, the statistical analysis gives an angle of incidence $\phi$ of 60.50° ± 0.02°, and a top layer

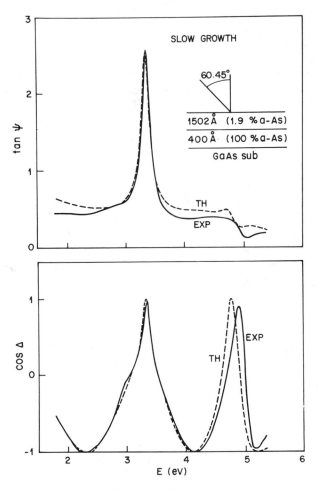

FIGURE 15.   Best-fit simulation of the data of Figure 12a, as discussed in the text (Theeten et al., 1980).

thickness $d_o$ of $1510 \pm 2$ Å, with a volume fraction $f_{ao}$ of $a$-As of $0.015 \pm 0.0005$. The unbiased estimator of the mean-squares deviation for this fit is 0.07. The comparison between this simulation and the data of Figure 12b is shown in Figure 14. As expected, the curves differ above 4.5 eV, indicating that the physical mixture assumption is no longer valid.

A similar five-parameter fit to the slow-growth data of Figure 12a is given in Figure 15. In this case, the fit is less satisfactory: The unbiased estimator is now 0.17. The following parameters are obtained: the angle of incidence of $60.45° \pm 0.02°$; a top layer thickness of $1502 \pm 210$ Å containing $0.019 \pm 0.00004$ $a$-As; an interface layer thickness of $400 \pm 210$ Å consisting of pure $a$-As. The fact that the tan $\psi$ data are not reproduced very well, even below 4.5 eV, in the energy region where the mixture approximation was found reasonable for the fast-grown sample, indicates that a four-phase model is not a good representation. In fact, the thickness and composition obtained for the interface indicate that this region is actually broad and very absorbent. A gradual variation of its composition from the top layer to the substrate is therefore more likely, in which case a more accurate representation would require more layers to be included in the model. Also, the top layer is itself more absorbent than that of the fast-grown oxide, making an interface analysis more difficult.

Although the four-phase model is not completely satisfactory, the relative values of the parameters obtained in the analysis of the two samples give some insight into the mechanism of plasma oxidation of GaAs. Whatever the rate of growth, some unoxidized material is present in the oxide, and this unoxidized material tends to accumulate at the interface between the oxide layer and the substrate. As already mentioned, the relative instability of $As_2O_3$ with respect to GaAs and $Ga_2O_3$ can be invoked to speculate that unoxidized arsenic is left in the oxide during the plasma oxidation.

If the process is very fast, the dynamics tend to favor a stoichiometric oxide. However, if the rate of oxidation is slow, more As atoms will be left in the metallic state. More precisely, the total amount of unoxidized As in the fast-grown sample is equivalent to a layer 42 Å thick, while that for the slow-grown sample is 410 Å. The fact that unoxidized material is mainly located at the oxide-substrate interface is probably due to segregation. To obtain a good quality oxide, the above analysis stresses the importance of a fast rate of oxidation.

B. HYDROGEN PLASMA ETCHING OF COMPOUND
   SEMICONDUCTOR SURFACES

In these experiments a hydrogen plasma was shown to be useful for cleaning and etching compound semiconductor surfaces (Chang and Darack,

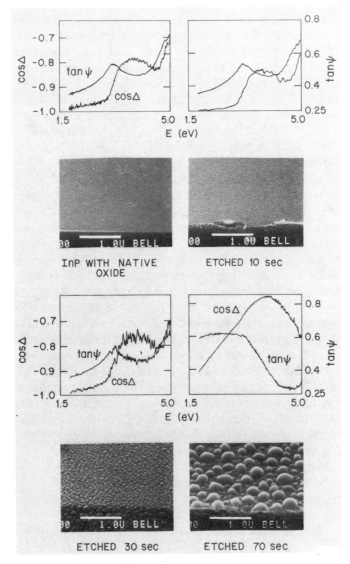

FIGURE 16.   Spectroscopic ellipsometric and associated scanning electron micrograph data for an InP surface after various etching durations in a hydrogen plasma (Chang et al., 1983).

1981), especially for removing surface carbon contamination (Tu et al., 1983). An extensive study was also performed for the hydrogen plasma etching of native oxides on GaAs, GaSb, and InP. Figure 16 shows spectroscopic ellipsometric data and associated scanning electron micrographs of InP surfaces etched in a hydrogen plasma. The rf power was 50 W and the hydrogen pressure was 200 $\mu$m. The etch rate for the oxide on InP is about 10 times that of the oxide on GaAs. The ellipsometric data show that after 10 seconds exposure a portion of the native oxide has been removed. However, after 30 seconds a thin surface film begins to form, causing the $\cos \Delta$ to increase again near 4.0 eV. Up to this point, independent Auger data showed that the P/In ratio was very similar to that of an air-cleaved InP surface. But after 70 seconds of etching, SEM micrographs showed hemispherical bumps starting to form. At this point the ellipsometric data are quite different from those for shorter exposures, probably due to the changing surface topography.

In order to interpret these data, a detailed model of the surface chemistry and geometry is needed. Auger measurements for the 70 second etched surface showed no gross surface segregation, although the P/In ratio within the first 30 Å of the surface was only $\frac{1}{4}$ that of the air-cleaved InP sample. The conclusion that surface segregation of P was minimal was drawn from the fact that the P/In ratio attained the bulk value after removal of less than 200 Å of the etched surface by ion milling. In fact, except for the first 50 Å, the Auger depth profile was indistinguishable from that taken on an air-cleaved InP surface. Therefore, the development of the large surface bumps seen in Figure 16 does not necessarily mean gross segregation of one element. As with GaAs, the mechanisms that cause the formation of these surface morphologies are not yet clearly known.

## C. PLASMA ENHANCED CHEMICAL VAPOR DEPOSITION OF $SiO_2$

The third application illustrates how a spectroscopic ellipsometer can be used to control the stoichiometry of $SiO_2$ during plasma-enhanced chemical vapor deposition (Chang et al., 1983). In this new deposition scheme $SiO_2$ films are formed by reacting a beam of atomic Si with atomic oxygen formed in a plasma discharge. By precisely controlling the flux ratio of the two source beams during deposition using the ellipsometer as a feedback monitor, $SiO_2$ films can be formed at 100°C with optical qualities similar to those of $SiO_2$ films grown thermally at 1000°C, as shown explicitly in Figure 17. Figure 18 shows a series of spectroscopic scans during a deposition run where the film stoichiometry was maintained as $SiO_2$ during

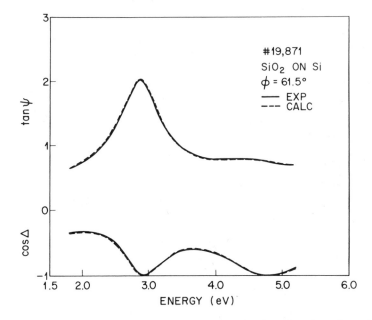

FIGURE 17. Comparison of spectroscopic ellipsometric data for $SiO_2$ films plasma-deposited at 100°C and thermally grown at 1000°C (Chang et al., 1983).

the first five minutes, with the Si evaporation current increased by 10% for the next 18 minutes. As a result, a Si-rich $SiO_2$ layer was formed. The final shapes of the ellipsometric curves can be interpreted by a simple three-layer model, as indicated by the solid curve and the drawing in the figure. This example demonstrates the power of spectroscopic ellipsometry for fabricating compositionally accurate, high quality thin films useful in VLSI technology.

## D.  PROCESSING DAMAGE

Energetic ionic species impinging on semiconductor surfaces cause damage that can be investigated by optical means. In principle, the surface left by plasma etching can be undamaged, but this process can be self-limiting if there is a pathway that forms a nonvolatile product. In this case a layer can accumulate on the surface and ultimately prevent further interactions from taking place. A combination of plasma etching and sputtering can be used

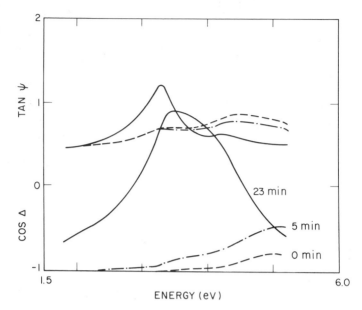

FIGURE 18.    Spectroscopic ellipsometric data for an $SiO_2$ film of variable stoichiometry, as discussed in the text (Chang et al., 1983).

to control this layer, but then damage can again result. The next example (Scherer et al., unpublished) considers this problem.

Figure 19 shows the pseudodielectric function of a GaAs sample whose surface has been ion milled by 500 eV Ar ions for five minutes. Segments of the dielectric function of nominally undamaged GaAs are also shown for comparison. The substantial reductions of the $E_1$ peak in $\langle \epsilon_1 \rangle$ and the $E_2$ peak in $\langle \epsilon_2 \rangle$ are clear evidence of an overlayer, while the noticeable shift of the $E_2$ peak of $\langle \epsilon_2 \rangle$ to lower energy shows that this overlayer absorbs light relatively strongly. Assuming that this overlayer can be represented as an aggregate physical mixture of amorphous GaAs, crystalline GaAs, and voids, and using least-squares regression to determine the best-fit values of the parameters of a three-phase model, we obtain the agreement shown, with an overlayer thickness of $47 \pm 13$ Å and an overlayer composition of $0.36 \pm 0.07$ a-GaAs and $0.14 \pm 0.04$ voids (leaving $0.50 \pm 0.11$ c-GaAs). The damage depth is consistent with that previously observed for single-crystal Ge bombarded with 500 eV Ar ions (Aspnes and Studna, 1980). Although the model spectrum approximates the data reasonably well, the discrepancies indicate that the overlayer is too complex to be completely represented by a single layer.

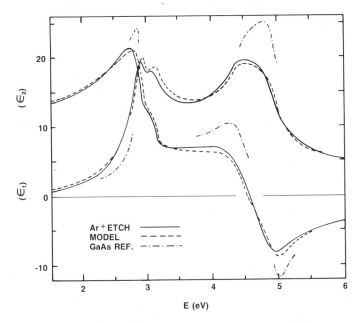

E (eV)

FIGURE 19.   Solid lines: pseudodielectric function spectrum of a GaAs sample $Ar^+$-ion-bombarded for five minutes at 500 eV. Dashed lines: spectrum calculated from model described in text. Dot-dashed lines: parts of the reference dielectric function spectrum of GaAs (Scherer et al., unpublished).

Figure 20 shows the $\langle \epsilon \rangle$ spectrum of a similar sample whose surface has been etched by a plasma containing both $Cl_2$ and Ar. A surface layer is still present, but the higher value of the $E_2$ peak indicates that it is thinner and probably much less absorbing than that which results from ion bombardment alone. Assuming the same model for this overlayer, a similar calculation shows that the film is $15.1 \pm 0.7$ Å thick and contains $-0.01 \pm 0.08$ $a$-GaAs and $0.52 \pm 0.03$ voids (leaving $0.49 \pm 0.11$ $c$-GaAs). The model spectrum in this case is in very good agreement with the data. The fact that the overlayer contains no $a$-GaAs indicates that material removal is now due primarily to reactive ion etching. The calculated void fraction is consistent with values typically encountered in the optical analysis of microscopically rough surfaces.

The above results might imply that the substrate material is substantially less damaged in the reactively ion etched sample, but what is really being established by this procedure is the nature of the residual layer that results

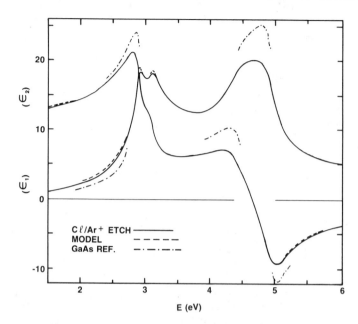

FIGURE 20. A reactively ion-etched GaAs sample, otherwise as Fig. 19. The $E_1$ and $E_1 + \Delta_1$ peaks are at 2.9 and 3.1 eV, respectively (Scherer et al., unpublished).

from plasma processing. The quality of the substrate is best examined by considering the $E_1$ and $E_1 + \Delta_1$ fine structure in $\epsilon_2$ near 3 eV. A close examination reveals that it is not as sharp as that given by model simulations. Thus in both experiments the substrates have been damaged. An estimate of the amount of this damage can be obtained by the method of Erman et al. (1983), who established a connection between impurity concentration and the $E_1 - E_1 + \Delta_1$ fine structure. They defined an "$s$-parameter" as the area between the measured tan $\psi$ spectrum and a line tangent to the spectrum drawn between the $E_1$ and $E_1 + \Delta_1$ peaks. This was shown to correlate with impurity concentration, showing noticeable deviation from its value for undoped material by impurity concentrations of the order of $3 \times 10^{17}$ cm$^{-3}$ and reaching its zero value at carrier concentrations near $3 \times 10^{18}$ cm$^{-3}$ and $3 \times 10^{19}$ cm$^{-3}$ for electrons and holes, respectively. As the $\langle \epsilon_2 \rangle$ and tan $\psi$ spectra are similar in form, the reduction by half of the dip between the $E_1$ and $E_1 + \Delta_1$ peaks seen in the $\langle \epsilon_2 \rangle$ spectra of both samples therefore suggests that both processes cause substrate disorder equivalent to that due to impurity concentrations in the $10^{18}$ cm$^{-3}$ range.

Of course, carrier lifetime measurements, not optical properties, must be used to critically assess material quality for device applications.

### E. SURFACE PREPARATION

Figure 21 shows ellipsometric data obtained during the cleaning of a Syton-polished (111) $c$-Si wafer that had been exposed to air for a two-year period prior to the study (Aspnes and Studna, 1981). The data are plotted in the complex $\langle \tilde{\epsilon} \rangle$ plane. Also shown is the trajectory calculated by assuming that the film is $SiO_2$, with increments of 5 Å indicated by fiducial marks. The data follow the trajectory very closely, beginning with an apparent thickness of 27 Å. However, methanol and water remove about 12 Å of this layer, which clearly could not have been $SiO_2$ but rather must have been organic contamination. A further reduction of 12 Å by hydrofluoric acid shows this to be the true limiting thickness of the oxide. The final 3 Å could only be removed by chemical polishing and preferential etching, indicating that this final layer was microscopic roughness. This example demonstrates that considerable insight as to the nature of over-

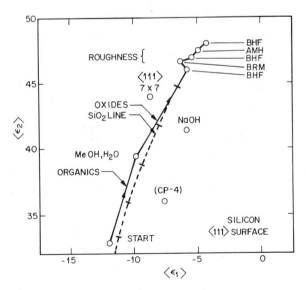

FIGURE 21.  Removal of an air-grown natural overlayer on a $c$-Si wafer, measured at 4.24 eV. Results for selected other surface treatments are also shown (Aspnes and Studna, 1981).

layer material can be achieved by examining its chemical reactivity, simply using the ellipsometer as a thickness monitor.

## F.   REAL-TIME MONITORING DURING CHEMICAL
## VAPOR DEPOSITION

Figure 22 shows an ellipsometrically measured $\Delta - \psi$ trajectory for an $a$-Si film prepared by chemical vapor deposition on a $Si_3N_4$-coated $c$-Si substrate at 580°C (Hottier and Theeten, 1980). Deposition followed the heating cycle, as indicated. As the thickness increased, optical contact to the substrate was gradually lost and the trajectory approached the location characteristic of $a$-Si. The dashed curve shows the trajectory expected if the material were being deposited uniformly. While the initial and final points agree, there is clearly a large discrepancy for intermediate coverages and

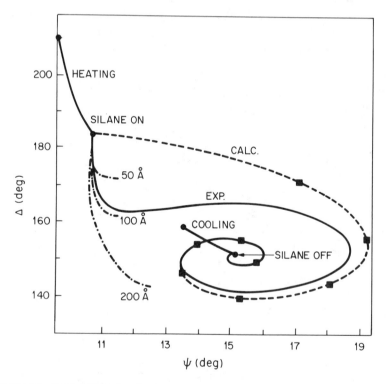

FIGURE 22.   Evolution of the ellipsometrically determined $\Delta - \psi$ trajectory for a film of $a$-Si deposited on a $Si_3N_4$-covered $c$-Si substrate (Hottier and Theeten, 1980).

particularly for the initial stages of deposition. The dot-dashed curves show the trajectories calculated assuming that initial growth does not proceed uniformly but by accretion about nucleation sites. The density of nucleation sites is an adjustable parameter that can be determined from the fact that the film coalesces at a thickness roughly equivalent to half the average separation of the nucleation centers. The best agreement using the Bruggeman effective medium model to approximate the dielectric response of the growing layer is found for an average separation of nucleation sites of about 90 Å. The interesting aspect of these data is that the nucleation growth mechanism can already be identified during the first Å of film growth.

## VI. Conclusion

The field of spectroscopic ellipsometry has undergone extensive development within the last decade, and possible applications are only beginning to be explored. The capability of optical probes to return real-time information about samples in reactive or high pressure ambients has not yet been used to any significant extent, yet as shown by the examples discussed here, such real-time feedback could have important and far-reaching consequences in plasma processing. As commercial spectroscopic ellipsometers have now become available, we anticipate a significant increase in their use in the future.

## References

Arwin, H. and Aspnes, D. E. (1984). *Thin Solid Films* **113**, 101.
Aspnes, D. E. (1976). Spectroscopic Ellipsometry of Solids. In: *Optical Properties of Solids: New Developments* (B. O. Seraphin, ed.). North-Holland, Amsterdam, p. 799.
Aspnes, D. E. (1982). *Thin Solid Films* **89**, 249.
Aspnes, D. E. (1985). *J. Mater. Education* **7**, 849.
Aspnes, D. E., Kelso, S. M., Lynch, D. W., and Olson, C. G. (1982). *Phys. Rev. Lett.* **48**, 1863.
Aspnes, D. E., Schwartz, B., Studna, A. A., Derick, L., and Koszi, L. A. (1977). *J. Appl. Phys.* **48**, 3510.
Aspnes, D. E. and Studna, A. A. (1980). *Surf. Sci.* **96**, 294.
Aspnes, D. E. and Studna, A. A. (1981). *SPIE Conf. Proc.* **276**, 227.
Aspnes, D. E., Studna, A. A., and Kinsbron, E. (1984). *Phys. Rev.* **B29**, 768.
Azzam, R. M. and Bashara, N. M. (1977). *Ellipsometry and Polarized Light.* North-Holland, Amsterdam.
Bagley, B. G., Aspnes, D. E., Adams, A. C., and Mogab, C. J. (1981). *Appl. Phys. Lett.* **38**, 56.
Bergman, D. J. (1980, 1982). *Phys. Rev. Lett.* **44**, 1285; *Ann. Phys.* **138**, 78.
Bruggeman, D. A. (1935). *Ann. Phys.* (Leipzig) **24**, 636.
Chang, R. P. H. (1977). *J. Vac. Sci. Technol.* **14**, 278.

Chang, R. P. H., to be published.

Chang, R. P. H. and Darack, S. (1981). *Appl. Phys. Lett.* **38**, 898.

Chang, R. P. H., Darack, S., Lane, E., Chang, C. C., Allara, D., and Ong, E. (1983). *J. Vac. Sci. Technol.* **B1**, 935.

Chang, R. P. H., Polak, A. J., Allara, D. L., Chang, C. C., and Landford, W. A. (1979). *J. Vac. Sci. Technol.* **16**, 888.

Erman, M., Theeten, J. B., Vodjdani, N., and Demay, Y. (1983). *J. Vac. Sci. Technol.* **B1**, 328.

Garnett, J. C. (1904, 1906). *Philos. Trans. R. Soc. London* **203**, 385; **205**, 237.

Greaves, G. N. and Davis, E. A. (1976). *Philos. Mag.* **34**, 265.

Hashin, Z. and Shtrikman, S. (1962). *J. Appl. Phys.* **33**, 3125.

Hauge, P. S. and Dill, F. H. (1973). *IBM J. Res. Dev.* **17**, 472.

Hottier, F. and Theeten, J. B. (1980). *J. Cryst. Growth* **48**, 644.

Jackson, J. D. (1975). *Classical Electrodynamics.* John Wiley & Sons, New York.

Keeping, E. S. (1962). *Introduction to Statistical Inference.* Van Nostrand, Princeton, Chapter 12.

Milton, G. W. (1980). *Appl. Phys. Lett.* **37**, 300.

————— (1981). *Optical Characterization Techniques in Semiconductor Technology.* SPIE Conf. Proc. **276**.

————— (1983). *Proceedings of the International Conference on Ellipsometry and Other Optical Methods for Surface and Thin Film Analysis.* J. Phys. Colloq. **C10**.

Rhinewine, M., Chang, R. P. H., and Slusher, R. P., unpublished.

Scherer, A., Craighead, H. G., and Aspnes, D. E., unpublished.

————— (1983). *Spectroscopic Characterization Techniques for Semiconductor Technology.* SPIE Conf. Proc. **476**.

Theeten, J. B., Chang, R. P. H., Aspnes, D. E., and Adams, T. E. (1980). *J. Electrochem. Soc.* **127**, 980.

Tu, C. W., Chang, R. P. H., and Schlier, A. R. (1983). *J. Vac. Sci. Technol.* **A1**, 637.

Vedam, K., McMarr, P. J., and Narayan, J. (1985). *Appl. Phys. Lett.* **47**, 339.

Viña, L. and Cardona, M. (1984). *Phys. Rev.* **B29**, 6739.

Wiener, O. (1912). *Abh. Math. Phys. Kl. Königl. Sächs. Ges.* **32**, 509.

Wilmsen, C. W. (1976). *Thin Solid Films* **39**, 105.

Yeh, P. (1979, 1980). *J. Opt. Soc. Am.*, **69**, 742; *Surf. Sci.*, **96**, 41.

# 4 Ion Beam Analysis of Plasma-Exposed Surfaces

B. L. Doyle

*Ion–Solid Interactions Division*
*Sandia National Laboratories*
*Albuquerque, New Mexico*

Wei-Kan Chu

*Department of Physics and Astronomy*
*University of North Carolina*
*Chapel Hill, North Carolina*

PLASMA DIAGNOSTICS
Surface Analysis and Interactions

109

## I.  Ion Beam Analysis

### A.  INTRODUCTION

Plasma research involves the study of the plasma itself as well as the investigation of the effects of plasmas on exposed surfaces or, more precisely, the analysis of surfaces exposed to and thereby modified by plasmas. This chapter concentrates on analysis methods utilizing MeV-energy ion beams for the study of plasma surface interaction effects on materials.

We will start with a general description of the various ion beam analysis techniques, including a description of the laboratory setup and instruments required. Each method will then be described in detail, followed by several ion-beam analysis examples which emphasize the effects experienced by surfaces that are exposed to various types of plasmas. Two areas of plasma-surface interactions are highlighted in this chapter—results which have been obtained in the plasma-induced modification of the near surfaces of 1) electronic materials and 2) first-wall or limiter materials used to define the plasma boundary in magnetic confinement fusion reactors.

### B.  ION BEAM METHODS OF SURFACE ANALYSIS

When the surface of a sample is bombarded with an energetic (several MeV) ion beam, several outcomes can result:

a) Most likely, the ion will move forward into the target and simply lose all of its energy to the sample atoms through the excitation of the electrons of the target atoms. When all of the ion's initial kinetic energy is expended, it will finally come to rest inside the sample just as in the case of *ion implantation.*

b) Occasionally, some of the ions will suffer an elastic collision with a target nucleus and radically change their direction of motion. When the collision is nearly head-on (i.e., when the impact parameter of the collision is close to zero) and the projectile ion is lighter than the target atoms, the projectile can scatter backward and re-emerge from the sample. This scattering scenario is the basis for *Rutherford backscattering spectrometry* (RBS).

c) During an elastic scattering event, a target atom will, at times, be recoiled into a forward direction. Some of these recoiled atoms will possess a velocity and direction which allow them to escape the sample, and this is the foundation for *elastic recoil detection* (ERD).

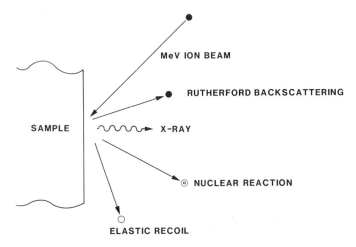

FIGURE 1.   Schematic of MeV-energy ion beam analysis methods.

d) During the excitation and ionization process discussed above, an inner shell electron of a target atom may be ejected, generating characteristic x-rays when the atom relaxes to its ground state. The spectral analysis of x-rays generated in such a manner is called *particle induced x-ray emission* (PIXE).

e) The incident projectile can cause nuclear reactions with the target nucleons when the Coulomb barrier is exceeded. When some of the reaction products such as protons, $\alpha$ particles, or $\gamma$ rays are detected, this ion beam analysis technique is called *nuclear reaction analysis* (NRA).

Figure 1 gives a schematic diagram of the above phenomena, revealing the information contained in the sample.

Regarding the samples highlighted in this chapter, they are typically flat surfaces of a bulk material which have been exposed to a certain plasma environment. The sample specimen could be a semiconductor exposed to sputtering or *reactive ion etching* (RIE) treatments, or it could be a low atomic number fusion material, such as graphite, used on a limiter or aperture subjected to various fusion plasma conditions. As indicated above, the individual examples will be given as the analytical ion-beam techniques are introduced in the next few sections. However, before we start describing these methods, we will first discuss the experimental setup required to perform ion beam analyses.

C. Ion Beam Instrumentation

The experimental setup needed to execute the experiments illustrated in Figure 1 requires: 1) a facility for producing MeV ions as projectiles, 2) a beam transport system capable of bringing the ion beam to a target, 3) a target chamber into which the samples to be analyzed are placed and oriented, and 4) detectors for the energy analysis of the atomic or nuclear reaction products which result when the ion beam strikes the target sample.

The simple setup which produced Rutherford's (1911) famous scattering experiment 75 years ago is illustrated in Figure 2. Figure 2a shows the conceptual layout of the scattering experiment, while the drawing in Figure 2b reproduces the apparatus used by Geiger and Marsden (1913) to test and confirm the new model of an atom as conceived by Rutherford. These early experiments used radioactive α emitters as their MeV ion beam source.

The ion beam accelerators producing variable but monoenergetic ion beams which were invented and developed in the 1930s were more versatile. These instruments were used primarily for research in atomic and nuclear physics. Numerous discoveries of modern physics were made through the use of accelerators, and the need to perform difficult experiments stimulated the accelerator technology into producing better and larger machines.

Today, the experimental systems for ion beam analysis use accelerators instead of natural α-particle sources and sophisticated detecting systems are used for analysis. Most of these accelerators are similar to those utilized for low-energy atomic or nuclear physics experiments. These accelerators are typically Van de Graaff (tandem or single-ended) style generators or other electrostatic ion accelerators. The floor plan of a very elaborate laboratory facility specializing in ion beam modification and analysis is illustrated in Figure 3. This system, located at Sandia National Laboratories in Albuquerque, New Mexico, is capable of generating a beam of virtually every element in the periodic table at energies in excess of 10 MeV. A tandem Van de Graaff generator with a terminal voltage of 5 MV is used, and a switching magnet diverts the ion beam to separate beam lines for different types of analyses.

The ion beam analysis community represents a group of new users emerging in the late 1960s, and the expansion of their activities has stimulated the development of compact-size medium-energy accelerators dedicated to materials analysis. For example, both the General Ionex and the National Electrostatic corporations supply such products for analytical purposes. Figure 4 shows a very small tandem accelerator (2 MeV) with beam line, target chamber, and electronics for data handling all confined within a volume of 3 m × 2 m × 2 m.

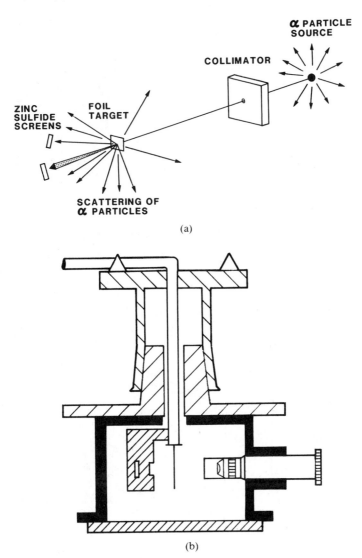

(a)

(b)

FIGURE 2.   (a) Layout of scattering experiment used by Geiger and Marsden. (b) Drawing of the apparatus used by Geiger and Marsden to test the new model of the atom.

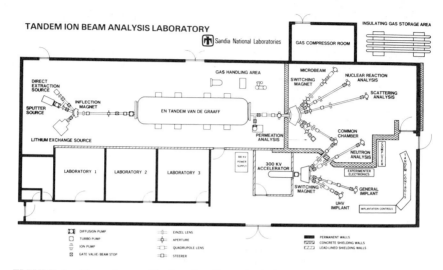

FIGURE 3.  Floor plan on Tandem Ion Beam Analysis Laboratory at Sandia National Laboratories, Albuquerque, NM.

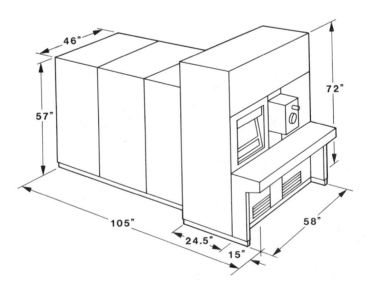

FIGURE 4.  Drawing of Small General Ionex fixed-energy tandem accelerator for RBS analysis.

The energy regime required for RBS, PIXE, and ERD is around 1–2 MeV, while nuclear reactions generally require higher energies. Protons and $^4$He ions are most useful for RBS, PIXE, and ERD, while some other light ions such as D or $^3$He can be used for nuclear reactions. Some reactions, such as $^{15}$N(p, $\alpha\gamma$), $^{19}$F(p, $\alpha\gamma$), can be inverted by using $^{15}$N or $^{19}$F beams for the detection of hydrogen. In general, an accelerator which can provide light ions with an energy of a few MeV is capable of performing most, probably 90%, of the ion beam analysis techniques which are in common use.

The ion beam experiments require a beam line, a switching or steering magnet, and a target chamber all under vacuum. A typical vacuum of $10^{-6}$ torr is sufficient for routine analysis.

Backscattered ions, recoil atoms, or nuclear reaction products from the sample can be detected and analyzed by solid state detectors. A silicon surface barrier detector 1 cm or less in diameter is commonly used for energy-analyzing the particles which scatter into a small solid angle defined by slits. This detector, in combination with a preamplifier and linear amplifier, generates a voltage signal proportional to the energy of the particle entering the detector. When photons need to be energy analyzed, as in the case of PIXE (x-rays) or resonant NRA-RNRA ($\gamma$-rays), either lithium drifted Si [Si(Li)], intrinsic Ge, or NaI detectors are commonly used.

A multichannel analyzer converts the analog input voltage signal into a digital form by sorting all signals according to detected energy during an experiment. At the end of a typical measurement, which takes about 10 to 30 minutes, the detectors receive about $10^6$ particles or photons with various energies. The multichannel analyzer yields an energy spectrum of counts versus energy-channel number. This spectrum represents the "raw" form of ion beam analysis data and as such contains information about the sample, such as the depth at which scattering events or nuclear reactions occur, in addition to the number of scattering event centers which exist within the sample. The extraction of this information in the form of a depth concentration profile or an areal density requires a quantitative understanding of the underlying atomic or nuclear physics responsible for the detected event. This basic understanding is outlined in Section II.

### D. NUCLEAR MICROPROBE

For most ion beam analysis experiments, the beam spot is about 1 mm$^2$ or larger. Since the development of strong focusing lenses for high-energy ions, beam spot sizes of $\mu$m$^2$ dimensions have been obtained. (A good background in this area can be obtained through the recent PIXE and Micro-beam Conferences, Nuclear Instruments and Methods, Vols. 165, 181, 197, and B23, 1987). Small beam diameters are particularly important in experi-ments where the sample is either very small or for experiments which require lateral information, as is the case for studies of microelectronic circuits or in metallurgical applications where individual crystalline grains are to be examined. Other areas which involve the need for analysis beams of small dimensions have been detailed in a review articles by Cookson (1979, 1981), Legge (1984), and Doyle (1985). Beam scanning capabilities have been incorporated in many nuclear microprobes to provide two-dimensional elemental imaging similar to that obtained on an electron microprobe. Beam resolution is not seriously affected by scanning the beam a few 100 $\mu$m. Three-dimensional profiling is possible by using lateral scanning while detecting one of the depth-sensitive nuclear signals (Doyle, 1983a, 1986).

Nuclear microprobe analysis, the name applied to the technique of performing ion beam analysis with a microfocused beam, is highly competi-tive with other elemental-imaging techniques and has clear advantages over most in the area of sensitivity and versatility. Because virtually all of the various nuclear scattering and ion induced x-ray production cross sections have been determined, direct quantitative measurements can be performed without the need of reference standards.

Another very attractive feature of this type of microanalysis is that it does not necessarily have to be performed in an evacuated sample chamber (Doyle, 1983). Such *ex vacuo* ion beam analysis (X-IBA) is usually per-formed by passing the high energy ion beam through a thin foil into either a He or air environment. The best results are obtained by minimizing the air path, and detection of the resultant nuclear or atomic signal can be performed either in air or, for RBS, back inside the vacuum chamber. X-IBA is at times the only analysis technique capable of studying samples which are either too large or otherwise unsuitable for insertion into a vacuum chamber. The fact that this type of analysis is nondestructive is also a benefit, allowing samples to be either reused or to undergo further processing.

The nuclear microprobe at Sandia National Laboratories is pictured in Figure 5. The various components of the probe, which is $\sim 3$ m in length, are indicated in the figure and caption. A magnetic quadrupole doublet lens is used which is capable of focusing ions with a $ME/q^2$ (mass-energy) product of $\simeq 100$ MeV amu. The high mass-energy product of this lens is necessary in order to focus the beams used for high-energy ERD (20 MeV $Si^{4+}$).

The smallest spot sizes obtained to date on the Sandia nuclear micro-probe have been $\simeq 1$ $\mu$m in diameter, which is the spherical aberration

## SANDIA NUCLEAR MICROPROBE

FIGURE 5. Photograph of nuclear microprobe at Sandia. The various beam forming elements are indicated.

limit, but as the beam current density is low ($\approx 1$ pA/$\mu$m$^2$), useful beams must have diameters of 2–3 $\mu$m or more in order to accomplish experiments in reasonable amounts of time. The Melbourne probe currently has the smallest spot size of only 0.25 $\mu$m (Legge et al., 1986).

Computer assisted data collection and reduction are essential for nuclear microanalysis, not only to process the vast amount of information gathered, but also because of the complexity of some of the techniques. The importance of this automation cannot be overstated; without sophisticated hardware and software, the operator quickly becomes overwhelmed by the amount of data collected. Several examples of multidimensional nuclear microprobe data are given in the following sections.

### E. PLASMA-MATERIALS EXAMPLES

There are two groups of problems involving materials exposed to plasmas. The first group uses plasma exposure as a means to enhance materials properties or to actually form materials. For example, plasma deposition and plasma etching are widely used in semiconducting materials processing. The second group involves the situation where the interaction between a plasma and surrounding materials is unavoidable. For example, magnetic-confinement fusion reactor research requires a minimum interaction between the plasma and the reactor wall or other in-vessel components. We will briefly discuss the two sets of problems here.

*1.  Semiconductor Materials Processing*

In silicon or compound semiconductor large scale integration, plasma etching techniques are essential for device fabrication. For example, to open a submicron contact, one must etch through the overlaying $SiO_2$ without broadening the contact area. The Si surface should also be left free of contamination and hopefully free of radiation damage as well. This process requires a good mask and the anisotropic etching in $SiO_2$, high etch ratio between $SiO_2$ and Si (selectivity), and a minimum deleterious effect between the plasma and the Si surface. Similar etching requirements are needed for various combinations of insulators/metals/semiconductors. Thin film etch rate, selectivity, contamination, and substrate damage are frequently investigated by ion beam analysis. In the Rutherford backscattering section we will give an example of Si surface damage by exposure to low-energy hydrogen or noble gas plasmas.

Plasma enhanced deposition is another important area where semiconductor device processing directly benefits from plasma-materials interactions (PMI). Deposition of $SiO_2$, $Si_3N_4$, Si-oxynitride, TiN, AlN, and many other elements or compounds can be achieved by coupling the plasma energy to a given chemical reaction on a hot substrate. Plasma assisted deposition and etching are topics covered in a recent book edited by Coburn et al. (1986).

*2.  Fusion Reactor Research*

In contrast to the many beneficial uses of plasmas to process/grow semiconductors or other materials, there are really no redeeming features of the PMI which occurs at the edge of an operating tokamak or any other controlled fusion device. Virtually every consequence of the PMI on the fusion plasma or surrounding components is deleterious. In fact, the only working examples of controlled fusion (i.e., stars) involve gravitationally confined plasmas which are maintained in a UHV system which does not have walls, and therefore no PMI results! Plasma scientists have seized on this clue from nature and expended considerable effort over the past three decades of fusion energy research to minimize the PMI in their plasma containment devices.

## II. Rutherford Backscattering Spectrometry (RBS)

In this and the next few sections, we will explain the RBS, ERD, PIXE, NRA, and RNRA methods individually. Their concepts and principles will be described first, then their applications will be illustrated with various examples related to plasma treated surfaces. Incorporation of the ion beam microprobe will also be given when measurements require lateral resolution. We start with the most common ion beam analysis technique of all, RBS.

Among all the ion beam analytical methods, RBS is probably the easiest to understand. During the past decade RBS has evolved into one of the major techniques of materials characterization. This is primarily due to the fact that quantitative information can be obtained by RBS without the need for standard samples, and the typical analysis time is only 10 to 20 minutes. Monographs and review articles on RBS are available (for example, Chu et al., 1973 and 1978). Nuclear and atomic physicists have realized the analytical power of the ion beam and frequently use the scattering technique to check for impurities, thickness, and composition of their targets. The basic concept of RBS is quite straightforward. In this section, we will first describe the concept and principle of RBS, then present applications on various samples subjected to plasma bombardment.

### A. Basic Concepts

Scattering is a two-body collision process. The basic information involved in an elastic scattering experiment, such as in Figure 6, are the energy of the scattered particle and the scattering probability at a given scattering angle. If the energy loss of the projectile due to ionization and excitation of the target electrons before and after the scattering event is understood, RBS provides the capability of depth profiling. We will elaborate on the basic concepts of scattering energy, probability, and slowing down.

### 1. Collision Kinematics (K), Scattering Energy

When a projectile of mass $M_1$ and energy $E_0$ is scattered from a target atom of mass $M_2$ ($M_1 < M_2$) and the impact is very close, the target atom

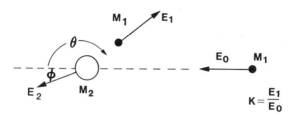

FIGURE 6.   Scattering geometry for RBS.

will recoil into the forward direction with energy $E_2$, while the projectile is scattered back with a scattered energy $E_1$, as shown in Figure 6. One can easily derive a scattering kinematic factor, which is defined as the ratio of $E_1$ to $E_0$, from the conservation of momentum and energy.

$$K = \frac{E_1}{E_0} = \left[ \frac{\left( M_2^2 - M_1^2 \sin^2 \theta \right)^{1/2} + M_1 \cos \theta}{M_2 + M_1} \right]^2.  \qquad (1)$$

One can see that Eq. (1) reduces to $(M_2 - M_1)/(M_2 + M_1)$, at $\theta = 90°$, and it reduces to $(M_2 - M_1)^2/(M_2 + M_1)^2$ at $\theta = 180°$.

One observes that the kinematic energy loss from a collision is largest for full backward scattering (180°). This is the reason why in backscattering spectrometry the detector is placed at the most backward angle feasible to yield the best mass resolution and to minimize the effects of kinematic energy broadening caused by finite detection solid angles.

Equation (1) states that by knowing the incident energy $E_0$, a measurement of the scattered energy $E_1$ determines the mass of the target atom where scattering occurs. In other words, the energy scale on a given backscattering spectrum can be translated into a mass scale. The mass resolution for RBS is directly related to the energy resolution of the detecting system. Typically for He ion scattering at 2 MeV, a solid-state detector with energy resolution of around 15 keV is sufficient to distinguish scattering from different Cl isotopes, but not able to separate Ta from Au and Pt.

## 2.   Scattering Cross Section σ

In the previous section we described the energy of the scattered projectile when a scattering event occurs. One should realize that the probability of a scattering event occurring is very low. The vast majority of the projectiles experience only small-angle deflections and eventually stop inside the solid.

Only a very small portion of the ion beam is backscattered and detected for
RBS analysis. The total number of detected particles, $A$, can be written as

$$A = \sigma \Omega Q (Nt)$$

$$\begin{pmatrix} \text{number of} \\ \text{detected} \\ \text{particles} \end{pmatrix} = \sigma \begin{pmatrix} \text{solid} \\ \text{angle} \end{pmatrix} \begin{pmatrix} \text{total number} \\ \text{of incident} \\ \text{projectiles} \end{pmatrix} \begin{pmatrix} \text{number of} \\ \text{target atoms} \\ \text{per unit area} \end{pmatrix}, \quad (2)$$

where $\sigma$ has the dimension of area ($cm^2$) and is called the differential
scattering cross section and abbreviated as scattering cross section. Equation (2) can be thought of as a general equation which relates detected
events to the concentration or areal density of target atoms, and this
equation is used in following sections.

One can calculate the scattering cross section for an elastic collision
where Coulomb repulsion is the force between the two nuclei. Based on
Rutherford's formula for the lab coordinates (Ziegler and Lever, 1973)

$$\sigma = \left( \frac{Z_1 Z_2 e^2}{4E} \right)^2 \left( \frac{4}{\sin^4 \theta} \right) \left( \frac{\left[ \left\{ 1 - \left( \left[ \frac{M_1}{M_2} \right] \sin \theta \right)^2 \right\}^{1/2} + \cos \theta \right]^2}{\left\{ 1 - \left( \left[ \frac{M_1}{M_2} \right] \sin \theta \right)^2 \right\}^{1/2}} \right). \quad (3)$$

This equation reveals that the significant functional dependences of
Rutherford differential scattering cross sections are:

a) proportional to $Z_1^2$; the higher the atomic number of the projectile, the
   larger the scattering cross section (for He ions $Z_1 = 2$);
b) proportional to $Z_2^2$; for a given projectile, it is more sensitive to detecting
   higher $Z$ elements than lower $Z$ elements; and
c) proportional to $E^{-2}$; the scattering yield increases with decreasing ion
   beam energy.

One can easily see from Eq. (2) that the calculation of $\sigma$ and the
measurement of the number of detected particles give a direct measure of
the number of target atoms per unit area ($Nt$), where $N$ is the number of
target atoms per unit volume and $t$ is the thickness of the target. The
scattering cross-section concept expressed in Eq. (2) therefore leads to the
capability of quantitative analysis of atomic composition by RBS. This
exploitation of atomic or nuclear physics ideas is common to all ion beam
analysis techniques. In fact, the experiments performed in atomic/nuclear
physics are virtually identical to those of ion beam analysis—the only
difference is that the emphasis of the former is on determining $\sigma$ from Eq.
(2), while for ion beam analysis the unknown in Eq. (2) is ($Nt$).

The Rutherford scattering cross section expressed in Eq. (3) is based on the assumption that both the projectile and target are point charges and their interaction follows the Coulomb law (i.e., $1/r$). The factor containing mass ratios in Eq. (3) comes from the laboratory/center of mass conversion. A correction of 1%–2% due to electron screening, which is most pronounced for high-$Z$ target atoms, has been studied by Macdonald et al. (1983).

## 3.  Energy Loss (dE/dx)

We have mentioned that an energetic ion impinging on a target will penetrate into it along the direction of incidence, while large angle scattering from a target atom is highly unlikely. The energy loss of the incident ion is overwhelmingly determined by the Coulomb interaction between the moving ion and electrons in the target material. The moving ion loses its energy by excitation of the target electrons. The numerous discrete interactions with electrons slow down the moving ion by transferring the ion energy to the target electrons in an almost continuous friction-like manner. This type of interaction does not significantly alter the direction of the ion beam. The energy loss per unit path length (i.e., the stopping power) $dE/dx$ can be measured experimentally by passing an ion beam through a thin foil of thickness $\Delta x$ and measuring the energy loss $\Delta E$; it can also be calculated theoretically. It is the energy loss process that produces the "depth perception" for RBS.

The value of $dE/dx$ is a function of the atomic number and mass of both the target and projectile atoms, together with the instantaneous energy of the projectile. A stopping power term which is used frequently in RBS analysis is called the stopping cross section, $\epsilon$, which is defined as

$$\epsilon = \frac{1}{N} \frac{dE}{dx},  \tag{4}$$

where $N$ is the atomic density and the target material, i.e., the number of target atoms per $cm^3$. $\epsilon$ carries the unit of eV-$cm^2$/atom, hence the term *stopping cross section*. This is very similar in name but quite different in meaning from the *scattering cross section*, which carries the unit of $cm^2$. One can comprehend the term $\epsilon$ as the energy loss on the atomic scale. For example, if a thin layer contains ($N\Delta x$), number of atoms per $cm^2$, then the energy loss of an ion beam passing through this thin layer becomes $\Delta E(eV)$ and $\Delta E = \epsilon(N\Delta x)$, where $\epsilon$ is the cross section of the given ion in the given elements. The concept of stopping cross section can be generalized for molecule-containing targets or for mixture samples by the Bragg's rule of additivity of stopping cross sections

$$\epsilon^{A_m B_n} = m\epsilon^A + n\epsilon^B, \qquad (5)$$

where $\epsilon^A$ and $\epsilon^B$ are stopping cross sections of elements $A$ and $B$, which are constituents of molecule $A_m B_n$ (or mixture of $A$ and $B$ with atomic concentration ratios of $m$ and $n$).

A number of reviews and reports on the subject of energy loss of charged particles in matter have been written over the years. For the purpose of RBS analysis using MeV He ions, tabulations of $\epsilon$ for all elements have been given by Ziegler and Chu (1974) and updated by Ziegler (1978).

## 4. Channeling Effect

When charged particles penetrate a single crystal along or near a major axis, they experience a collective string potential produced by the rows of atoms along that axis. The string potential will steer the charged particle back and forth within a "channel" as it moves through the crystal. A critical angle is defined as the limiting angle between the incident direction and row of atoms such that the steering effect exists. When charged particles are incident in a direction exceeding the critical angle, those particles have a transverse kinetic energy exceeding that provided by the collective string potential. The collective steering effect subsequently disappears. The axial steering effect is called axial channeling; planar channeling is similarly defined.

Ion channeling in conjunction with ion backscattering measurements have been used extensively for the study of the near surface of single crystals. Most of the studies by channeling involve backscattering of light particles such as protons or helium ions with energies of a few hundred keV to a few MeV. A highly precise ($< .01°$) multi-axis goniometer is required for target manipulation and crystal alignment.

An excellent monograph on the theory of channeling and the experimental study of channeling effects with emphasis on materials analysis is given by Feldman et al. (1982). An example of damage measurement by RBS channeling analysis on reactive ion etched surface will be given in the next section.

## B. PLASMA ETCHING

Reactive ion etching (RIE) is a preferred dry process for the fabrication of micron as well as submicron structures of very large scale integrated (VLSI)

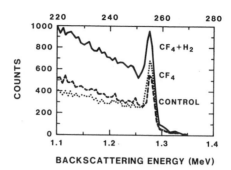

FIGURE 7.   RBS-channeling spectra of 2.3 MeV He ions from three Si surfaces exposed to reactive ion etching. Greater damage is observed when $H_2$ is present.

circuits. A high selectivity for the etching of $SiO_2$ as compared to Si is a necessity in many processing steps. Such selectivity can be obtained readily by the addition of hydrogen to a $CF_4$ plasma (Heinecke, 1976) or by using a hydrogen containing gas such as $CHF_3$. The addition of $H_2$ to a $CF_4$ plasma reduces the etching rate of Si by deposition of a polymeric material on Si. It does not affect the etching rate of $SiO_2$ significantly (Coburn, 1979).

However, wafers exposed to any hydrogen containing plasma were observed to have excessively high leakage current, thus poor retention times (Frieser et al., 1983). They have also observed by RBS/channeling that hydrogen containing plasmas ($CF_4 + H_2$) produce greater damage at the surface. Figure 7 illustrates their RBS/channeling studies of $CF_4$ and ($CF_4 + H_2$) etched samples. The channeling spectra clearly indicate that $CF_4 + H_2$ gives a higher surface damage of Si. For the control sample, the surface peak is due to the scattering of the probing projectile 2.3 MeV $He^4$ from the top layer of atoms and any residual surface oxide on Si. They attribute the excessive damage caused by the $CF_4 + H^2$ plasma to the hydrogen bombardment which results from the H component of the plasma.

Vitkavage et al. (1986) have studied the Si surface damage due to the bombardment of low-energy Ar, Ne, and H as a function of dose and energy using a 98° scattering angle geometry. Figure 8 shows four RBS/channeling spectra of the 1.6 MeV $He^4$ ion scattering from Si $\langle 100 \rangle$ bombarded by 1 keV H, Ne, and Ar with a dose of $1.5 \times 10^{17}$ cm$^2$. The analysis of this figure along with other energies and doses are listed in Table 1. The above examples show that RBS/channeling is a powerful method which yields quantitative results on surface damage due to various sputtering or etching conditions.

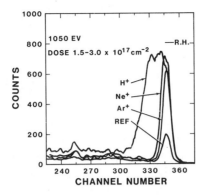

FIGURE 8. Ion channeling spectra for different mass ions implanted into Si.

**Table 1.** Analysis of H, Ne, and Ar Implantation Damage Distributions in Si. (Vitkavage et al., 1986)

| Ion | Energy (eV) | Dose (×10 cm) | Total Damage (Å) | Oxide (Å) | 50% Width (Å) | 15% Width (Å) |
|-----|-------------|---------------|------------------|-----------|---------------|---------------|
| Ar  | 250  | 19.00  | 23  | 28  | —   | —   |
| Ar  | 1050 | 30.00  | 52  | 35  | —   | —   |
| Ne  | 260  | 24.00  | 25  | 24  | —   | —   |
| Ne  | 1060 | 16.00  | 67  | 37  | 93  | 106 |
| H   | 260  | 25.00  | 130 | 36  | 170 | 203 |
| H   | 560  | 23.00  | 171 | 52  | 220 | 248 |
| H   | 1060 | 0.87   | 45  | 36  | 105 | 148 |
| H   | 1060 | 4.80   | 265 | 45  | 310 | 334 |
| H   | 1060 | 9.20   | 254 | 46  | 309 | 370 |
| H   | 1060 | 16.00  | 245 | 40  | 328 | 378 |
| H   | 1060 | 130.00 | 349 | 105 | 385 | 472 |

Widths have been corrected for detector broadening

## C. Fusion Materials Research

This section deals with one of the special areas where RBS analysis in conjunction with a nuclear microprobe enjoys a certain uniqueness, namely, its use to obtain multidimensional concentration profiles of elements in solids for cases in which one of the dimensions is depth. The application of a scanning nuclear microprobe to measure the concentration of elements in three dimensions was first suggested by Martin (1973) (just following the introduction of the initial probe by Cookson and Pilling (1972)), and it remains the only instrument capable of providing 3-D elemental concentration images nondestructively.

The subject of this example was the moveable limiter in the Tokamak Fusion Test Reactor (TFTR), which is located at the Princeton Plasma Physics Laboratory in Princeton, New Jersey (Dylla et al., 1985, Cecchi et al., 1984). The plasma contacting surfaces of the moveable limiter consisted of an array of titanium carbide (TiC) coated graphite tiles. The TiC coating was deposited as a nominally 20 $\mu$m thick coating on POCO™ graphite (grade AXF-5Q) substrates using a CVD process described by Trester et al. (1981). The primary purpose of the TiC coating was to decrease the anticipated plasma-induced and chemically-enhanced sputtering which these components would have suffered if the POCO™ graphite substrate had been exposed to the tokamak plasma (Mattox et al., 1979). Reductions in the eventual tritium inventory of these tiles was also expected through the use of these thin TiC coatings, as were beneficial decreases in limiter conditioning time (Doyle et al., 1984). *Conditioning* is a term applied to the apparent physical and chemical changes that occur in the near surface region of limiter materials as a result of high fluence plasma exposure.

This example focuses on the microbeam RBS profiling analysis of the limiter tiles following exposure to 800 high-power ohmically heated discharges in TFTR. The primary reason for this analysis was that during the plasma operations period a failure of the TiC coating was evident from visual observations and plasma contamination observed during discharges. After the tiles were removed from the TFTR, subsequent examination revealed that approximately 30% of the tile surface area had been stripped of the TiC coating. The failure appeared to be due to an uncontrolled variable in the TiC deposition process, judging from the observations that the coating failure was only weakly coupled to the power deposition profile on the limiter and occurred at power loadings that were insignificant in comparison to the design values ($\sim$ 2 kW/cm$^2$). It was hoped that detailed ion microbeam analysis would help shed light on the cause and lead to a remedy for the coating detachment process. After removal of the tiles from TFTR, several tiles were chosen for detailed surface measurements. The chosen tiles came from an area of the limiter that experienced the peak heat flux. Two of the tiles had almost complete removal of the TiC coating, and one of the tiles had the TiC coating virtually intact.

Two RBS spectra are plotted in Figure 9; the top spectrum was taken at a position on the TiC coated graphite tile (tile #28) where the coating had remained intact, while the bottom spectrum shows the results of an RBS measurement of a control TiC coated graphite standard. For both of these

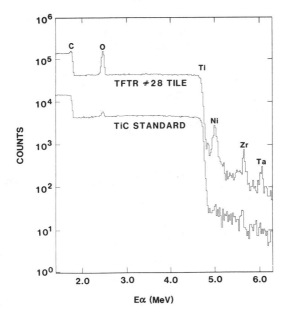

FIGURE 9. RBS spectra taken on a TiC coated graphite standard (lower) and a similarly coated tile used in the TFTR tokamak at Princeton. A deposition layer comprised of Ta, Zr, Ni, and O is observed on the TFTR tile.

experiments, 6.5 MeV $\alpha$ particle projectiles were used in the RBS analysis. For the control sample, only the Ti and C ledges, together with a thin surface oxide peak, are observed. The depth profiled for the element Ti by the 6.5 MeV $\alpha$ particle beam is approximately 4 $\mu$m between the Ti surface ledge at 4.8 MeV and at a backscatter energy just above the O peak at 2.4 MeV. For the TFTR tile sample, the same Ti and C ledges are observed, but, in addition to evidence for a thicker oxide, an added impurity layer consisting of Ni, Zr, and Ta is detected. Analysis of this data determined that the TiC coating of both the standard and TFTR tile was stoichiometric to within the uncertainty of the measurement (5%), and therefore the coating detachment could not be blamed on an improper composition. Further details concerning this impurity layer is given in Section IV on PIXE.

Figure 10 shows a 3-D profile of Ti on the TiC coated graphite tile at a position near the maximum power flux of the plasma. The top three panels were taken using 6.5 MeV $\alpha$ particles, while the remaining five panels, which represent deeper depths, were collected using 6.5 MeV protons. The two beams were selected in order to highlight details near the surface and

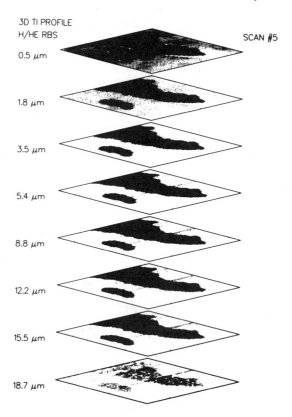

FIGURE 10.   A 3-D profile of the concentration of Ti for a TiC coated graphite tile used in TFTR. Each panel represents a different depth and darkness indicating increasing concentrations.

still be able to profile through the entire thickness of the TiC coating, which was nominally 20 μm thick. As indicated above, the 6.5 MeV α particles can profile only ~ 4 μm in Ti before the O surface peak appears. For 6.5 MeV protons, this depth is extended to ~ 25 μm at the expense of loss of resolution at the surface. The results plotted in Figure 10 are interpreted as follows: The darker regions in each panel indicate a greater local concentration of Ti. When an area is fully black, this indicates that the concentration of Ti at this position is equal to that found in TiC or $5.0 \times 10^{22}/\text{cm}^3$. Each panel represents a single depth slice of the sample and the mean depth of each panel is indicated to the right. The area scanned was 6.4 × 6.4 mm, and therefore this experiment could be categorized as more of a milliprobe than microprobe analysis. These data clearly show that the coating in this

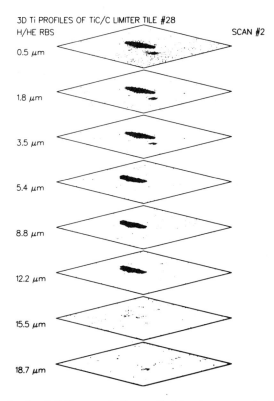

3D Ti PROFILES OF TiC/C LIMITER TILE #28
H/HE RBS                                    SCAN #2

0.5 μm

1.8 μm

3.5 μm

5.4 μm

8.8 μm

12.2 μm

15.5 μm

18.7 μm

FIGURE 11.   Another 3-D Ti concentration profile taken in an area of extreme TiC coating exfoliation.

region is uniform in thickness. This conclusion can be drawn because of the grayness uniformity of the island and peninsula of TiC displayed by the slice at 18.7 μm. The area adjacent to these two features suffered coating detachment, and the only Ti remaining is located in the very near surface. It is thought that the Ti in this region was transported there following the removal of the coating by an erosion/redeposition process which was operating on regions of the tile where the coating remained.

Another 3-D Ti profile taken on the same tile but in a region of more severe coating loss is shown in Figure 11. Two small islands of TiC are observed, one which is only 3.5 μm thick, the other which is 12.2 μm thick. Very little near-surface deposition of Ti is observed at this location, indicating either a low deposition rate, high erosion rate, or that the coating detachment in this vicinity happened just before the removal of this tile from the tokamak. The thinness of the smaller island of TiC, together with its proximity to the thicker island, suggests that it may have suffered an intracoating fracture, which could have played a role in the detachment of the rest of the coating.

These and other two- and three-dimensional concentration profiles generated by the nuclear microprobe verified that large sheets of the coating had indeed been removed, probably in single events, and that for most cases the coating had become detached at or near the Ti interface. More data obtained on this sample is presented in the PIXE section.

## III. Elastic Recoil Detection (ERD)

The elastic recoil detection (ERD) technique was first used by L'Ecuyer et al. (1976) and is very similar to RBS. Both techniques utilize the elastic collision, collision cross section, and energy loss concepts developed in the previous section. The only major difference between ERD and RBS is that with ERD recoil atoms from the sample are analyzed, while in RBS the backscattered projectiles are analyzed. Because a projectile can only be backscattered ($\theta > 90°$) from target atoms heavier than itself, helium ion scattering cannot detect hydrogen or helium atoms. No such restriction applies to ERD, and this makes ERD a very powerful tool for light atom detection. In this section we will describe the ERD technique and give examples which illustrate the method.

### A. Basic Concepts

The basic concept and experimental setup of ERD are schematically illustrated in Figure 12. The scattering geometry given in this figure is very similar to that of RBS given in Figure 6. The only difference is that in Figure 12, the recoiled particle (mass $M_2$ with $M_2 < M_1$) will only make a forward scattering angle $\theta$ of less than 90° and the recoil will also be in the forward direction. Some of the recoil atoms with recoil energy $E_2$ will escape the target surface, penetrate a thin foil particle filter (typically 10 $\mu$m of mylar), and be energy analyzed by a solid state detector. By conservation of both energy and momentum, it can be shown that the recoil kinematic factor is given by

$$K_R = \frac{E_2}{E} = \frac{4 M_1 M_2}{(M_1 + M_2)^2} \cos^2 \phi. \qquad (6)$$

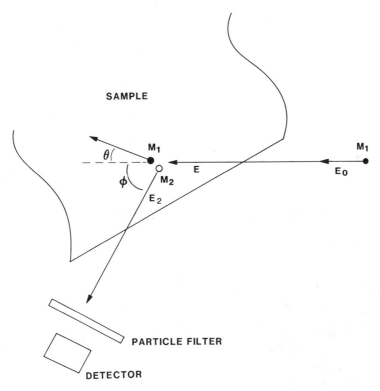

FIGURE 12. Scattering geometry for performing ERD.

Equation (6) gives the ratio of the energy of the recoil atom to that of the incident projectile and is derived in the same manner as that for the backscattered particle in Eq. (1).

The recoiling cross section for the case of simple Coulomb collisions is given by

$$\sigma = \left( \frac{Z_1 Z_2 e^2 (M_1 + M_2)}{2 M_2 E} \right)^2 \frac{1}{\cos^3 \phi}. \quad (7)$$

Equation 2 can be used to relate the number of detected recoils to the areal density of the constituents recoiled from the sample, just as in the case of RBS. The depth scale can also be established by knowledge of the stopping powers (or stopping cross sections) of both the incoming projectile and outgoing recoil atom in the target matrix.

The particle filter in Figure 12 is used to sort the recoil atoms from the forward-scattered projectiles. Through careful selection of this foil's thickness, the recoil atoms can be transmitted with minimal loss of energy while the scattered projectiles are completely stopped. Doyle and Peercy (1979a) showed that this even works for incident He projectiles for the profiling of H isotopes. This is accomplished because of the fact that the recoil atoms normally have both greater velocities and lower atomic numbers than the scattered projectiles; both effects lower the stopping power of the recoil as compared to the projectile. Standard beam-target-detector geometries use both incident and exit angles of the incident and recoiled nucleons of 15° with respect to the surface plane of the sample. For example, with this geometry a 10 $\mu$m mylar film will stop all the 2 MeV helium ions scattered from any sample, while protons recoiled by the helium will penetrate through the mylar with an energy loss of only .3 MeV and be detected by the solid state detector.

## B.  HYDROGEN ISOTOPE PROFILING IN GRAPHITE

The accumulation of ion-implanted hydrogen isotopes (hydrogen, deuterium, and tritium (H, D, T)) in the near surface region of solids has recently received much study by the plasma-surface interaction community. This interest has been generated largely by the ramification of this accumulation in components such as limiters or other first-wall surfaces of magnetic confinement fusion reactors (eg., tokamaks). In particular, the near surface buildup of H-isotopes strongly affects the recycling process and may lead to unacceptable levels of T. ERD offers an attractive way to study H, D, and T buildup and replacement in materials because it is the only nondestructive ion-beam analysis technique capable of detecting all of the isotopes of hydrogen simultaneously.

The objective of the present example (Doyle et al., 1980) was to determine the saturation and isotopic replacement of hydrogen implanted into a low-$Z$ material, namely, C. The main benefit of utilizing low-$Z$ materials in a tokamak is the reduction of the plasma's $Z_{eff}$ and the corresponding increase in core plasma temperature.

The hydrogen retention and replacement properties of graphite was studied by implanting H or D at an energy of 1.5 keV/ion, chosen to correspond to the upper end of typical particle energies at the plasma edge. All implants were carried out at ambient temperatures. The total amount ($\#/cm^2$) of H and D detained within the first ~ 500 nm of the C samples was measured using 2.5 MeV He ERD, and the results of both a saturation D implant followed by an isotope exchange implant of H are shown in

**ERD SPECTRA**

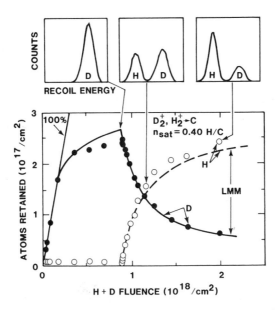

FIGURE 13. Areal density of implanted H and D into graphite using ERD analysis. The top three panels show the raw ERD spectra for the implant conditions indicated. Saturation and isotope exchange are observed.

Figure 13. The three panels in the upper part of this figure show the raw ERD spectra taken at three points during the course of the D and H implants.

For graphite it was found that at low fluences ($< 10^{17}$ cm$^{-2}$) all the incident H, or D, was retained in the sample, apart from a small fraction (typically 5%–10%) which was kinematically reflected. Depth profiling showed that the hydrogen was retained at the end of its implant trajectory in the material, and therefore the deposition probability is totally determined by the range and straggling of the hydrogen, both of which can be theoretically predicted. This immobility of implanted hydrogen is found in many non-hydride-forming materials at temperatures where normal diffusion is suppressed.

As the incident fluence was increased ($> 10^{17}$ cm$^{-2}$), a maximum hydrogen concentration (saturation) was eventually reached, as is shown in Figure 13. To characterize the isotopic exchange behavior, the D saturated graphite ($\sim 9.0 \times 10^{17}$ cm$^{-2}$) was implanted with H, and the D and H content of the samples was monitored as the H displaced the D.

**Table 2.** Room Temperature Hydrogen Implantation-Induced Saturation Concentrations and Release Temperatures for Selected Low-$Z$ Materials. (Wampler et al., 1981)

| Material | Saturation Concentration at Room Temperature (H/Host) | Temperature Range for H Release (°C) |
|----------|----------------------------------|---------------------------|
| C        | .40  | 350–550 |
| Si       | .50  | 300–600 |
| B        | .45  | 100–525 |
| $B_4C$   | .57  | 400–700 |
| TiC      | .26  | 100–600 |
| $VB_2$   | .16  | 100–500 |
| $TiB_2$  | .16  | 100–500 |
| 316 SS   | 1.00 | −200    |

\* − 120°C

It was found that for projectile energies in the range 0.5–14 keV, both the saturation behavior and the isotopic exchange process can be described by a simple macroscopic model called the local mixing model (LMM) (Doyle et al., 1980). The solid and dashed lines in Figure 13 result from LMM calculations. The LMM has as its single characteristic parameter the atomic density of hydrogen at saturation, $N_s$, which can be determined experimentally by analysis of data such as that shown in Figure 13. This hydrogen isotope saturation concentration is a material characteristic which depends on temperature and for graphite equals 0.4 H/C. The saturation concentrations for several other low $Z$ elements and compounds are listed in Table 2.

The LMM treats the saturation and isotopic exchange processes differentially in the depth variable $x$. That is to say, it is assumed that at a depth where saturation has not been reached, the incident particles are trapped near the end of their trajectory. This behavior implies 1) a high probability that a hydrogen ion coming to rest near an unoccupied trap will be captured by the trap and 2) a high probability for a recapture by a trap should a hydrogen atom be dislodged by a subsequently implanted projectile.

When saturation is reached, additional hydrogen coming to rest in the region will find no available sites unless some of the previously trapped hydrogen is detrapped. It is assumed in the model that a hydrogen atom coming to test in an already saturated region competes on an equal basis with the local hydrogen atoms already present for the available trap sites

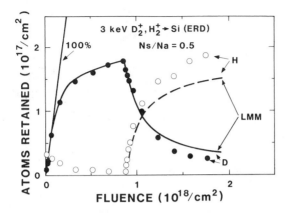

FIGURE 14.   Areal density of H and D implanted in Si using ERD.

and that the single atom which does not find a trap is then lost from the material.

These results have been used to show that graphite and the other materials listed in Table 2 do not pose a serious near-surface T buildup threat if used in TFTR. It was also found that hydrogen isotopes trapped in graphite can be replaced by subsequent implantation with a different isotopic species so that recovery of tritium by replacement with hydrogen or deuterium should be feasible. The LMM agrees very well with the data on graphite and other low-$Z$ materials and provides predictions of retention and exchange behavior in a tokamak environment. This result is important not only from the tritium inventory standpoint but also for including wall effects in computing D-T ratios in the plasma.


C.  ANALYSIS OF HYDROGEN IN SILICON

The presence of hydrogen in semiconducting materials such as Si (Peercy, 1981 and Picraux, 1987) is also of great concern, just as in the case of H isotopes in fusion materials such as the graphite example described above. Figure 14 shows the buildup and isotope exchange of D and subsequence H implantation into Si. As with the graphite experiment, 2.5 MeV He ERD was used to collect this data, and the results are quite similar. The LMM agrees very well with the data for a saturation concentration of 0.5 D/Si.

The validity of the concepts in the LMM are born out further by the results of another saturation-isotope exchange experiment shown in Figures 15 and 16. In this case the implantation of the D was performed at 60 keV,

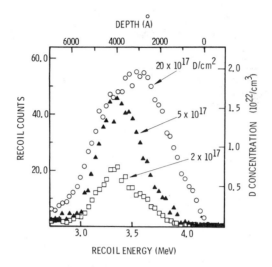

FIGURE 15. 24 MeV Si ERD depth profiles of D implanted into Si to three different fluences. The local D concentration is observed to saturate.

and the H was implanted at 75 keV. The ERD analysis beam in this case was 24 MeV $Si^{+5}$. For this experiment, the depth of the implant is great enough for the depth concentration profiles of both the D and H to be measured together with the areal densities. In Figure 15 the saturation of D implanted into Si is displayed. For this data, this recoil counts and energies are converted to concentration and depth, respectively, using the concepts developed in Section A above. The three sets of data plotted in this figure result from implants of D to fluences of 2, 5, and $20 \times 10^{17}/cm^2$, as indicated. It is obvious from this figure that the D has yet to saturate the Si for fluences less than $5 \times 10^{17}/cm^2$, and that the D concentration reaches $\sim 2.0 \times 10^{22}/cm^3$ when the Si does become saturated. This concentration corresponds to a D/Si ratio of $\sim 0.4$, which is close to that observed for low-energy D implants. As the D saturates the Si, the D profile is observed to move toward the surface and does not extend significantly deeper into the bulk.

The results of an isotope exchange implant of H into the D saturated Si sample is shown in Figure 16. The upper panel shows the evolution of the newly implanted H profiles, while the bottom panel plots the corresponding profiles of the D which is in the process of being replaced. The respect H and D profiles are plotted for H fluences of 0.5, 2, 5, and $20 \times 10^{17}/cm^2$, as indicated. It is clear from this figure that the LLM correctly describes the

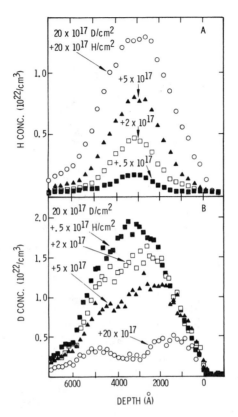

FIGURE 16. Depth profiles of D (A-lower) and H (B-upper) for the H implantation of the D-saturated Si sample of Figure 15. Isotope exchange is observed, and, for the $20 \times 10^{17}$ $H/cm^2$ implant, the D profile displays a trough at the same position as the peak in the H distribution.

physics of the isotope exchange process. Even the bimodel distribution predicted by the LMM is observed for the highest H implant fluence.

Hydrogen isotopes (and other light elements) can also be profiled in 3-D using ERD (Doyle et al., 1983a and 1986) by employing a nuclear micro-probe and methods analogous to those used to profile heavier elements in 3-D by RBS. An example of such a 3-D hydrogen profile using 20 MeV Si ERD analysis of laser-assisted plasma deposition of crystalline Si on a Si substrate is shown in Figure 17. This sample was prepared by depositing a thin Si layer onto a Si substrate using a $SiH_4$ plasma. The deposition was enhanced by shining the light of a KrF laser (2 mm diameter) onto the

**3D H PROFILE**

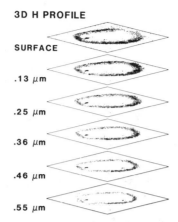

FIGURE 17. 20 MeV Si ERD profile of H in a laser-assisted deposit of Si on Si. The H concentration was measured in 3-D; where each panel is at a different depth and darkness corresponds to the presence of H. Full black represents 1 at. %.

sample. As with the 3-D RBS data shown above, darker regions in this figure indicate an increase in the abundance of the hydrogen. Full black in this figure corresponds to an H concentration of 1 at. %. The lateral region scanned was fairly large, 3.2 mm on a side, while the maximum depth was only 0.5 $\mu$m. A ring of H is observed which decreases in width as a function of depth. Such a 3-D profile is consistent with the interpretation that at the central region of the laser spot the temperature was high enough to promote the liquid phase epitaxy of single crystalline Si on the Si substrate and thereby eliminate the presence of H in this region by zone refinement; whereas, at the perimeter of the spot, the temperature decreases as a function of radial position, which results in the growth of polycrystalline Si where H can exist in abundance at grain boundaries. The rate of growth of this polycrystalline layer decreases to zero at the very edge of the laser spot and hence neither Si nor H is deposited here. This tapered growth pattern results in the 3-D H profile which is observed in Figure 17. Millibeam channeling of He in the central H-free region of this sample confirmed that it was indeed single crystalline material.

## IV. Particle Induced X-Ray Emission (PIXE)

The detection and energy-analysis of characteristic x-rays which are generated by the interaction of MeV-energy particles with matter is called particle induced x-ray emission or PIXE. An excellent recent treatment of

PIXE analysis has been given by Campbell and Cookson (1984). This type of analysis is almost completely insensitive to depth and therefore is most applicable to problems where areal densities of thin deposits or the determination of the composition of homogeneous samples is important. Any MeV ion is capable of generating even the highest energy x-rays; however, protons are favored because of analysis complications resulting from multiple inner-shell ionizations by high-Z projectiles and the fact that x-ray generation cross sections generally scale with velocity. In the following, we describe the basic concepts of PIXE and present several fusion-reactor examples of its use.

## A. Basic Concepts

Although it is another ion beam analysis technique, PIXE is actually quite different from RBS or ERD because it exploits atomic physics, not nuclear physics, to generate a detectable signal characteristic of the material being studied. The following events must occur before a characteristic x-ray can be detected: 1) At a certain depth within the sample the incident projectile (assume it is a proton here) deposits a small fraction of its energy to an inner shell electron (assume that this is a K-shell electron) through an ion-electron collision; 2) if the energy is great enough, this electron will be ejected from the atom and an inner-shell vacancy produced; 3) this inner-shell vacancy places the atom into an excited state which can relax one of two ways—radiatively or nonradiatively; 4) if the atom relaxes radiatively, an electron from a higher shell (L, M, N, etc.) falls into the vacancy and a K x-ray with an energy equal to the energy difference between these electron states is simultaneously emitted; 5) in order to be detected, this x-ray must propagate in the direction of the detector and must avoid being absorbed by the sample material located between its generation and penetration points; and, finally, 6) as opposed to the 100% efficient Si surface barrier detectors which are used to energy-analyze particles, x-ray detectors have efficiencies less than 100%, which are quite dependent on energy, and, therefore, the detection of this K x-ray is subject to chance.

Thick target x-ray yield analysis (Campbell and Cookson, 1984) will not be covered here because of its complexity and the fact that the examples we give involve the determination of areal densities of thin deposits and not concentrations of elements in a homogeneous solid. Equation (2) in the RBS section can be used to describe the relationship between the yield of x-rays and the areal density of target atoms when the layer is so thin that the attenuation of x-rays passing through this layer is negligible. The parameter $\sigma$ in Eq. (2) is then interpreted as an x-ray generation/detection

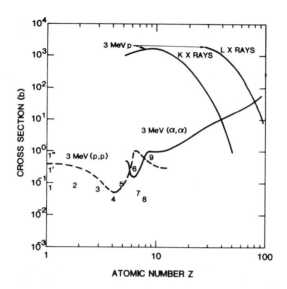

FIGURE 18.   Cross sections for various ion-atom collision events as a function of target atomic number. The $(\alpha, \alpha)$ and $(p, p)$ cross sections correspond to RBS at 165°. The PIXE cross sections are for 3 MeV $p$ for $K\alpha$ and $L\alpha$ x-rays. The numbered cross sections are for NRA and RNRA analyses of light elements were:

$1 - {}^2D({}^3He, p)$     $1' - {}^1H({}^{15}N, \alpha\gamma)$     $1'' - {}^1H({}^{19}F, \alpha\gamma)$
$2 - {}^3He(d, p)$
$3 - {}^6Li(p, \alpha)$
$4 - {}^9Be(d, \alpha)$
$5 - {}^{11}B(p, \alpha)$
$6 - {}^{12}C(d, p)$
$7 - {}^{14}N(d, \alpha)$
$8 - {}^{16}O(d, p)$
$9 - {}^{19}F(p, \alpha\gamma)$

cross section and can either be calculated or determined through the use of standards. This x-ray detection cross section can be expressed as

$$\sigma_{det} = \epsilon\sigma_x, \qquad (8)$$

where $\sigma_x$ is the x-ray generation cross section and $\epsilon$ is the detector efficiency. A plot of both K-shell and L-shell x-ray generation cross sections versus target atomic number for incident protons at 3 MeV is plotted together with the RBS cross sections for 3 MeV $\alpha$ particles and protons in Figure 18. Several cross sections for selected nuclear reactions useful for the detection of low atomic number materials are also indicated in this figure. From Figure 18 it is apparent that K x-ray PIXE is much more sensitive than RBS for $Z < 30$ and L x-ray PIXE is more sensitive for the elements $30 < Z < 80$.

## B. FUSION-PLASMA LIMITER STUDIES

An important aspect of materials testing for magnetic fusion energy development is the determination of the composition and total amount of impurities deposited on in-vessel components such as limiters during plasma exposure. Impurity deposition on limiters is of particular interest because the plasma-surface interaction is most intense in this region. A quantitative knowledge of these deposits is necessary to understand plasma impurity transport, impurity generation sources, and the limiter conditioning process. One of the primary goals of current fusion technology is the reduction of impurities in the plasma core because of their deleterious effects (e.g., dilution and radiative cooling) on plasma performance. Because of the importance of impurities, deposits on limiters or other wall elements have been studied recently in virtually every tokamak in the world. (Numerous examples of fusion component analysis can be found in the proceedings of the Sixth International Conference on Plasma Surface Interactions in Controlled Fusion Devices, *J. Nucl. Mater.* **128 / 129** (1984).)

PIXE is an ideal analytic technique for this type of investigation because of: 1) its high sensitivity for the impurity elements of primary interest (structural metals), 2) the low backgrounds which result because of the current widespread use of low-$Z$ materials (e.g., C) in the construction of first-wall components, 3) the possibility of using scanned proton microbeams to study areal distributions, and 4) the capability of performing PIXE analyses external to a vacuum chamber (X-PIXE) so that extremely large wall components can be analyzed nondestructively. For these and other reasons, PIXE has been widely used in the study of impurity deposits on fusion reactor components (Behrish et al., 1984; Doyle et al., 1984; Amemiya et al., 1984). In this section we describe PIXE analyses of the TiC coated graphite moveable limiter tiles described in the RBS section (Dylla et al., 1985).

Figure 19 shows a 4.5 MeV PIXE spectrum taken on the TiC coated graphite limiter tile #28 used in TFTR and plotted together with the spectrum from a control standard. This analysis was performed in the same region as the RBS data shown in Figure 9. For this particular limiter tile, most of the TiC was removed, supposedly by a spallation process which took place during discharges. The reason for studying this limiter with PIXE was to determine whether impurities, either preexistent or deposited on the surface, played a role in the coating detachment process. The

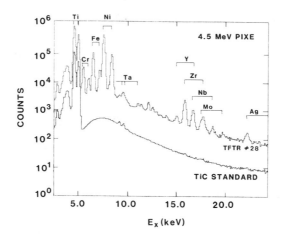

FIGURE 19.   4.5 MeV PIXE on (lower) TiC coated graphite standard and (upper) TFTR coated tile. A deposition layer is apparent.

interpretation of this data is that most of the layer deposited on this tile during its use in TFTR consists of Cr, Fe, and Ni (1000 Å thickness), which comes from the walls, and that the other higher $Z$ elements identified in this spectrum are at levels indicative of thicknesses less than 1–10 Å. All of these higher $Z$ elements also come from components in the interior of TFTR, except the Mo which may have been deposited during a preliminary baking step. PIXE spectra have also been taken on limiters from ALCATOR-C, PLT, PDX, and TEXTOR, with each displaying a slightly different "fingerprint" of elements which are characteristic of each device. The only element found to diffuse into the TiC coating was Zr, but the levels detected ($< 0.1\%$) were thought to be insufficient to cause the large-scale spallation which was observed.

Figures 20, 21, and 22 show 2-D element maps taken on the TFTR limiter tile discussed in the RBS section. The C and O maps were determined using 4.5 MeV $\alpha$ RBS, while the Ti, Ni, and Zr maps were measured using 4.5 MeV $\alpha$ PIXE. These data demonstrate the utility of simultaneous use of both RBS and PIXE to study a material. Note that logarithmic gray scales are used in this figure, where, as with the 3-D RBS data, darker regions indicate greater abundance of elements. The same peninsula and island of TiC still coating the graphite can be seen in Figure 20, as is observed in the 3-D RBS profile of Ti shown in Figure 10. Likewise, the two islands of TiC shown in the 3-D RBS profile in Figure 11 are easily identified in the 2-D PIXE maps plotted in Figure 21. Figure 22

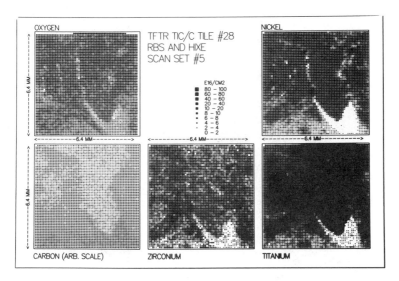

FIGURE 20.   2-D PIXE (for Ti, Ni, and Zr) and RBS (for C and O) maps of a TiC coated graphite tile used in TFTR. This is the same area displayed in 3-D in Figure 10.

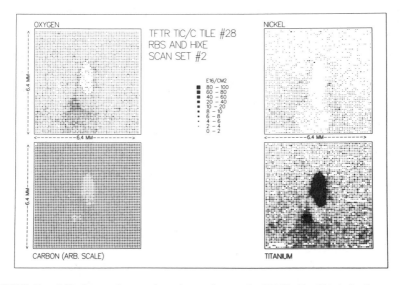

FIGURE 21.   2-D elemental map of another region on the TFTR tile. This is in the same area as that of Figure 11.

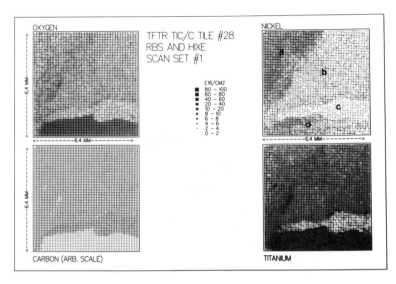

FIGURE 22.   2-D elemental map of a different area on the TFTR TiC coated graphite tile.
The amount of Ni preserves the history of the TiC coating detachment process.


displays a different region—one which graphically illuminates the coating
detachment process.

For each of these three figures, the TiC areas which have not been
removed are easily identified by noting regions which have higher Ti (or
lower C) than their surroundings. The carbon increases in regions of
coating detachment because the graphite substrate becomes exposed.

The Ni areal density is especially useful in determining the evolution of
the loss of coating. Areas with either no coating loss or those which have
suffered very early spallation have a greater Ni overlayer than those which
experienced coating detachment at a later time. This is because the Ni
deposition (from Inconel erosion) is nearly continuous. The amount of
deposited Ni therefore records the history of the coating spallation process
by providing an indirect measurement of the time difference between
coating detachment and the removal of the tile for analysis. For example,
the light Ni region at the bottom of the TiC peninsula in Figure 20 defines a
region of quite recent coating removal. The Ti, Zr, and O are also lower in
this region, but the amount of detected C increases here as expected
because more of the graphite substrate is exposed. Similar interpretations
can be made for the area displayed in Figure 21, and in the case of Figure
22, four distinct areas of coating loss are dramatically preserved. At the
bottom of the area scanned in Figure 22, a Ti rich and C poor region
labeled "d" is observed. This is clearly a small chip of the original TiC

coating which was not removed. Notice that the amount of Ni deposited on the TiC coating is approximately equal to that found in a triangular region a at the upper left-hand region of the scanned area; however, in region a the TiC coating has obviously detached. The similarity of the Ni overlayers in these two regions indicates that the region a suffered coating detachment at a very early time of this tile's use in TFTR. Region b has a thinner Ni deposit than region a, and c has even less Ni than b. The time history of the TiC coating detachment in this region can now be reconstructed: Region a was removed first, probably during the first few discharges of TFTR following the insertion of this tile; region b detached at a later time; region c suffered its coating removal just before this tile was removed from TFTR for analysis; and in region d the TiC coating still remains. The uniformity of these Ni overlayers in each region suggests that the TiC removal was not caused by a gradual, continuous erosion of the coating but that the coating detached in large sheets.

The fact that large sheets of TiC coating were removed in single events strongly indicated that a problem existed with the bond between the coating and substrate. It was later determined that a process step involving the ultrasonic cleaning of the graphite prior to coating deposition was omitted during the manufacture of these tiles. This processing error easily explains the poor coating-substrate bond, which was apparently responsible for the inferior performance of these TiC coated tiles in TFTR.

## V. Nuclear Reaction Analysis (NRA)

Two types of nuclear reaction analyses exist: resonant and nonresonant (RNRA and NRA). Both of these techniques utilize nuclear reactions which occur between the MeV-energy incident projectile and target nucleons. Both RNRA and NRA are most useful in the analysis of light elements because of the existence of large cross-section reactions resulting from low Coulomb barriers. In this section, the basics of both NRA and RNRA are given followed by several examples which were taken from fusion-plasma research.

### A. BASIC CONCEPTS

#### 1. NRA

NRA is quite similar to RBS and ERD, except for the fact that inelastic scattering—not elastic scattering—is used to indicate the presence of specific nucleons in the sample. These reactions are usually quite exothermic and result in nuclear reaction products which are very energetic.

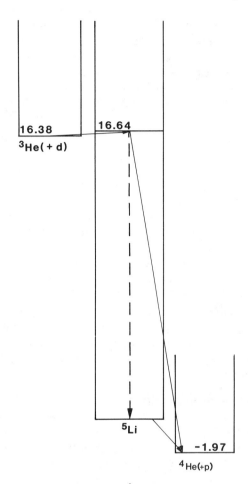

FIGURE 23. Energy level diagram of the $^5$Li compound nucleus system. This diagram presents the NRA use of the highly exothermic $^3$He + $d$ → $^4$He + $p$ reaction to detect either $^3$He or $d$.

Figure 23 shows an energy level diagram of a very useful nuclear reaction to detect deuterium (d). In this figure the total energy of the labeled systems ($^3$He + $d$, $^5$Li, and $^4$He + $p$), including rest masses, are indicated. This energy scale has been normalized to zero for the compound nucleus of $^5$Li. A $^3$He beam is used for the analysis. When the kinetic energy of $^3$He increases, the total energy of the $^3$He + d system increases above the rest mass line at 16.38 MeV. When this total energy equals that of the first excited state of $^5$Li at 16.64, the cross section for formation of this compound nucleus reaches a maximum. This occurs at 0.26 MeV in the center of mass frame or at 0.65 MeV in the rest frame of the d. This excited state in $^5$Li exists only momentarily and quickly decays into $^4$He $+ p$. This system has a combined rest mass of only $-1.97$ MeV and therefore 18.61 MeV of kinetic energy is shared between the $^4$He and the $p$ where most of this energy is provided to the $p$. Energy-analysis by a surface barrier Si diode of the high-energy $p$ resulting from this nuclear reaction provides the signal used to analyze the concentration or areal density of $d$ in the sample being studied. Equation (2) can be used to relate the yield of detected high-energy protons to the areal density of $d$.

Several other light-ion low-energy nuclear reactions can be used to study material, and a few selected nuclear reaction cross sections which are especially useful are plotted in Figure 18. McMillan et al. (1982) have recently reviewed the most useful nuclear reactions for ion beam analysis applications.

## 2. RNRA

As the name implies, *resonant nuclear reaction analysis* (RNRA) involves the use of nuclear scattering resonances to measure the composition of solids. A simplified schematic of the use of the $^{15}$N $+ p$ reactions is shown in Figure 24. Plotted in this figure is an energy level diagram where the total energy (including rest masses) of the system increases vertically. These total energies are normalized to the compound nucleus of the scattering system, $^{16}$O, which is set to zero. The very bottom of each box, from right to left, indicates the rest mass of $^{15}$N $+ p$, $^{16}$O, and $^{12}$C $+ \alpha$, where $p$ and $\alpha$ indicate a proton and alpha particle, respectively. The energy levels positioned within each box are for the $^{15}$N, $^{16}$O, and $^{12}$C nucleons. Only the lowest excited states of $^{12}$C and $^{15}$N are indicated; whereas, for $^{16}$O just two energy levels (out of dozens which have been measured) at 12.53 and 12.79 MeV are highlighted to avoid confusion. Again, the total energy of each of the three systems increases as the center-of-mass kinetic energy of each nucleon is increased. A simplified description of the nuclear physics which occurs during the $^{15}$N $+ p$ collision can now be given with the use of Figure 24.

First assume that the $^{15}$N is at rest in a target material being bombarded by protons ($p$). As the center-of-mass kinetic energy of the $p$ is increased, the total energy of the $^{15}$N $+ p$ system eventually equals that of the 12.44 MeV excited state in the $^{16}$O compound nucleus. At this resonance energy (0.40 MeV in the center-of-mass system, 0.43 MeV in the rest frame of the $^{15}$N), there exists an enhanced cross section for the $p$ and $^{15}$N to unite and

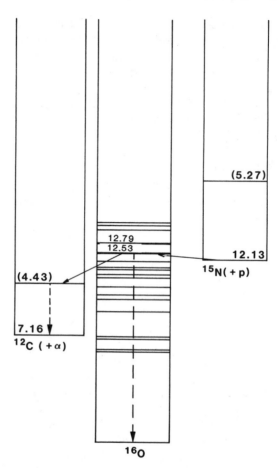

FIGURE 24.   Energy level diagram of the $^{16}O$ compound nuclear system. The sharp 12.53 MeV resonant excited state at 12.53 MeV in the $^{16}O$ system can be used to depth profile either $^{15}N$ or $p$ using RNRA.

form a compound nucleus of $^{16}O$. The energy dependence of this cross section is given by the Breit-Wigner relationship

$$\sigma(E) = \frac{C}{(E - E_r)^2 + \dfrac{\Gamma^2}{4}}, \tag{9}$$

where $E$ is the lab energy of the incident particle (i.e., the proton in this case), $E_r$ is the resonance energy of the reaction in the lab frame, $\Gamma$ is the resonance width in the lab frame, and $C$ is an energy-independent constant.

$\Gamma$ is related to the lifetime, $\tau$, of the excited state through the Heisenberg uncertainty principle

$$\Gamma\tau = h, \tag{10}$$

where $h$ is Planck's constant. Ordinarily, these excited compound nuclear states simply decay by the emission of a gamma ray, as indicated with the downward pointing arrow in Figure 1. However, in this case, the system has so much angular momentum that particle emission is favored, and, as shown in the figure, the 12.53 MeV excited state in $^{16}$O normally decays into an $\alpha$ particle and $^{12}$C in its 4.43 MeV excited state, followed by the emission of a 4.43 MeV gamma ray. As a consequence, the lifetime, $\tau$, of this excited state is quite long, resulting in remarkably small values for $\Gamma$. For example, the resonance in the $^{15}$N + $p$ system at a proton energy of 0.43 MeV is only 0.9 keV wide and may be much smaller based on recent measurements by Amsel and Maurel (1983 and 1986). At a higher center of mass energy, the next compound nuclear resonance at 12.79 MeV comes into play.

There are two ways by which a nuclear resonance reaction can be used as an ion beam analysis tool. Using the $^{15}$N + $p$ example, either the $^{15}$N can be analyzed using a $p$ beam, or the reverse reaction can be employed where a $^{15}$N beam is used to detect $p$ (i.e., hydrogen) in a target. In either case, the 4.43 MeV gamma ray from $^{12}$C provides the detection signal, and the relationship between concentration and gamma ray yield is expressed as

$$c_x\left(^{15}\text{N}, p\right) = CY_\gamma S\left(p, ^{15}\text{N}\right) \tag{11}$$

$$x = \frac{\left(E - E_r\right)}{S\left(p, ^{15}\text{N}\right)},$$

and the depth resolution is

$$\delta x = \frac{\Gamma}{S\left(p, ^{15}\text{N}\right)}, \tag{12}$$

where $c$ is the concentration of $p$ or $^{15}$N at a depth $x$, $Y_\gamma$ is the gamma ray yield per incident $p$ or $^{15}$N, $E$ is the incident energy, and $S$ is the stopping power (assumed constant) of the $p$ or $^{15}$N in the sample material.

## B. Tritium Inventory in Fusion-Plasma Reactors

The amount of tritium ($t$) that will be retained in graphite and related components exposed to $t$ plasmas is an important issue for tokamaks like TFTR because of the large amount of graphite in TFTR (about 2000 kg) and the limited on-site inventory of $t$ (5 grams) (Baskes et al., 1984). $t$ retained in the graphite will be unavailable for experiments. Therefore it is important to know how much of the $t$ might be retained and to investigate ways of recovering $t$ from the graphite. NRA—$d({}^3\text{He}, p)$—measurements of the amount of $d$ present in the TFTR TiC coated graphite limiters after prolonged exposure to $d$ plasmas are reported in this section and provide the most reliable indication presently available of the amount of $t$ remaining in the limiters after exposure to $t$ plasmas.

The amount of $d$ in the limiters was determined from the yield of energetic protons produced by nuclear reactions between a ${}^3\text{He}$ analyzing beam and $d$ in the target. Details of this technique were given above. An analysis beam energy of 700 keV was used to scan the near-surface areal density of $d$ as a function of position. This measures the amount of $d$ within about 1 micron of the surface. The analysis beam did not cause any significant loss of $d$ from the targets.

At selected locations, the depth profile of the $d$ was determined by measuring proton yields as a function of beam energy analysis from 300 to 2000 keV. Since the nuclear reaction cross section and the stopping power are known functions of energy, the depth profile can be obtained from a deconvolution of the measured proton yields. This method gives the $d$ concentration versus depth down to about 5 microns, with a depth resolution of a few tenths of a micron. Most of these depth profiles showed that the major fraction of the $d$ was within 1 micron of the surface and was therefore included in the 700 keV scans of near-surface areal density versus position.

The measurements were made of $d$ retained in a TiC coated graphite tile from the TFTR limiter (Wampler et al., 1987). This tile was in the same set discussed in the PIXE and RBS sections. The limiter was exposed to about 800 ohmically heated $d$ discharges. At the beginning of this period all of the tiles had a 20 micron coating of TiC, however failure of the coating during operation resulted in removal of the TiC from about 30% of the limiter surface area by the end of the period.

Figure 25 shows a scan of the near-surface areal density of $d$ over part of this tile. The amount of $d$ retained in the TiC remaining near the apex is only slightly less than the amount retained in carbon from which the TiC was lost. Depth profiles were measured at the three positions indicated in Figure 25. In the carbon near the boundary with the TiC, the $d$ profile extends several microns below the surface, with an additional $4 \times 10^{17}$ $d/\text{cm}^2$ not seen in the initial scan. At the other two locations most of the $d$ was within 0.5 $\mu$m of the surface. Measurements of the concentration of $d$ in the bulk of this tile gave $12 \pm 6$ at. ppm.

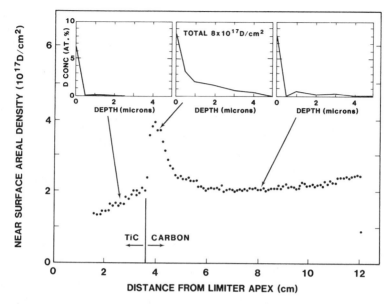

FIGURE 25.   Areal density of D (i.e., $d$) as a function of distance from the limiter apex (i.e., point of contact with the plasma) for a TiC coated graphite limiter used in TFTR. D($^3$He, $p$) NRA was used. D is implanted into this component during D discharges to TFTR. The upper three panels show depth profiles taken at the positions indicated.

Uncoated graphite tiles from the TFTR limiter had an average near-surface areal density of $d$ about $10^{17}$ $d/cm^2$. Multiplying this by the area of the limiter gives about $2 \times 10^{21}$ $d$ in the moveable limiter and $2 \times 10^{22}$ $d$ in the inner bumper limiter. The limiter therefore contains more $d$ than the amount in a typical TFTR plasma estimates from the volume ($4 \times 10^7 cm^3$) times the average density ($2 \times 10^{13}/cm^2$). The quantity of $d$ in the limiter is thus large enough to affect the density and isotopic composition of the plasma.

The high concentration of H (i.e., $p$) (1 at. %) found in the bulk of the graphite suggests that the previously reported long changeover from $p$ to $d$ plasmas in TFTR might have been due to release of $p$ from the graphite during initial conditioning of the limiter. This mechanism could be tested by observing the changeover from $d$ back to $p$ plasmas, which should take fewer discharges. If this is found to be the case, then the changeover from $d$ to $t$ plasmas should also be fairly rapid.

The amount of $t$ that will be retained in graphite components exposed to $t$ plasmas in TFTR can be estimated from the $d$ retention measurements reported here. Using an average near-surface retention of $10^{17}/cm^2$, the inventory of $t$ is estimated to be 0.04 gm in the moveable limiter and 0.4 gm in the bumper limiter. This retention is, therefore, not expected to be a large fraction of the allowed inventory. Furthermore, some of this near-surface $t$ will be recoverable by isotope exchange during subsequent operation with H or $d$ plasmas.

$t$ retention in the bulk graphite is potentially more serious. A concentration of only 10 at. ppm of $t$ throughout the 2000 kg of graphite in TFTR would consume nearly the entire allowed 5 gram inventory of $t$. This concentration is comparable to the bulk concentration of $d$ ($9 \pm 4$) found in the TFTR tiles using RNRA.


C.  STUDY OF HYDROGEN IMPLANTED IN
    LOW-$Z$ MATERIALS

Knowledge of the temperature at which hydrogen is released from a fusion first-wall material is necessary to permit prediction of recycling behavior and to anticipate problems relating to the control of tritium inventories. In this example of RNRA, the temperature at which hydrogen is released from $TiB_2$ was measured (Vook et al., 1979; Doyle and Vook, 1979b and 1979c). $^1H$ (i.e., $p$) was first implanted at 60 keV to depths of 3000–4000 Å into the $TiB_2$ and the samples were annealed isochronally for 20 minutes at increasing temperatures. $TiB_2$ is a refractory material which, because of physical properties, seems very well suited for coating applications to first-wall elements (e.g., limiters, armor plating, channel walls, etc.) of tokamak fusion reactors. Also the low-$Z$ constituents minimize the possibility of poisoning the plasma during a discharge.

$^1H$ profiles were measured using RNRA (see Figure 26) between each anneal, and these profiles, along with plots of H retention as a function of anneal temperature, are shown in Figure 27. H is clearly retained in $TiB_2$. One half of the H in $TiB_2$ is released by 400°C. The relative constancy of the width of the H peaks as a function of temperature indicates that the release is trap-limited in both materials. This observation indicates that the implanted H is trapped in its own ion-induced damage. The release temperature of implanted H or D for other low $Z$ materials are listed in Table 2.

The direct exposure of samples in an operating tokamak provides a rigorous check on conclusions drawn from implantation studies to simulate plasma-wall interactions. Si, C, and stainless steel 304 surfaces have been exposed and analyzed for H profile and H saturation characteristics, and these results are discussed in the section on NRA and ERD. In addition, these measurements have led to the development of a diagnostic technique

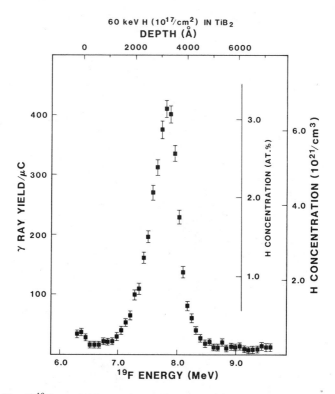

FIGURE 26. H($^{19}$F, $\alpha\gamma$) RNRA profile of H implanted into a TiB$_2$ coating on graphite. Both the raw ($\gamma$-ray yield vs. $^{19}$F energy) and analyzed data (concentration vs. depth) are given.

by Wampler (1982) to determine the fluence and energy spectra of escaping H isotopes which relies on changes in resistance to analyze plasma flux and energy.

## VI. Conclusions

In this chapter, we have tried to provide a very general description of the various types of MeV-energy ion beam analysis techniques which have been successfully applied to plasma-materials interaction/processing research. The emphasis has been on providing the reader with sufficient information in the form of description and theory of analysis, in conjunction with examples, so that an educated decision can be made as to whether ion beam analysis should be used in plasma based research studies.

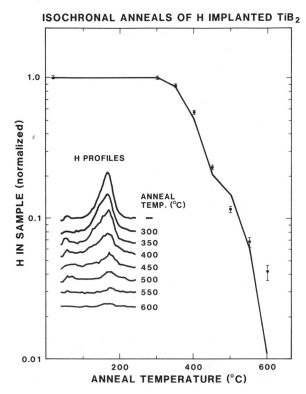

FIGURE 27.   Areal density of H implanted into $TiB_2$ coated graphite vs. isochronal annealing temperature. The release at ~ 400 °C indicates traps with activation energies of ~ 2 eV.

The arsenal of techniques which are available through the use of MeV-
energy ions can provide information about samples which cannot be
obtained by other methods. This fact is particularly true in the case of
light-atom ($Z < 6$) detection, trace element analysis, isotope selectivity, and
the possibility of ex vacuo analysis.

Another important attribute of ion beam analysis is that it can be used to
measure depth concentration profiles of elements in solids nondestructively.
For example, it is desirable at times to perform analyses at depth solely to
avoid surface contamination or interference effects—this is particularly true
for the analysis of common elements such as H, C, or O. By combining the
depth profiling capability with the lateral scanning available on a nuclear
microprobe, elemental profiles can be determined in three dimensions. No
other existing technique can provide this type of data without destroying
the sample.

Ion beam analysis is not, however, without its drawbacks, and while it enjoys a certain degree of superiority over many other quantitative analysis techniques, it clearly is no panacea. The cost of setting up such a laboratory is rather high and there are currently very few "turn-key" systems for either acquiring or analyzing data. Nevertheless, it should be clear from our description of ion beam analysis that this technique is rapidly evolving into a major analysis tool in the area of plasma-materials interaction research and that this will continue to be the case for the foreseeable future.

One of the authors (W. K. Chu) acknowledges partial support from the Semiconductor Research Corporation and the Microelectronics Center of North Carolina. B. L. Doyle's work performed at Sandia National Laboratories supported by the U. S. Department of Energy under contract DE-AC04-76DP00789.

# References

Amemiya, S., Asawa, A., Tanaka, K., Tsurita, Y., Masuda, T., Katoh, T., Mohri, M., and Yamashina, T. (1984). *J. Nucl. Instr. Meth.* **B3**, 549.

Amsel, G. and Maurel, B. (1983). *J. Nucl. Instrum. & Meth.* **218**, 183.

Amsel, G. and Maurel, B. (1986). *J. Nucl. Instrum. & Meth.* **B14**, 226.

Baskes, M. I., Brice, D. K., Heifetz, D. B., Dylla, H. F., Wilson, K. L., Doyle, B. L., Wampler, W. R., and Cecchia, J. L. (1984). *J. Nucl. Mater.* **128 & 129**, 629–635.

Behrish, R., Borgesen, P., Ehrenberg, J., Scherzer, B. M., Sawicka, B. D., Sawicki, J. A., and the ASDEX Team (1984). *J. Nucl. Mater.* **128 & 129**, 470.

Campbell, J. L. and Cookson, J. A. (1984). *J. Nucl. Instrum. & Meth.* **B3**, 185–187.

Cecchi, J. L., Bell, M. G., Bitter, M., Blanchard, W. R., Bretz, N., Bush, C., Cohen, S., Coonrod, J., Davis, S. L., Dimock, D., Doyle, B., Dylla, H. F., Efthimion, P. C., Fonck, R., Goldston, R. J., Von Goeler, S., Grek, B., Grove, D. J., Hawryluk, R. J., Heifetz, D., Hendel, H., Hill, K. W., Hulse, R., Isaacson, J., Johnson, D., Johnson, L. C., Kaita, R., Kaye, S. M., Kilpatrick, S., Kiraly, J., Knize, R. J., Little, R., McCarthy, D., Manos, D., McCune, D. C., McGuire, K., Meade, D. M., Medley, S. S., Mikkelsen, D., Mueller, D., Murakami, M., Nieschmidt, E., Owens, D. K., Ramsey, A. T., Roquemore, A. L., Sauthoff, N., Stangeby, P., Schivell, J., Scott, S., Sesnic, S., Sinnis, J., Sredniawskim, J., Strachan, J., Tait, G. D., Taylor, G., Tenney, F., Thomas, C. E., Timberlake, J., Towner, H. H., Ulrickson, M., and Young, K. M. (1984). *J. Nucl. Mater.* **128 & 129**, 1–9.

Chu, W. K., Mayer, J. W., Nicolet, M-A., Buck, T. M., Amsel, G., and Eisen, F. (1973). *Thin Solid Films* **17**, 1.

Chu, W. K., Mayer, J. W., and Nicolet, M-A. (1978). *Backscattering Spectrometry*. Academic Press.

Coburn, J. W. (1979). *J. Appl. Phys.* **50**, 5210.

Coburn, J. W., Gottscho, R. A., and Hess, D. W. (1986). *Plasma Processing Materials Research Society Sym. Proc.* Vol. 68.

Cookson, J. A. and Pilling, F. D. (1972). AERE Report R6300.

Cookson, J. A. (1979, 1981). *J. Nucl. Instrum. Meth.* **165**, 477; **181**, 115.

Doyle, B. L. and Peercy, P. S. (1979a). *Appl. Phys. Lett.* **34**, 811.

Doyle, B. L. and Vook, F. L. (1979b). *IEEE Trans. on Nucl. Science* **NS-26** (1), 1305.

Doyle, B. L. and Vook, F. L. (1979c). *J. Nucl. Materials* **85 & 86**, 1019–1023.

Doyle, B. L., Wampler, W. R., Brice, D. K., and Picraux, S. T. (1980). *J. Nucl. Mater.* **93 & 94**, 551–557.

Doyle, B. L. and Wing, N. D. (1983a). *IEEE Trans. Nucl. Sci.* **NS-30**, 1214.

Doyle, B. L. (1983a). *Nucl. Instrum. & Meth.* **218**, 29.

Doyle, B. L., Wampler, W. R., Dylla, H. F., Owens, D. K., and Ulrickson, M. L. (1984). *J. Nucl. Mater.* **128 & 129**, 955.

Doyle, B. L. (1985). *J. Vac. Sci. Technol.* **A3**, 1374.

Doyle, B. L. (1986). *J. Nucl. Instrum. & Meth.* **B15**, 654–660.

L'Ecuyer, J., Brassard, C., Cardinal, C., Chabbal, J., Deschenes, L., Labrie, J. P., Terreault, B., Martel, J. G., and St.-Jacques, R. (1976). *J. Apply. Phys.* **47**, 881.

Feldman, L. C., Mayer, J. W., and Picraux, S. T. (1982). *Materials Analysis by Ion Channeling.* Academic Press, New York.

Frieser, R. G., Montillo, F. J., Zingerman, N. B., Chu, W. K., and Mader, S. R. (1982). *International Dry Process Symposium*, 57, Sendai, Japan.

Geiger, H. and Marsden, E. (1913). *Phil. Mag.* **25**, 606.

Heinecke, R. (1976). U.S. Patent 3940506.

Legge, G. J. (1984). *J. Nucl. Instrum. Meth.* **B2**, 561.

Legge, G. J., McKenzie, C. D., Mazzolini, A. P., Sealock, R. M., Jamieson, D. N., O'Brien, P. M., McCallum, J. C., Allan, G. L., Brown, R. A., Colman, R. A., Kirby, B. J., Lucas, M. A., Zhu, J., and Cerini, J. (1986). *J. Nucl. Instrum. & Meth.* **B15**, 669–674.

Macdonald, J. R., Davies, J. A., Jackman, T. E., and Feldman, L. C. (1983). *J. Appl. Phys.* **54**, 1800.

Martin, F. W. (1973). *Science* **12**, 173.

Mattox, D. M., Mullendore, A. W., Pierson, H. O., and Sharp, D. J. (1979). *J. Nucl. Mater.* **85 & 86**, 1127.

McMillan, J. W., Pummery, F. C., and Pollard, P. M. (1982). *J. Nucl. Instrum. & Meth.* **197**, 171.

Peercy, P. S. (1981). *Nucl. Instrum. & Meth.* **182 & 183**, 337.

Picraux, S. T., Vook, F. L., and Stein, H. J. (1979). *The Institute of Physics Conference*, Ser. No. 46, Chapter 1.

Rutherford, E. (1911). *Phil. Mag.* **21**, 669.

Trester, P. W., Sevier, D. L., Chin, J., Horner, M. H., Staley, H. G., and Kaplan, J. (1981). *J. Nucl. Mater.* **103 & 104**, 193.

Vitkavage, D. J., Dale, C. J., Chu, W. K., Finstad, T. G., and Mayer, T. M. (1986). *J. Nucl. Instrum. & Meth.* **B13**, 313.

Vook, F. L., Doyle, B. L., and Picraux, S. T. (1979). *IEEE Trans. on Nucl. Science* **NS-26** (1), 1272.

Wampler, W. R., Doyle, B. L., Brice, D. K., and Picraux, S. T. (1981). Sandia National Laboratories Internal Report, SAND80-1184.

Wampler, W. R. (1982). *Appl. Phys. Letts.* **41**, 4.

Wampler, W. R., Doyle, B. L., and A. E. Partan (1987). *J. Nucl. Materials* **145–147**, 353–356.

Ziegler, J. F. and Chu, W. K. (1974). *At. Data Nucl. Data Tables* **13**, 463.

Ziegler, J. F. (1978). *Helium: Stopping Powers and Ranges in All Elemental Matter.* Pergammon Press.

Ziegler, J. F. and Lever, R. F. (1983). *Thin Solid Films* **19**, 291.

# 5 The Interpretation of Plasma Probes for Fusion Experiments

P. C. Stangeby

*Institute for Aerospace Studies*
*University of Toronto*
*4925 Dufferin Street*
*Downsview, Ontario, Canada*

## I. Introduction

Perhaps the simplest of all plasma diagnostic techniques involves the insertion of a solid object into the plasma and the measurement of the particle and energy fluxes to the object, i.e., a probe. The interpretation of the probe signals, however, is notoriously difficult, even for the simplest plasmas. For the case of a strongly magnetized plasma, the problem is particularly difficult, and progress has been modest despite decades of fusion research employing probes. The field remains in a developing state, and this chapter necessarily constitutes a progress report rather than a review of a completed body of knowledge.

It is a common observation that probe diagnosis is experimentally straightforward, but there is a commensurate penalty in interpretive difficulty. For the diagnosis of fusion plasmas—and undoubtedly others as well—even the experimental aspects have grown to be highly challenging. This aspect of the topic has been well covered elsewhere, particularly in the recent review by Manos and McCracken (1986). It will therefore not be the focus of this chapter, which will instead outline the present understanding concerning *interpretation*. This chapter does not include a review of fusion probe literature per se, since most probe work does not address questions of interpretation. Likewise, the very extensive body of knowledge, theoretical and experimental, covering the use of probes in nonmagnetic plasmas is well covered elsewhere, including a chapter of the present book; it will only be briefly referenced here.

Although now two decades old, the classical probe reviews by F. F. Chen (1965) and by Swift and Schwar (1969) remain relevant and highly useful for nonmagnetic plasmas. More recent reviews are also available (Chung et al., 1975; Chen, 1985; Smy, 1976; Clements, 1978; Manos, 1985; Lipschultz et al., 1986), which relate particularly to fusion plasmas.

## II. Survey of Probe Techniques

For a complete coverage of this topic the reader is referred to Manos and McCracken (1986). *Probe* herein refers to the full range of techniques and not just to Langmuir probes. The field can be divided into active probes and passive probes.

*Active Probes:*
1. Langmuir probes (Manos and McCracken, 1986; Chen, 1965; Swift and Schwar, 1969) which record the electrical current $I$ to the probe as a function of the voltage $V$ applied to the probe. This most important of probe techniques yields, in principle, the plasma density $n_e$ and electron temperature $T_e$. Single and triple Langmuir probes also measure the floating potential $V_f$ from which it is possible, in principle, to infer the plasma potential, $V_{p\ell}$, thus the electric field in the plasma, $\vec{E}$. Double and triple Langmuir probes require no reference electrode. Triple probes provide continuous records (no voltage sweep required) of $n_e$, $T_e$, and $V_f$.

2. Finite ion Larmor radius probes (Manos and McCracken, 1986) such as the Katsumata probe (Katsumata and Okazaki, 1967) can, under certain circumstances, provide a measure of $T_i$.

3. Heat flux probes (Manos et al., 1982; Stangeby et al., 1982; Ertl, 1984) measure the heat flux from the plasma deposited on the sensing element via its rate of temperature rise. Assuming the energy reflection coefficient is known, then the incident heat flux is inferrable. The latter information, if combined with information on $T_e$ and $n_e$, e.g., from a Langmuir probe, can provide a measure of $T_i$ (Stangeby, 1982a; Stangeby, 1982b). In principle, combined Langmuir heat flux probes can be used to infer information about the presence of non-Maxwellian components, electron and/or ion (Stangeby, 1984c).

4. Rotating calorimeter probes (Manos et al., 1982; Manos et al., 1986) combine the finite ion Larmor radius effects of very fast ions together with a heat detecting element to deduce the flux and particle energy of very fast ions present as a plasma minority.

5. Mach number probes (Harbour and Proudfoot, 1984; Stangeby, 1984e) consist of two Langmuir probe sensors, facing in opposite directions along $\vec{B}$. Under certain circumstances the ratio of the ion saturation currents collected provides a measure of the drift velocity of the plasma past the probe.

6. The carbon resistance probe (Wampler, 1982) records the ion-impact damage caused to a carbon substrate as a change of its electrical resistance, thus providing a measure of ion flux density and energy.

7. Emissive probes (Hershkowitz et al., 1983) strongly emit electrons due, for example, to thermionic emission, and under some circumstances (Chen, 1985) this causes the probe to float at, or very near, the local plasma potential, thus providing a *direct* measure of $V_{p\ell}$.

8. "Sniffer" probes (Poschenrieder et al., 1982) consist of a small duct which draws in a particle flux sample at one end and detects it, for example, with a mass spectrometer, at the other. A permeation membrane may be used to distinguish atomic from molecular fluxes (Shmayda et al., 1986).

9. Gridded energy analyzers (Simpson, 1986; Matthews, 1984) provide a measure of the energy distribution and flux of ions and electrons.

10. $\vec{E} \times \vec{B}$ probes (Staib, 1980) exploit the strong $\vec{B}$ field present in the machine and can, under some circumstances, (Matthews, 1984) provide a measure of the ion energy distribution.

*Passive Probes:*

1. Hydrogen trapping probes (Staib, 1982; Zuhr et al., 1984) consist of substrates such as carbon which trap ions during plasma exposure for subsequent analysis after withdrawal from the machine. Analysis can

be by thermal desorption, SIMS, nuclear reaction analysis, etc. In principle flux and energy $(T_i)$ can be inferred.

2. Impurity deposition probes (Zuhr et al., 1984) operate similarly for impurities which adhere to the surface of the substrate or penetrate it.

3. Erosion probes (Manos and McCracken, 1986; Zuhr et al., 1984) contain markers, for example, ion implanted impurities in a substrate. Analysis of the substrate subsequent to plasma exposure reveals erosion.

Clearly, the interpretation of data from each of these techniques requires substantial effort. For details, the reader is referred to the primary sources and reviews indicated. Much of the interpretation in each case is the same whether the probe is inserted into a nonmagnetic plasma or a high-field fusion plasma. The focus of this present review is on the *special* interpretative problems associated with use in plasmas with very high magnetic fields. In this regard, it may be noted that virtually all of the probes mentioned encounter the same problems—principally ones arising from the unusual sampling volume for probes in high-$\vec{B}$ plasmas, i.e., very long and thin, and the anomalous nature of cross-field transport. These problems are present in all of their manifestations for Langmuir probes, and since this type of probe is the most widely used, it is appropriate to investigate these general probe questions in the context of this specific technique. Accordingly, the bulk of the following discussion is on Langmuir probes; a brief section is included on the other techniques which indicates how, if at all, their interpretative problems for high-$\vec{B}$ plasmas differ.

The general task then may be stated to be one of relating what is observed at a probe inserted in a high-$\vec{B}$ plasma to the properties of the undisturbed plasma, e.g., $n_e$, $T_e$, $T_i$, impurity density, plasma drift speed, plasma potential, fluctuation frequencies and amplitudes of $n_e$, etc.

## III. Theoretical Basis for Interpretation of Probes in Very Strong Magnetic Fields

### A. The Basic Model

We consider first the definition (Chung et al., 1975; Brown et al., 1971) of "very strong magnetic field" on the basis of the ion and electron Larmor radii:

$$\ell_{e,i} = \frac{v_{e,i} m_{e,i}}{eB}, \tag{1}$$

which are to be compared with the characteristic size of the probe $d$.

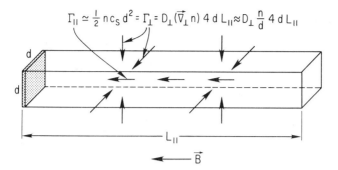

FIGURE 1.   The probe collection or disturbance length $L_\parallel = L_{col}$ is defined by the balance between cross-field diffusion into the flux tube of the probe and the plasma flow parallel to $\vec{B}$ to the probe. Ambipolar collection.

"Strong magnetic field" has been defined to be the case

$$\ell_e < d < \ell_i, \tag{2}$$

while for " very strong magnetic field"

$$d > \ell_{e,i}. \tag{3}$$

Typically $B = 1\text{--}10\text{T}$ in fusion devices, while probe sizes are typically $d = 1\text{--}30$ mm. As plasma pulse lengths have grown longer in fusion devices, now reaching tens of seconds in some tokamaks, and plasma energy densities have increased, probe sizes have tended to increase for heat dissipation reasons. Typically, probes on the JET and TFTR tokamaks, for example, have $d \simeq 10$ mm (Erents et al. 1986; Kilpatrick et al. 1986). Example $T_i = 100$ eV, $\therefore$ $v_i \simeq 10^5$ m/s $(D^+)$, $B = 4T$, $\therefore$ $\ell_i \simeq 0.5$ mm, and thus the " very strong magnetic field" regime generally obtains.

In a magnetic plasma, the probe disturbs the plasma over a very great length (Bohm et al., 1949; Cohen, 1978) $L_{col}$—the probe disturbance or collection length, $L_{col} \gg d$—extending along $\vec{B}$ away from the probe in each direction (Figure 1). The sink action of the probe reduces $n_e$ in the disturbed zone, and cross-field transport draws particles into this zone, achieving a steady state in which cross-field rates equal parallel rates. The key aspect of the very strong magnetic field regime is that the magnetic field totally dominates cross-field transport to the probe, even for the ions, and individual particle trajectory effects are not important. As shown in Figure 1, the value of $L_{col}$ is given (Stangeby, 1982; Cohen, 1978) approximately by the flux balance expression

$$\Gamma_{col} \simeq \tfrac{1}{2} n c_s d^2 = \Gamma_\perp \left( \vec{\nabla}_\perp n \right) 4 d L_{col} \simeq D_\perp \frac{n}{d} 4 d L_{col}. \tag{4}$$

Thus

$$L_{col} \simeq \frac{d^2 c_s}{D_\perp}. \tag{5}$$

Equation (4) assumes that parallel flow to the probe in the collection flux tube involves acceleration of the ions to the sound speed

$$c_s = \left[ \frac{k(T_e + T_i)}{m_i} \right]^{1/2}, \tag{6}$$

while the density drops to half the far-field value, see Section III.B; it is assumed that cross-field transport is diffusive and proportional to the density gradient caused by the probe's presence, $\approx n/d$. Example: $d = 10$ mm, $T_e = T_i = 100$ eV $\therefore$ $c_s = 10^5$ m/s ($D^+$), $D_\perp = 1$ m$^2$/s $\therefore$ $L_{col} = 10$ m, i.e., the disturbed zone is extremely long and narrow. One may note one practical consequence of this: The probe does not measure local conditions but ones averaged over length $L_{col}$.

In one sense, the very strong field regime involves a simplification in that finite ion larmor radius effects can be neglected. Cross-field motion is presumably "bulk" or "fluid-like." An additional advantage is that plasma flow to the probe is presumably, therefore, of the same nature as to the very large wall structures contacting the plasma, e.g., limiters. Thus the extensive body of experimental knowledge (Post and Lackner, 1986) concerning plasma-limiter interactions should be applicable to probes in fusion devices. This similarity, however, points up a critical problem: Cross-field transport is not understood in high-$\vec{B}$ plasmas, it is anomalous (Bohm et al., 1949). So long as this situation persists, there can be no complete probe theory for the very strong field case.

There is considerable evidence that cross-field transport to limiters behaves as *if* it were diffusive, that is, particle flux density is proportional to the local density gradient (Post and Lackner, 1986); recently, experimental measurements of plasma near a probe have provided similar evidence (Matthews et al., 1986). The measured diffusion coefficient $D_\perp$, however, is much too large to be explained by classical cross-field diffusion (Chen, 1984), i.e., resulting from particle–particle collisions. The latter is $D_\perp \simeq D_\parallel (\ell_i/\lambda_{ie})^2$ for a fully ionized plasma, where $\lambda_{ie}$ is the mean free path and $D_\perp$ is orders of magnitude smaller than the generally observed value, which often obeys the purely empirical Bohm expression (Bohm et al., 1949):

$$D_\perp^{Bohm} \simeq 0.06 \frac{T_e}{B} \quad [\text{m}^2/\text{s}],$$

where $T_e$ is in [eV], $B$ in [T]. That is, $D_\perp \approx 1$ m$^2$/s for edge plasma conditions in fusion devices. There is strong evidence (Zweben and Gould,

1983; Ritz et al., 1984; Robinson and Rusbridge, 1969) that cross-field transport in high $\vec{\mathbf{B}}$ devices is due to fluctuations in density $\tilde{n}$ and plasma potential, $\tilde{V}_{p\ell}$. Experimentally, it is found that in fusion edge plasmas $\tilde{n}/\bar{n}$ and $\tilde{V}_{p\ell}/V_{p\ell}$ range up to 100%. The cross-field flux appears to be due to the $\vec{\mathbf{E}} \times \vec{\mathbf{B}}$ plasma drift associated with $\tilde{V}_{p\ell}$ and the applied magnetic field $\vec{\mathbf{B}}_{app}$. For net outward flow to occur the $\tilde{n}$ and $\tilde{V}_{p\ell}$ fluctuations must be in the appropriate phase

$$\Gamma_r = \langle \tilde{n}\tilde{E}_\theta \rangle / B_{app},$$

where $\tilde{E}_\theta$ is the fluctuation electrical field in the plasma in the direction normal to $\vec{\mathbf{B}}$ and to $\vec{\mathbf{r}}$. The fluctuation level appears to be caused by the existence of density gradients (Callen et al., 1982), thus explaining why transport appears to be diffusive, i.e., $\Gamma_r \propto dn/dr$.

Because the cross-field transport is nonclassical, the probe theories which have been developed assuming classical transport (Sanmartin, 1970)— including those for space applications (Rubinstein and Laframboise, 1982) —cannot be expected to reproduce the experimental situation in fusion devices.

There is also the difficulty of explaining nonambipolar cross-field transport. Such flows occur, under some circumstances, to limiters (Erents et al., 1986) and always to probes which draw net current. Classically, cross-field transport is automatically ambipolar for a fully ionized plasma (Chen, 1984). The simplest picture of fluctuation-induced cross-field transport also implies ambipolar flow. Thus major questions remain unanswered, see Section VIII.

All is not lost, however, as it appears that certain aspects of probe interpretation are insensitive to the uncertainties in the mechanism of cross-field transport. Happily, these aspects govern the most important practical uses of probes, e.g., measuring $n_e$. This is discussed in more detail in Section IV.B. Here we may note that the very strong field case enjoys the simplification that flow to the probe is only quasi-two-dimensional. To a good approximation, one can treat the cross-field and parallel-field motion separately, principally because the scale lengths for the two motions are so enormously different. Thus, so far as the parallel-motion is concerned, one can model the flow as one-dimensional, nonmagnetic flow with a distributed particle source term. The latter is given by the cross-field flow. One-dimensional, zero-$\vec{\mathbf{B}}$ plasma motion has been much studied experimentally and theoretically, and this well established body of knowledge may be appropriated for probe interpretation. This work is briefly reviewed in Section III.B. One of the findings of this work is that, independent of the source distribution (i.e., of the cross-field transport in the present case), the flow speed reaches $c_s$ at the probe while the density diminishes to about half the distant value. Thus the inference of $n_e$, for example, from the probe measurements should be insensitive to uncertainties in the cross-field mechanism. Other aspects of probe interpretation, unfortunately, are less immune, e.g., inferring $T_e$, see Section IV.C.

One may note, in passing, one additional simplification for probe interpretation in typical fusion conditions—the sheath thickness is generally vanishingly small compared to the probe size. The sheath thickness is on the order of the Debye length,

$$\lambda_D = \left( \frac{\epsilon_0 k T_e}{n_e e^2} \right)^{1/2} \tag{7}$$

and unless $\lambda_D \ll d$, strong electric fields extend far from the probe, making interpretation complex even for $\vec{B} = 0$. Typically, $n_e = 10^{17}-10^{19}$ m$^{-3}$, $T_e = 5-100$ eV, thus $\lambda_D = 10^{-3}-10^{-1}$ mm.

We thus arrive at the following simple model for probes in magnetic fusion devices: The effective collection area of the probe is simply its *projected* area perpendicular to $\vec{B}$. The complexities of finite $\ell_i$ and $\lambda_D$ are usually (but not always) absent. The flow is essentially the result of two quasi-independent one-dimensional motions. The parallel flow is classical, simple, and understood. Cross-field transport is not understood at the present time, despite almost half a century of very substantial research effort. For the present, an important task is, therefore, isolating and identifying the influence of this uncertain aspect on probe interpretation.

As indicated, objects in contact with plasmas spontaneously assume a potential (if allowed to float) which is negative relative to the plasma (Figure 2). The object may also be biased to more or less than this potential. So long as the probe-plasma potential drop is negative, this drop is accommodated in a very thin sheath, thickness O($\lambda_D$). The problem thus divides naturally into two parts: the sheath and the presheath (or more simply, plasma) of length $L_{col}$. The plasma density, for example, varies in the presheath by an amount which must be known if probe measurements are to be interpreted. This aspect is addressed in Section III.B. Separately one can consider the relation between the potential drop in the sheath and the density, etc., just at the sheath/presheath interface. This aspect is addressed in Section III.C. The actual probe characteristic results from the combination of the two parts.

## B. PLASMA FLOW TO SURFACES IN PLANAR GEOMETRY

We proceed with the basic assumption of the model that the parallel and cross-field flow are quasi-independent and that the parallel flow can be treated as 1-$D$, $\vec{B} = 0$ flow to a surface. Such plasma motion has been

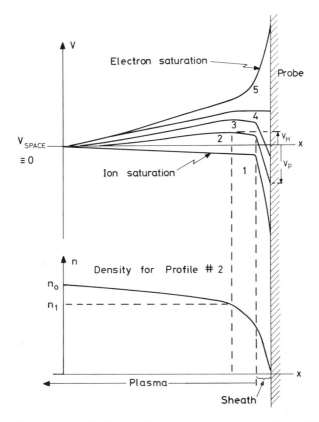

FIGURE 2. Potential and density profiles near a biased probe. Curve 1/5 describes ion/electron saturation. For intermediate conditions (net electron collection), a positive potential ($V_H$) can develop in the plasma as a result of friction between ions and electrons flowing to the probe. Floating conditions closely conform to curve 1.

widely studied and theoretically analyzed since the pioneering work of Irving Langmuir (Langmuir, 1929) in the 1920s. We consider first the case of ambipolar flow of a fully ionized plasma, thus neither electrons nor ions suffer momentum-loss collisions. Collisions may nevertheless occur, e.g., ion–ion, and one might think it important to distinguish collision-less (kinetic) flow and collisional (fluid) flow. In fact, this is found to make little difference to the predictions of interest.

The ion–ion collision length

$$\lambda_{\mathrm{ii}} \simeq \frac{10^{17} T_i^2}{n_i \ell n \Lambda},$$

(8)

with $T_i$ in [eV], $n$ in [m$^{-3}$], $\lambda$ in [m]. Example: $T_i = 100$ eV, $n_i = 3 \times 10^{18}$ m$^{-3}$, $\ell n \Lambda = 20$ $\therefore$ $\lambda_{ii} \simeq 10$ m, which can be comparable to $L_{col}$.

We consider first the collisional (fluid) case. Assume isothermal flow ($T_e$, $T_i$ constant) and that the electrons satisfy the Boltzmann relation

$$n_e = n_0 \exp\left[\frac{eV}{kT_e}\right],\qquad (9)$$

where $V(x)$, the plasma potential varies with $x$ (along $\vec{B}$) and $n_0$ is the far-field, undisturbed plasma density. The electrons are in a retarding field as they approach the solid, owing to the negative charge on the (floating) surface which built up in the first instants after plasma formation, owing to the higher random velocity of the electrons than the ions. The electrons still drift toward the probe, $v_e = v_i$ (ambipolar flow, $Z = 1$ ions) despite the electrostatic repulsion, due to a pressure (i.e., density) gradient which is established along $x$, $n_e(x)$. There is almost exact balance between these two forces on the electrons; precise equality gives the Boltzmann relation precisely. (The relation is in error by $0(M_e)$, where $M_e = v_e/\bar{c}_e$ = electron Mach number, however, $M_e \ll 1$). The ions, on the other hand, experience two nearly-equal accelerating forces: that due to $dV/dx \equiv -E$ and that due to $dp_i/dx$. The conservation of mass and momentum for the ions gives

$$\frac{d}{dx}(nv) = S \qquad (10)$$

$$nm_i v\frac{dv}{dx} = -\frac{dp_i}{dx} + eE - Svm_i, \qquad (11)$$

where $S$ is the source strength due to cross-field transport:

$$S \approx \frac{D_\perp n_0}{d^2}. \qquad (12)$$

Combining Eqs. (10) and (11) gives for the ion Mach number $M \equiv v/c_s$ the equation (Stangeby, 1984e)

$$\frac{dM}{dx} = \frac{S}{nc_s}\frac{1 + M^2}{1 - M^2}. \qquad (13)$$

One may note that once the flow has accelerated to the point where $M = 1$ the plasma solution "blows up," i.e., $dM/dx$, $dV/dx \to \infty$, and one has a natural transition to the electrostatic sheath just in front of the solid surface with its very large electric field. By analyzing the nature of the equations of motion *within* the sheath it can be shown (Bohm et al., 1949; Stangeby, 1986; Chodura, 1986) that the ion velocity must be *at least* $c_s$, thus one arrives at the result that the flow velocity at the sheath edge is exactly sonic, i.e., the Bohm criterion (Bohm et al., 1949; Chen, 1965). The equations may also be rewritten to give the result

$$\frac{n(M)}{n_0} = \frac{1}{1 + M^2}.$$ (14)

Thus at the sheath edge $n = \frac{1}{2}n_0$. We have thus arrived at the surprising and critically important result that the ambipolar flux density to the probe is

$$\Gamma = \frac{1}{2}n_0 c_s$$ (15)

*independent* of $S$, its absolute magnitude or spatial distribution.

We consider next collisionless ambipolar flow to a surface. This was first analyzed by Tonks and Langmuir in 1929 for the case of cold ions, $T_i = 0$. They obtained $\Gamma = 0.487\, n_0 c_s$. Emmert et al. (1980) have extended the analysis to cover $T_i \neq 0$ and, using a particular assumption about the velocity distribution of the ions as they enter the flow, found that $\Gamma = f_E(T_i/T_e)n_0 c_s$, with $f_E$ varying monotonically from 0.487 at $T_i/T_e = 0$ to 0.798 as $T_i/T_e \to \infty$. Emmert et al. assumed that the source ion distribution is such that if $E = 0$, the ion *fluxes* will be Maxwellian, i.e., there are no ions with zero random velocity. Bissell and Johnson (1986) have employed a slightly different assumption, namely, that the ions are created, for example, by the ionization of a Maxwellian neutral distribution, i.e., the ion *density* will be Maxwellian (if $E = 0$), and therefore there are some ions with zero random velocity at their time of entry to the flow. Bissell obtains the result $\Gamma = f_B(T_i/T_e)n_0 c_s$, with $f_B$ decreasing monotonically from 0.487 at $T_i/T_e = 0$ to 0.41 to $T_i/T_e = 4$.

We thus find that so far as the principal measurable quantities of interest are concerned, e.g., particle fluxes, it makes little difference whether the plasma flow is collisionless or (self) collisional. We may therefore proceed to consider the case of nonambipolar flow from the fluid viewpoint. This is deferred, however, until Section IV.C.

## C. Sheath Transmission Factors

In this section $V_{ps}$ refers to the plasma potential just at the plasma/sheath interface, while the plasma density at the interface is $n_{ps}$. So long as $V_{probe} < V_{ps}$ the ion flux is independent of $V_{pr}$ and is

$$\Gamma_i = n_{ps} c_s.$$ (16)

The electron flux, allowing for secondary electrons, is

$$\Gamma_e = \tfrac{1}{4} n_{ps} \overline{c_e} (1 - \gamma_{se}) \exp\left[\frac{e(V_{pr} - V_{ps})}{kT_e}\right]. \tag{17}$$

Thus for ambipolar, i.e., floating, conditions and $Z = 1$ ions, one has the probe floating potential $V_f$ from $\Gamma_i = \Gamma_e$:

$$V_f = V_{ps} + \frac{kT_e}{2e} \ell n\left[\left(2\pi\frac{m_e}{m_i}\right)\left(1 + \frac{T_i}{T_e}\right)(1 - \gamma_{se})^{-2}\right]. \tag{18}$$

(Note that the second term is generally negative.)

The *sheath* (although not necessarily the *probe*) I–V characteristic is then

$$I = I_s^+\left[1 - \exp\left(\frac{eV_{pr}'}{kT_e}\right)\right], \tag{19}$$

where $I_s^+ \equiv en_{ps}c_s$, the ion saturation current ($Z = 1$ ions)

$V_{pr}' \equiv V_{pr} - V_f$, the probe potential relative to the floating potential.

Note that secondary electron emission generally does not affect the I–V characteristic. If the $\gamma_{se}$ is large enough, however, then $V_f \to V_{ps}$ and the assumptions employed above fail. Hobbs and Wesson (1967) showed that for $\gamma_{se} > 0.8$ (hydrogenic plasma) the sheath potential becomes nonmonotonic in space with a negative space charge sublayer developing adjacent to the surface, effectively suppressing further release of secondary electrons back into the plasma (Figure 3). Thus the *effective* value of $\gamma_{se}$ has an upper limit of $\sim 0.8$. The matter is not entirely clear (Chen, 1985), however, for the case of *very* strong electron emission from the surface. In this case, evidently the sublayer can be "washed out" and the probe then floats at $\simeq V_{ps}$ (or slightly above), with no conventional sheath being

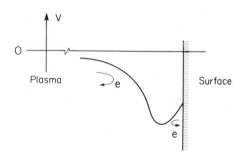

FIGURE 3. Double sheath can develop for $\gamma_{se} \gtrsim 0.8$ which limits the number of secondary electrons reaching the plasma.

present at all. The latter, of course, describes the emitting probe (Hershkowitz et al., 1983) which is used to directly measure plasma potential. The transition between the two regimes for $\gamma_{se}$ in the $\sim 0.8$ to $\sim 1$ range requires further study, see Section VIII.

Turning to the matter of heat transmission through the sheath, if the electrons have a Maxwellian distribution far from the probe, then the mere act of retarding them in the presheath and sheath potential drops does not change this distribution. Thus the electron heat flux density reaching a floating surface is

$$Q_e = \frac{2kT_e}{1 - \gamma_{se}} \frac{I_s^+}{e}, \tag{20}$$

where the factor 2 derives from the nature of the Maxwellian distribution and the $1 - \gamma_{se}$ factor accounts for the fact that each secondary electron removes little heat from the surface but permits an additional plasma electron to reach the surface.

The ions are accelerated by the presheath field, and thus their velocity distribution at the plasma/sheath interface is not simple. This matter has been examined, and it is concluded (Stangeby, 1986) that to a first approximation one can take

$$Q_i = \left[2kT_i + \left|\frac{e(V_f - V_{ps})}{kT_e}\right|\right]\frac{I_s^+}{e}, \tag{21}$$

where the second term represents the energy gained by the ions in falling through the sheath.

The total heat transmission factor is defined to be

$$\delta \equiv \frac{(Q_e + Q_i)}{(kT_e I_s^+ / e)}, \tag{22}$$

thus

$$\delta = \frac{2T_i}{T_e} + \frac{2}{1 - \gamma_{se}} - 0.5\,\ell n\left[\left(\frac{2\pi m_e}{m_i}\right)\left(1 + \frac{T_i}{T_e}\right)(1 - \gamma_{se})^{-2}\right]. \tag{23}$$

For the case of a biased probe it can be shown (Stangeby, 1984d) that

$$\delta = \left|\frac{e(V_{pr} - V_{ps})}{kT_e}\right| + \frac{2T_i}{T_e}$$
$$+ 2\left[\left(1 + \frac{T_i}{T_e}\right)\left(\frac{2\pi m_e}{m_i}\right)\right]^{-1/2} \exp\left[\frac{e(V_{pr} - V_{ps})}{kT_e}\right]. \tag{24}$$

See Figure 4. Note that here $\delta$ is independent of $\gamma_{se}$. One may also note

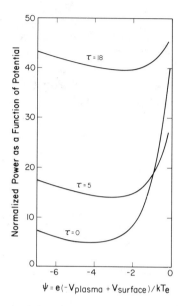

FIGURE 4.   The heat transmission coefficient as a function of surface potential shows a minimum at $V_{\text{surface}} \approx V_{\text{floating}}$. $\tau \equiv T_i/T_e$.

that the heat load is a minimum, approximately, at floating conditions (Kimura et al., 1978).

For these heat transmission coefficients, a number of refinements may be appropriate in certain operating conditions of fusion devices:

(a) When $T_e$ is very low, $\lesssim 5$ eV, as in high recycling divertors, then the heat deposition associated with ion–electron and atom–atom surface recombination can dominate the kinetic energy fluxes; the energy of secondary electrons should not be neglected.

(b) Ion back-scattering reduces the energy deposited by the ions.

(c) The electron and/or ion distribution far from the probe may be non-Maxwellian.

(d) Impurity ion effects may be significant, see Section VII.B.

## IV. Langmuir Probe Interpretation

Undoubtedly the most extensively used probes in fusion devices are Langmuir probes. Theoretical understanding is also more advanced for this than for other probe methods.

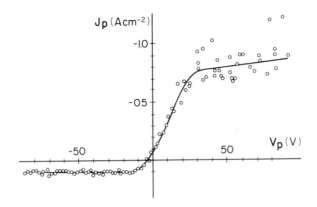

FIGURE 5. Langmuir probe characteristic taken in the DITE tokamak with plasma current = 120 kA, toroidal field = 2T and the probe 35 mm behind the limiter.

## A. GROSS FEATURES OF THE LANGMUIR PROBE I–V CHARACTERISTIC

Figure 5 is a typical experimental I–V characteristic (Manos and McCracken, 1986) from a tokamak plasma. One may note certain features:

(a) The ion saturation portion is usually flat, yielding a well defined value of $I_s^+$. This contrasts with the nonmagnetic plasma result, where the ion current usually never actually saturates. This saturation is presumably attributable to the small values of $\lambda_D/d$ typical of fusion experiments.

(b) An electron saturation current $I_s^-$ is relatively well defined, but more noisy than $I_s^+$. The most remarkable feature, however, is the small ratio of $I_s^-/I_s^+$, which sometimes approaches unity and is typically $\sim 10$. In a nonmagnetic plasma one generally observes the simple relation (Langmuir, 1929)

$$\frac{I_s^-}{I_s^+} = \frac{\frac{1}{4}n\bar{c}_e}{nc_s} \approx \sqrt{\frac{m_i}{m_e}} \simeq 60 \qquad \text{for } D^+. \tag{25}$$

It is believed that for very strongly magnetized plasmas the value of $I_s^+$ is about as expected on simple considerations (see the next section) and that the complications relate to electron collection (Bohm et al., 1949; Cohen, 1978; Stangeby, 1982a). The latter matter and the explanation of the $I_s^-/I_s^+$ ratio is discussed in Section IV.C.

## B. MEASURING $n_e$ WITH A LANGMUIR PROBE

On the basis of the deductions about ambipolar flow in Section III.B and the sheath properties in Section III.C one anticipates that for a floating probe the so-called *Bohm formula*,

$$\Gamma^+ \simeq 0.5 n_0 c_s, \tag{26}$$

should hold *independent* of the uncertainties in the cross-field transport mechanism (or, equivalently, uncertainties in $L_{col}$). Consider next the situation when the probe is biased more negatively so as to repel electrons, thus making it possible to actually measure $\Gamma^+$ (as $I_s^+$). While this change of bias may (?) cause a change in the cross-field transport mechanism, if we follow the same logic as before, then such a hypothetical change should not in itself alter relation (26). One must consider, however, the frictional drag over length $L_{col}$ on the ions as they drift toward the probe through a background of nondrifting electrons (probe highly biased negatively). Fortunately, the great mass disparity between ions and electrons means that the ions are not retarded significantly by the stationary electrons. Thus, in a pure, fully ionized plasma one is led to expect that

$$I_s^+ = 0.5 e n_0 c_s, \tag{27}$$

and so a simple measurement of the distant, unperturbed plasma density, $n_0$, is possible.

In reality, other particle collisions can cause momentum loss to the ions over length $L_{col}$. It can be shown (Franklin, 1976) that $n_{ps}/n_0$ is a monotonically increasing function of $\lambda_m/L_{col}$, where $\lambda_m$ is the mean free path for momentum loss by the ions; therefore significant errors in deducing $n_0$ are conceivable. Consider first neutral hydrogen-ion collisions. Typically the neutral atom density in the edge region is $\sim 10^{-2}$ of the plasma density (Heifetz, 1986), say, $10^{16}$ m$^{-3}$. The hydrogenic charge-exchange cross section is (Harrison, 1986) $\sim 3 \times 10^{-19}$ m$^2$, hence $\lambda_m^{cx} > 100$ m, and so one can generally neglect this process. Impurity ions are also present, but unless their densities are extremely high, they will also not retard the ions significantly.

One thus concludes that the unperturbed plasma density should be given by

$$n_0 \simeq \frac{2 i_s^+}{A_\perp c_s e}, \tag{28}$$

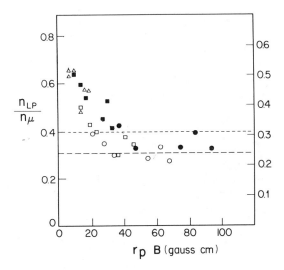

FIGURE 6. Transition from strong magnetic field regime, $\ell_i > r_p > \ell_e$, to very strong magnetic field regime $r_p > \ell_{e,i}$. Ratio of plasma density measured by a Langmuir probe, $n_{LP}$, and a microwave interferometer, $n_\mu$, as a function of $r_p B$ in a non-fusion magnetized plasma. Probe radius $r_p = 0.24$ mm (●), 0.18 mm (○), 0.12 mm (□), 0.085 mm (■), 0.045 mm (△). Left-hand scale for $n_{LP}$ calculated using Laframboise theory, right-hand scale for $n_{LP}$ calculated using Bohm expression, Eq. (28) and taking total circumferential area of cylindrical probe for both cases. Thus as $r_p B \to \infty$ one expects ratio $\to \pi^{-1}$ (dotted lines).

where $i_s^+$ is the saturation ion current in [Amps] and $A_\perp$ is the projected area of the probe collecting element.

This is perhaps the most important single prediction for probe diagnosis, and it is appropriate to consider in some detail what experimental evidence there is to support it. Conceptually, the simplest test is to compare values of $n_e$ deduced using a Langmuir probe and Eq. (28) with ones measured by a different diagnostic technique, e.g., microwave interferometry.

An early and important experiment of this nature was carried out by Brown et al. (1971) employing a nonfusion plasma facility. Because $\lambda_D \approx d$ in their system, the value of $I_s^+$ was not clearly defined; nevertheless, using the simplest definition of $I_s^+$ (namely, the point where the ion "saturation" curve departed from linearity, i.e., showed electron contributions), very good agreement was obtained between microwave measurements and Eq. (28) (see Figure 6), using $A$ = projected probe area. By varying the magnetic field strength, Brown et al. were able to span from the strong magnetic field case to the very strong case (break point at $r_p B \approx 0.018 T$ mm, where

$r_p$ = cylindrical probe radius). Also shown in Figure 6 is a comparison with $n_e$ deduced from the Langmuir probe using Laframboise's probe theory (Laframboise, 1966). The latter theory gives a value of $n_e$ which is smaller than the Bohm value by $\sqrt{2/\pi} = 0.798$, a variation which is probably not distinguishable within experimental error. In the nonfusion open-ended test facility used, $T_i \ll T_e$, also end effects caused uncertainties. Comparison of microwave and Langmuir probe values of $n_e$ in a tokamak are therefore of more direct relevance. Such comparisons have been made on a number of machines.

A very detailed comparison was made by Proudfoot (to be published) on the DITE tokamak using an array of four Langmuir probes mounted on a rake which could be moved horizontally into the plasma. Thus a large number of spatially distributed points were measured along a chord. This permitted direct comparison with microwave measurements which, of course, yield values of $n_e$ integrated along a chord (see Figure 7). As can be seen, the agreement with the Langmuir probe (Bohm) value is excellent. Unfortunately, the cylindrical probe diameter, 0.5 mm, was comparable to $\ell_i$, and thus there is some uncertainty in the definition of the ion collection area. The results in Figure 7 were obtained assuming the strong magnetic field regime, and thus the full surface (rather than projected) area of the probe was used. It was also assumed that $T_e = T_i$, although no direct measurement was made of $T_i$.

Uncertainties in the effective ion collection area were evidenced in other DITE probe experiments (Stangeby et al., 1984) in which a large Langmuir probe (plate, 5 × 10 mm) was compared with a small one (cylinder, 5 × 1 mm) in the same plasma (Figure 8). The cylinder appeared to collect ions over its entire surface area, although this varied with plasma position, the collection being more effective at smaller densities, apparently indicating finite $\lambda_D$ effects.

Other comparisons of microwave and Langmuir probe values of $n_e$ in tokamaks have been reported, e.g., Ditte and Grave in ASDEX (1985). These authors used cylindrical probes (2 × 2 mm) and, assuming the very strong field regime (projected area collection) and $T_e = T_i$, found that the Langmuir probe value of $n_e$ was low by a factor of 1.78 compared with the microwave value.

We may conclude, therefore, that the present experimental evidence confirms the validity of the Bohm formula, Eq. (28), for $n_e$ to within a factor of two and probably better. The definitive test remains to be done, however. This test would employ probes clearly operating in the very strong field regime ($d \gg \ell_i$) with very thin sheath ($d \gg \lambda_D$), and with $T_i$ measured, e.g., using a gridded energy analyzer and with numerous probe measurements along the line of sight of the microwave interferometer.

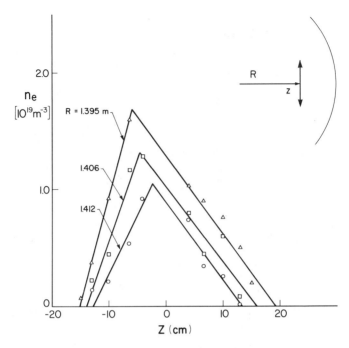

FIGURE 7. Langmuir probe and microwave interferometry measurements of $n_e$ made on DITE. Probe configuration described in *J. Nucl. Mat.* **111** & **112** (1982), 87, with interpretation based on Bohm formula, Eq. (28), and A = total surface area of collector. Microwave integrates along Z. Density fitted by straight lines.

| | R | $\dfrac{(\bar{n}_e \ell)_{LP}}{(\bar{n}_e \ell)_\mu}$ | |
|---|---|---|---|
| | [m] | $[10^{18} \text{ m}^{-2}/10^{18} \text{ m}^{-2}]$ | |
| △ | 1.395 | $\dfrac{2.8}{2.2}$ | = 1.3 |
| □ | 1.406 | $\dfrac{2.0}{2.0}$ | = 1.0 |
| ○ | 1.412 | $\dfrac{1.4}{1.6}$ | = 0.88 |

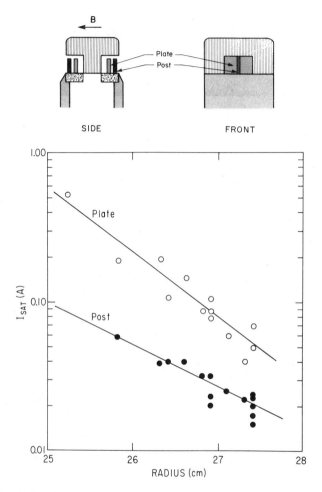

FIGURE 8.   Probe size effects shown with a probe including a 1 mm × 5 mm post collector in front of a 10 mm × 5 mm plate collector. Values of $I_s^+$ as a function of radial position in DITE, limiter radius 26 cm. Ratio of $I_{s\ plate}^+/I_{s\ post}^+$ is clearly less than ratio of the projected areas (9) indicating finite ion larmor radius collection by the post.


## C.   MEASURING $T_e$ USING A LANGMUIR PROBE

For a Maxwellian electron distribution, Eq. (19) gives the I–V characteristic of the *sheath*. If this is also the Langmuir *probe* characteristic then $T_e$ is simply deduced by first finding $I_s^+$, then plotting $\ell n(I^-)$ vs. $V_{\text{probe}}$.

Unfortunately, the sheath and probe characteristics are not identical in general. The reference electrode for the (single) probe is typically the limiter, and the complete probe circuit involves a long plasma path. For net electron collection by the probe, the electrons suffer momentum-loss collisions as they are drawn down the long flux tube, length $L_{col}^-$, moving necessarily with a higher drift velocity than the ions. An electric field thus arises along the flux tube to draw the electrons to the probe against friction. This field repels ions which, conforming to a Boltzmann factor (based on $T_i$), gives a density depression just in front of the probe. The value of $I_s^-$ is thus reduced from the $B = 0$ value of $1/4 n_0 \bar{c}_e e$, thereby explaining the depressed ratio $I_s^- / I_s^+$ generally observed.

Bohm (1949) first quantified these ideas, and his result is derived here in a simplified form. Bohm's analysis was applied to the case of electron saturation, curve E of Figure 2. The electron parallel flux density

$$\Gamma_\parallel^- = -D_\parallel^- \frac{dn}{dx} + n\mu_\parallel^- E \tag{29}$$

where $x$ is measured parallel to $\vec{B}$ and where it is assumed that the electron motion is diffusive (due to $e - i$ collisions) rather than free-fall.

One assumes the Einstein relation

$$\frac{D_\parallel^-}{\mu_\parallel^-} = -\frac{kT_e}{e} \tag{30}$$

and that the repelled ions satisfy the Boltzmann relation

$$n_i = n_0 \exp\left[\frac{eV}{kT_i}\right], \tag{31}$$

where $V$ is the plasma potential, $E = -dV/dx$ and $V \to 0$ far from the probe. Thus

$$\Gamma_\parallel^- = -D_\parallel^- (1 + \tau) \frac{dn}{dx}, \tag{32}$$

where $\tau \equiv T_i/T_e$.

Bohm assumed a similar relation for the cross-field transport into the flux tube, Figure 1, i.e.,

$$\Gamma_\perp^- = -D_\perp^- (1 + \tau) \frac{dn}{dz}, \tag{33}$$

where $z$ is measured perpendicular to $\vec{B}$. This assumption is examined below.

By assuming that $\Gamma_\perp A_\perp = \Gamma_\parallel A_\parallel$ with $A_\perp \simeq 4 d L_{col}^-$ and $A_\parallel = d^2$ where $L_{col}^-$ is the probe's electron collection length and the approximations

$$\frac{dn}{dz} \simeq -\frac{n_0 - n_1}{d} \tag{34}$$

$$\frac{dn}{dx} \simeq \frac{n_0 - n_1}{L_{col}^-}, \tag{35}$$

where $n_1$ is the density at the end of the long flux tube near the probe, one obtains for $L_{col}^-$

$$L_{col}^- = \frac{d}{2\sqrt{\alpha}}, \tag{36}$$

where $\alpha \equiv D_\perp^- / D_\parallel^-$. If we assume for the moment that $D_\perp^-$ is about the same as the experimentally measured ambipolar diffusion coefficient characteristic of fusion devices, i.e., $D_\perp \simeq 1 \text{ m}^2/\text{s}$, then using the classical value for $D_\parallel^-$:

$$D_\parallel^- \approx \lambda_{ei} \bar{c}_e, \tag{37}$$

one finds that $\alpha \simeq 10^6$ for typical edge plasma conditions. Therefore, $L_{col}^- \approx 10^3 d$, again indicating a very long, narrow disturbance zone.

One may also combine Eqs. (32)–(35) to give

$$\Gamma_\parallel^- = 2D_\parallel^-(1 + \tau)(n_0 - n_1)\sqrt{\alpha}/d, \tag{38}$$

from which one may eliminate $n_1$ by assuming that the electrons travel collisionlessly to the probe over the last mean free path from the point where $n = n_1$ near the end of the tube, i.e., we also have the relation

$$\Gamma_\parallel^- = \tfrac{1}{4} n_1 \bar{c}_e. \tag{39}$$

Thus

$$\Gamma_\parallel^- = \tfrac{1}{4} n_0 \bar{c}_e \frac{2\lambda_{ei}\sqrt{\alpha}\,\dfrac{(1 + \tau)}{d}}{1 + 2\lambda_{ei}\sqrt{\alpha}\,\dfrac{(1 + \tau)}{d}}, \tag{40}$$

where $D_\parallel^- = 1/4\lambda_{ei}\bar{c}_e$ is used.

We may rewrite as

$$\Gamma_\parallel^- = \tfrac{1}{4} n_0 \bar{c}_e \frac{r}{1 + r}, \tag{41}$$

where the reduction factor

$$r \equiv 2\lambda_{ei}\sqrt{\alpha}\,\frac{(1 + \tau)}{d}. \tag{42}$$

[Aside: an earlier derivation (Stangeby, 1982a) of equation (42) differed by a factor $8/\pi$. The latter allows for collection by *both* sides of the probe, as in Bohm's original derivation (1949), and also assumes Bohm's analogy between the solution of the diffusion equation for the probe and the electrostatic equation for a capacitor. The latter analogy is doubtful since it requires that the density gradient normal to the probe surface be constant across the face of the probe, in analogy to $\vec{E}$ for a conducting surface. There is no valid reason to assume this for the probe case.]

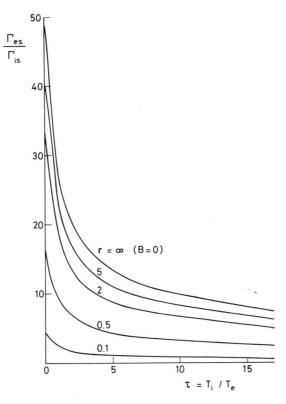

FIGURE 9.   Calculated ratio of electron to ion saturation currents as a function of $T_i/T_e$, with reduction parameter $r$, Eq. (42). As $B$ increases $r$ decreases. $D^+$ ions.

For strong magnetic fields $r \ll 1$, and thus one has a quantitative explanation for the reduction of the ratio $I_s^- / I_s^+$ (Figure 9).

The foregoing analysis, which just applies to electron saturation, has been extended (Stangeby, 1982a) to all values of electron collection, with the result that a nonexponential behavior is predicted (Figure 10). In this theory the electron current density to the probe, as a function of probe potential $V_{pr}$ relative to the distant plasma potential, is given by

$$I^- = \tfrac{1}{4} n_0 \bar{c}_e \left[ r^{-1} + \exp\left[ e\left(V_H - V_{pr}\right)/kT_e \right] \right]^{-1}, \qquad (43)$$

where $V_H$ is obtained from

$$\frac{eV_{pr}}{kT_e} = \frac{eV_H}{kT_e} + \ell n \left[ r\left(\exp(eV_H/kT_i) - 1\right) \right]. \qquad (44)$$

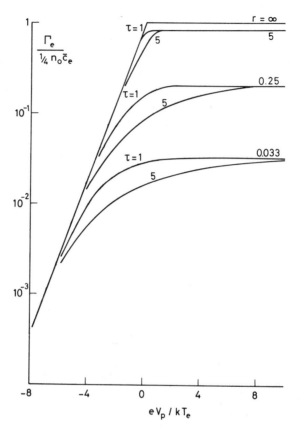

FIGURE 10. Normalized electron current density to a probe as a function of its potential relative to plasma potential with $\tau = T_i/T_e$ and $r$ (reduction) parameters, Eq. (42).

Independently, Ditte (Ditte and Muller, 1983; Ditte, 1983) has developed a similar theory for electrons and has also extended the analysis to ions in the case where neutrals control parallel ion diffusion.

In order to compare theory such as Eq. (44) with experiment, it is necessary to know $D_\perp^-$ and, more generally, to justify the nature of the assumed nonambipolar net-electron cross-field transport, Eq. (33). This is not readily done. According to *classical* cross-field transport theory (Chen, 1984), for fully ionized plasmas transport is automatically ambipolar for the simple probe geometry considered; also the concept of $\mu_\perp^-$ is invalid, since an applied electric field does not lead to transport in the direction of $\vec{E}$, but rather to flow perpendicular to $\vec{E}$ and $\vec{B}$. Fluctuation-induced transport, in its most simple formulation, is also inherently ambipolar since it is a type of

$\vec{E} \times \vec{B}$ drift. It is thus not evident how there can be any net electron (or ion) collection by a probe at all! In light of this fundamental conundrum, it is obviously impossible to assign any value to $D_\perp^-$ with confidence. Indeed, one may question the basic assumption of Eq. (33) entirely. Nevertheless, it is instructive to compare the predicted I–V characteristic, Figure 10, with experiment, assuming for convenience that $D_\perp^- = D_\perp^{\mathrm{amb}} = D_\perp^{\mathrm{Bohm}} = 0.06 T_e\, B^{-1}$, say. This has been done for a number of tokamaks (Bundy and Manos, 1984; LaBombard and Lipschultz; Erents; Tagle et al.), generally with the same results: The characteristic is explained qualitatively but the value of $r$ required to explain the observed characteristics is smaller than obtained using $D_\perp^{\mathrm{Bohm}}$ and $D_\parallel^-$ from classical theory (for a pure, fully ionized hydrogenic plasma). Figure 11 shows an example from measurements on JET (Tagle et al., 1987).

Various explanations for this disparity are evident, the simplest being that $D_\perp^-$ is smaller than $D_\perp^{\mathrm{amb}}$. This possibility can be directly tested, see Section V.D.

Tests of certain aspects of electron collection are indicated by the single Langmuir probe characteristics in Figure 12 taken on the DITE tokamak using the post-in-front-of-plate probe, Figure 8. In Figure 12a the post and plate were biased at the same (swept) potential, while their currents were separately recorded. The post current has been multiplied by 10 (i.e., virtually the ratio of the plate/post projected areas = 9). As can be seen, the electron current collected is shared between the post and plate almost in the exact ratio of their projected areas (while the $I_s^+$ values are not in this ratio, as noted earlier). This is perhaps expected since $\ell_e \ll d$. Note, however, that the post is located in the middle of the plate. Thus if there is a significant density gradient across the collection tube *near* the plate (one expects one further away, e.g., Figure 14), then the post would collect fewer particles than its area-based share. The result of Figure 12a implies that such a gradient is weak or absent, justifying the simple probe modelling used to derive Eqs. (43), etc., which assume the absence of any such spatial variations along the plasma/sheath interface.

In Figure 12b either the plate is left floating while the post is swept in bias or vice versa. In this situation the post, acting alone, collects more electrons than when it is effectively just part of a large probe as it was in Figure 12a. This is in accord with the calculation, Eq. (42), that the reduction parameter $r$ is smaller for larger probes.

From the practical point of view, one must ask the question: What should a diagnostician do? There is good reason to believe that the electron current collection for probe voltages less than floating should follow the simple exponential law, Eq. (19). When $V_{\mathrm{pr}}' < 0$, the electrons are not obliged to drift through a slower ion background. Indeed, the electrons

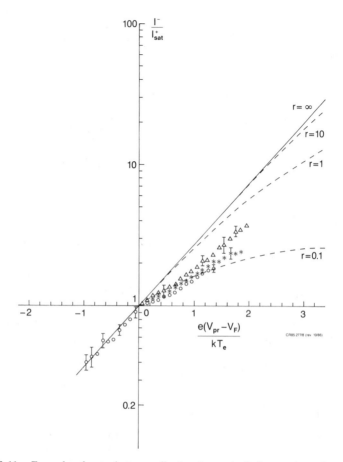

FIGURE 11. Example of net electron collection by a single Langmuir probe on JET, collection area 0.5 cm². Current normalized to measured value of $I_s^+$. Probe voltage relative to measured floating potential, normalized by $T_e$ measured using only data for $V_{pr} < V_F$. Latter data follows exponential in contrast with data for $V_{pr} > V_F$. For comparison dotted lines show theoretical characteristics, Eq. (44), for different values of reduction parameter $r$, Eq. (42). Value of $r$ calculated using Eq. (42) and JET edge data indicates $r \simeq 1$. Different symbols indicate different discharge conditions.

receive some *forward* momentum from $e - i$ collisions. (It can be shown, however, that such collisions have only a slightly distorting effect on the electron collection.) In practice, therefore, it would seem advisable to employ only data for $V_{pr}' \leq 0$ in deducing $T_e$. This point is demonstrated in Figure 13 using probe data taken in the JET tokamak (Tagle et al., 1987). Here $T_e$ is calculated as a function of $V_c$; the (upper) cutoff voltage,

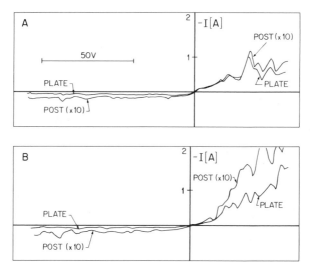

FIGURE 12. Langmuir probe characteristics taken on the DITE tokamak with the post-in-front-of-plate probe, Figure 8. A, the bias of post and plate is swept together, B, post floats while plate is swept and vice-versa. For explanation, see text.

reworking the same data sets. As can be seen the inferred, $T_e$ is indeed constant for $V_c \lesssim V_f$, but it then rises for $V_c > V_f$. This rise is, of course, a reflection of the tendency of $I^-$ to fall below the classical exponential characteristic (Figure 11).

It appears then that values of $T_e$ inferred classically from single probes in high-$\vec{B}$ plasmas may be valid provided that only data for $V'_{pr} < 0$ is used. Since double and triple probes automatically restrict themselves to almost this range (although going slightly positive, to $V'_{pr} \simeq +0.7kT_e/e$), they should be reasonably insensitive to this problem. The price paid, however, is evident: One is inferring $T_e$ from the tail of the electron distribution, which may not be representative of the bulk (de Chambrier et al., 1984). This could be a matter of concern in the present situation, where the cross-field transport mechanism is not understood and where non-average electrons may therefore dominate the sample reaching the probe. Fortunately, $\lambda_{ee} \gg L^-_{col}$ generally, and so good thermalization of those electrons actually present in the collection flux tube will exist; also because the cross-field electron heat conduction coefficient is known (experimentally) and is known to be large (Post and Lackner, 1986) ($X^-_\perp \simeq 1\text{--}10\, D^{amb}_\perp$), one can generally assume good thermal coupling to the main reservoir of electrons in the plasma.

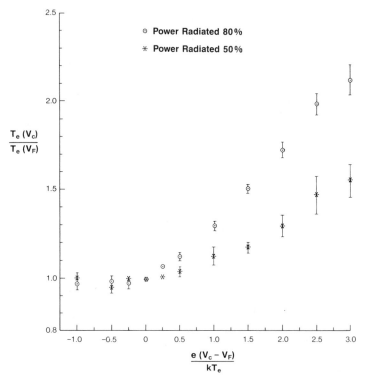

FIGURE 13.   Use of data from Figure 11 to infer $T_e$ by employing different voltage (upper) cut-offs, $V_c$, to the data. For $V_c \lesssim V_F$ the electron current is exponential and gives a value of $T_e$ independent of $V_c$. For $V_c \gtrsim V_F$ the best exponential fit to the data gives spuriously high values of $T_e$. Distortion appears to be worst for highly radiating, i.e., impure, plasmas.

As with measurements of $n_e$, one would like to have comparisons with $T_e$ measured using non-probe techniques. On a number of magnetic fusion devices, $T_e$ has been measured using gridded energy analyzers and Langmuir probes. Unfortunately, this is not a significant test: The orifice of the GEA floats, thus only electrons from the high-energy tail enter the grid system—as to the LP for $V'_{pr} < 0$. Comparisons with Thompson scattering measurements of $T_e$ are difficult, owing to the spatial resolution of this optical technique and the fact that edge conditions vary rapidly with minor radius of the plasma. Improved spatial resolution for special systems (Johnson et al., 1986) down to $\sim 1$ cm should change this situation. Nevertheless agreement in such studies can be to within a factor of two or better (Ditte and Grave, 1985; Harbour and Proudfoot, 1981). As with the case of $n_e$, it would be desirable to carry out more definitive tests in fusion devices and for a wide range of probe and plasma parameters.

## D. Other Probe Techniques

As indicated earlier, each probe technique requires substantial interpretative effort, even for use in nonmagnetic plasmas. Such material is not reviewed here. Only the interpretative problems associated with use in very high $\vec{B}$ fields are examined.

Fortunately, most probe techniques, Section B, involve electrically floating probes. As indicated in the discussion of Langmuir probes, this is the best understood mode of operation for probes. The problems associated with finite probe size must be allowed for, Section V, of course, particularly since these probes are almost always large, owing to reasons of construction.

Various probe techniques employ a sampling orifice which generally floats electrically, e.g., gridded energy analyzers, $\vec{E} \times \vec{B}$ probes, etc. It is usually necessary to keep the orifice width small compared to the sheath thickness in order that plasma does not penetrate into the interior region, confounding the workings of the diagnostic. In this case only the particles passing through the sheath potential drop are analyzed. Thus only the high-energy tail of the electron distribution is monitored, and one encounters the same problem as with Langmuir probes, viz., the bulk of the electrons may not be represented by the tail. The ion distribution is also rather distorted from its shape far from the probe, owing to the presheath and sheath accelerations; fortunately, the more energetic part of the ion distribution passing through the orifice is simply an accelerated Maxwellian (Emmert et al., 1980), assuming the ion distribution is Maxwellian far from the probe.

Thus the interpretative problems of most probes used in very high $\vec{B}$ plasmas, which are related specifically to this application, are essentially the same, viz., ones involving finite probe size, i.e., collection length effects, and the influence of the uncertainty in the cross-field transport mechanism. Fortunately, most probe techniques should be insensitive to the latter problem.

One class of probes requiring further consideration are the impurity ion probes. This matter is discussed in Section VII.

## V. Probe-Size and Limiter-Shadow Effects

## A. Introduction

In contrast with most other plasma diagnostic techniques, probes strongly interact with the plasma being measured and, for the case of high-$\vec{B}$ plasmas, the disturbance or collection length, $L_{col}$, extends considerable distances along $\vec{B}$ into the plasma. To further complicate matters, it will sometimes occur that $L_{col}$ exceeds the distance to the next solid object present in the plasma, typically a limiter. This interrupts the natural particle collection by the probe and necessitates additional interpretative effort (Stangeby, 1984a, 1985a, 1985b).

Since $L_{col}$ increases with probe size $d$, one possibly simplifying approach is to use very small probes. This, unfortunately, encounters two significant problems:

a) The heat removal capacity of large probes is greater than for small ones. This feature is of growing importance in fusion research as the pulse length of devices and the energy density in the edge plasma both increase.

b) As indicated above, there are interpretative advantages in operating a probe in the "very high $\vec{B}$" regime, $d \gg \ell_{i,e}$: 1) the collection areas are well defined and constant, 2) the cross-field transport mechanism, although not presently understood as to its fundamentals, is presumably identical to that occurring to the principal edge structures, e.g., limiters.

The latter point, which requires confirmation (see below) is quite important. If true, it means that one's confidence in probe modelling can be almost as good as scrape-off layer modelling. The latter field (Post and Lackner, 1986) has been quite successful at "explaining" observed results, albeit, by invoking anomalous cross-field transport coefficients, $D_\perp, \chi_\perp$, which are effectively fitting parameters. One can presumably, see below, invoke these same mechanisms and coefficients for probe modelling. By contrast, if one operates a probe in the merely "high $\vec{B}$" regime, $\ell_i > d > \ell_e$, then one has no comparable body of knowledge to draw on. Presumably nonclassical cross-field transport processes also characterize this regime, but one has little to work with at this time. Looking to the future, one may be optimistic that the enormous, ongoing effort (Hugill, 1983) directed at understanding cross-field transport in fusion devices will eventually successfully unravel this decades-old problem. At that time, it will be possible to set down a complete theory for probes in very high magnetic fields. (Indeed, one may also invert the preceding logic: experiments with $d > \ell_{ei}$ probes may help to unravel the mechanism of cross-field transport.)

## B. THE PROBE COLLECTION LENGTH

The concept of a precise cutoff to collection is an approximation—the disturbance extends to infinity—but of diminishing magnitude. One may see this more quantitatively by considering a simple 2-D model of flow to a

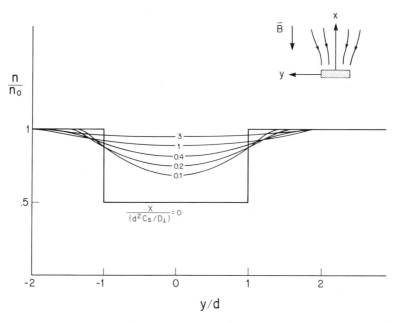

FIGURE 14.   Two-dimensional plasma transport to a semi-infinite plane probe in a mag-
netic field. Probe extends to $\pm \infty$ in Z-direction. Disturbance extends a distance $\simeq d^2 c_s / D_\perp$
in the direction parallel to $\vec{B}$.

probe. Consider a semi-infinite rectangular probe collection element, Figure
14, of width $2d$ in the $y$-direction and extending without limit in the
$z$-direction. $\vec{B}$ and plasma fluid flow are in the $x$-direction.

Flux balance gives

$$D_\perp \frac{\partial^2 n}{\partial y^2} = -\frac{\partial}{\partial x}(nv),    \tag{45}$$

where $v$ is the plasma drift velocity in the $x$-direction. We relate $v$ to $n$
using equation (14),

$$\frac{n}{n_0} = \frac{1}{1 + M^2} = \frac{1}{1 + \dfrac{v^2}{c_s^2}}.    \tag{46}$$

(Note: Eq. (45) assumes that fluid enters each flux tube by cross-field
transport, with zero velocity in the $x$-direction. Here, in fact, momentum as
well as particles are transported cross-field. This error does not, however,
significantly alter the result.)

Thus one obtains a simple equation for $N(x, y)$, where $N \equiv n/n_0$:

$$\frac{\partial^2 N}{\partial Y^2} = -\frac{\partial}{\partial X}(N - N^2)^{1/2}, \tag{47}$$

with the natural normalizations

$$Y \equiv \frac{y}{d} \tag{48}$$

$$X \equiv \frac{x}{\left(\dfrac{d^2 c_s}{D_\perp}\right)}. \tag{49}$$

$N(x, y)$ is plotted in Figure 14. Therefore, we see that indeed the disturbance is only significant over a length $L_{col} \simeq d^2 c_s/D_\perp$.

The first point to emphasize is that since the probe draws particles from the plasma over length $L_{col}$, the measurement is effectively an average of conditions over this length. Therefore, a probe cannot be used to reliably measure, for example, density variations where the density gradient scale length is shorter than $L_{col}$.

A second point relates to limiter-shadowing (Stangeby, 1984a, 1985a, 1985b). We may define a probe to be "large" in this sense, if $L_{col} > L_{con}$, where $L_{con}$ is the probe-limiter, connection length (along $\vec{B}$). A large probe acts effectively as a limiter and a new scrape-off layer (sol) (with its own scale length) is set up between the probe and limiter. The interpretative implications of large probes have been examined and four general cases identified (Stangeby, 1985a), (Figure 15).

(a) Configuration 1. The probe sensor is located adjacent to an existing limiter, facing away from it, capable of moving radially and with or without a probe housing. This is the simplest case to analyze since the probe, regardless of its size, does not disturb the existing scrape-off plasma, but merely measures plasma conditions utilizing the fluxes which would arrive at the limiter anyway.

(b) Configuration 2. The probe housing is "large" and fixed, as a "back stop" with its leading edge at the last closed flux surface (lcfs) defined by the limiters; the sensor moves independently. In this case the "large" housing defines a new sol, but with properties which are simply related to the unprobed sol. The scrape-off length measured by the probe $\lambda_p$ is related to the unprobed value $\lambda_{up}$ by

$$\frac{\lambda_p}{\lambda_{up}} = \left(\frac{L_p}{L_{up}}\right)^{1/2} \tag{50}$$

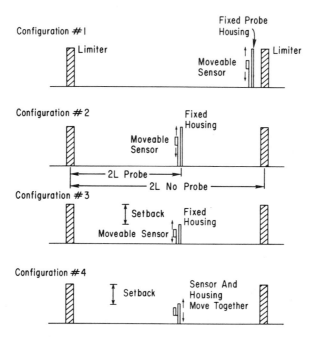

FIGURE 15.   Different probe-limiter configurations for a "large" (limiter-simulating) probe in a scrape-off layer. Configuration # 4 is used in practice, however, other configurations would facilitate interpretation. Explanation in text.

where $L_p$ = probe–limiter connection length and $L_{up}$ = limiter–limiter connection length.

(c) Configuration 3. As in configuration 2, but with the housing set back from the lcfs. This results in a two-zone (radially) sol but again with properties in the probed, outer sol 2, which relate simply to the inner sol 1; in fact, Eq. (50) also applies.

(d) Configuration 4 is the one generally used: sensor and housing move together. The relation between a measured quantity $f_{meas}$, e.g., $n_e$, and the value of $f = f(0)$ at the lcfs is then given by

$$f_{meas}(r_h + \Delta) = f(0) \exp\left(-\frac{r_h}{\lambda_{up}}\right) \exp\left(-\frac{\Delta}{\lambda_p}\right), \qquad (51)$$

where $r_h$ = radial position of the leading edge of the housing, measured outward from the lcfs, $\Delta$ = distance of setback of the sensor from the tip of the housing, and $\lambda_p/\lambda_{up}$ is given by Eq. 50.

Thus as $r_h$ is varied, a log plot of $f_{meas}$ will give the *undisturbed* scrape-off length $\lambda_{up}$.

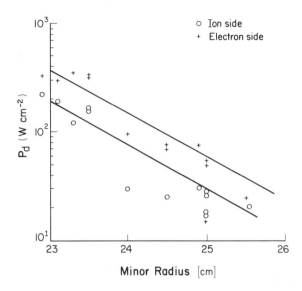

FIGURE 16.   Profiles of deposited heat flux outboard of the separatrix on DITE, measured using a large limiter-simulating probe, Figure 8. Probe has sensors facing in electron and ion drift directions. DITE operated here with a divertor with the separatrix at $r \simeq 21$ cm.

An illustration of the foregoing is shown in Figure 16, which shows radial profiles of heat flux measured by a heat flux probe in DITE. The probe is similar to the one in Figure 8 and is described by configuration 4, where the sol is defined by divertor target plates, rather than limiters. The probe-divertor $L_{con} \simeq 2\pi R$ in the ion side direction and $\simeq 8\pi R$ in the electron side direction. As may be seen, the measured scrape-off lengths are the same in each direction, $\sim 18$ mm, as expected theoretically. The short connection side absolute flux level is lower by a factor of $\sim 1/2$. This is also as predicted from Eq. (51) from which the ratio is expected to be

$$R = \frac{\exp(-\Delta/\lambda_i)}{\exp(-\Delta/\lambda_e)} \tag{52}$$

and where $\Delta$ = sensor set-back = 12 mm and $\lambda_{e,i}$ are the (unmeasured) scrape-off lengths which can be calculated from Eq. (50). Thus Eq. (52) gives

$$R = \frac{\exp(-12/(\sqrt{0.2}\,18))}{\exp(-12/(\sqrt{0.8}\,18))} = 0.47 \tag{53}$$

The foregoing applies to ambipolar (floating) probe operation with $D_\perp = D_\perp^{amb}$, the measured ambipolar value inferred, for example, from measurements of the sol thickness $\lambda_{sol}$, where

$$\lambda_{sol} = \left( \frac{2 D_\perp L_{con}}{c_s} \right)^{1/2} \tag{54}$$

is found to hold widely (Stangeby, in press) for the purely empirical value

$$D_\perp = D_\perp^{Bohm} = 0.06 T_e / B \quad [m^2/s]. \tag{55}$$

While it is not always the case that plasma flow to limiters *is* locally ambipolar, the departure from ambipolarity is often not great, i.e., $|i^- - i^+|/i^+ \lesssim 1$. (The latter can be inferred from measurements of the plasma potential (Erents et al.) near limiters, which, together with the known potential of the limiter and the theoretical relation for a floating sheath potential drop, Eq. (18), imply that local floating conditions are often absent.)

For net ion or electron collection, different collection lengths may apply, but here information is slight, see below. From a theoretical viewpoint one might expect that for net ion collection

$$L_{col}^+ \approx \frac{c_s d^2}{D_\perp^+}, \tag{56}$$

where $D_\perp^+$ is not necessarily equal to $D_\perp^{amb}$.

For net electron collection one anticipates that a distinction between collisional and collisionless ($e - i$ collisions) should be made. For the collisional case one has Eq. (36)

$$L_{col,c}^- \simeq d \left( \frac{D_\parallel^-}{D_\perp^-} \right)^{1/2}, \tag{57}$$

where it appears unlikely that $D_\perp^- = D_\perp^{amb}$. For the noncollisional case, one expects

$$L_{col,nc}^- \simeq \frac{\bar{c}_e d^2}{D_\perp^-}. \tag{58}$$

## C. The Simple and Complex SOL

Most probe applications in fusion devices occur within a scrape-off layer, sol. In this case the insertion of a "large" probe, i.e., limiter-simulating, can change more than the scrape-off length $\lambda_{sol}$, which is the assumption of the last section and which gives the simple relation, Eq. (50). Insertion of a

**Table 1.**  (McCracken and Stangeby, 1985) Criteria indicating transition from simple SOL (sheath sink dominates heat and particle balance of SOL) to complex SOL (distributed processes dominate)

| | | |
|---|---|---|
| 1. | Parallel field temperature gradients | $\dfrac{n_e L_{con}}{T_e^2} \geq 10^{17} \, \mathrm{m}^{-2} \, \mathrm{eV}^{-2}$ |
| 2. | Electron-ion equipartition (sheath dominant) | $\dfrac{n_e L_{con}}{T_e^2} \geq 10^{17} \, \mathrm{m}^{-2} \, \mathrm{eV}^{-2}$ |
| 3. | Electron-ion equipartition (radiation dominant) | $\dfrac{n_{imp}}{n_e} \geq \dfrac{0.1}{T_e^{1/2}}$ |
| 4. | Impurity radiation dominant (oxygen and carbon) | $\dfrac{n_{imp} L_{con}}{T_e^{3/2}} \geq 5 \times 10^{16} \, \mathrm{m}^{-2} \, \mathrm{eV}^{-3/2}$ |
| 5. | Ionization dominates particle balance | $n_{neut} L_{con} \geq 3 \times 10^{18} \, \mathrm{m}^{-2}$ |

large, limiter-simulating probe can change the sol from one in which processes such as local ionization, longitudinal $T_e$-gradients, impurity radiation, etc., are important factors in the sol energy and particle balance (which control $\lambda_{sol}$) to one in which these processes are unimportant (Stangeby, 1985b, 1984b, in press; McCracken and Stangeby, 1985).

The "simple SOL" can be defined to be one in which such *distributed* processes are unimportant, and the sheath sink dominates energy and particle balance. Insertion of a limiter-simulating probe into a sol which is already simple does not change this property and the simple relation, Eq. (50), should hold.

It can be shown that for $L_{up}$ long enough the various distributed processes become important. The criterion for these effects to be important are given in Table 1 (McCracken and Stangeby, 1985). One may note that most of these criteria depend on $L_{con}$. Thus insertion of a limiter simulating probe can change a complex SOL into a simple SOL. This change must be allowed for in probe interpretation and the relation, Eq. (50), should not then be used.

## D. EXPERIMENTS ON PROBES

To date there have been few experiments *on*, as distinct from *with*, probes in very high $\vec{B}$ plasmas. Such experiments are useful since they

(a) constitute a test of the various assumptions used in probe interpretation,
(b) provide unique opportunities to study cross-field transport since an extra "knob" is available, viz., probe bias.

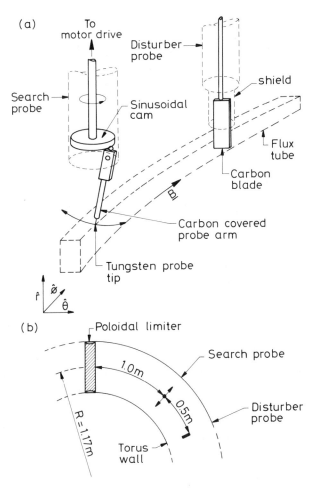

FIGURE 17.   Two-probe experiment on DITE in which a small search probe is scanned
through the collection tube of a large disturber probe on the same field line.

One such probe experiment (Matthews et al., 1986), carried out in the
DITE tokamak, is shown in Figure 17. The disturbance zone of a relatively
large disturber probe, DP, was probed by a small, moveable search probe,
SP. The SP in fact scans through a zone between the DP and a nearby
limiter. For the experiment described here the DP was 20 mm wide
(poloidally), extending radially into a plasma minor radius $r = 260$ mm (the
lcfs on DITE with limiters, as employed here), and extending outward,
effectively to infinity (i.e., further than $\lambda_{sol}$). The SP consisted of a Langmuir
single probe, 1 mm diameter, 2 mm long.

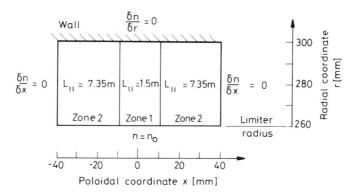

FIGURE 18. Boundary conditions assumed to analyze data from two-probe experiment on DITE, Figure 17.

Since the DP geometry does not conform to the simple case of a square plate in an infinite plasma, the particle balance equation

$$D_{\perp,r} \frac{\partial^2 n}{\partial r^2} + D_{\perp,x} \frac{\partial^2 n}{\partial x^2} = \frac{nc_s}{L_{con}} \qquad (59)$$

was solved numerically for the particular boundary conditions (Figure 18). The *RHS* of Eq. (59) is the average sink strength in the sol. One may note that cross-field diffusion in both the radial, $r$, direction and the poloidal, $x$, direction is assumed here. This is, in fact, not a trivial assumption. If cross-field transport is due to fluctuations and the fluctuations are caused by density gradients, then two possibilities may be considered:

(a) The fluctuation field may be the result of processes in the sol as a whole and may thus reflect the sol gradient $\lambda_{sol}$. This situation might lead to cross-field transport in the radial direction only and at a rate proportional to $n/\lambda_{sol}$, rather than having anything to do with the presence of the probe and $n/d$.

(b) Alternatively, the fluctuation level may be the result of local conditions, in which case one would anticipate that insertion of the disturber probe would create its own gradient and fluctuation level, with cross-field transport occurring in both radial and poloidal directions.

Assuming the latter, Eq. (59) was solved for $D_{\perp,r} = D_{\perp,x}$.

With regard to the boundary conditions, Figure 18: Over the poloidal extent of the DP ($-10 \leq x \leq +10$ mm), the probe-limiter connection length $L_{con} = 1.5$ m, while outside this zone $L_{con} = 7.35$ m for the limiter–limiter connection length.

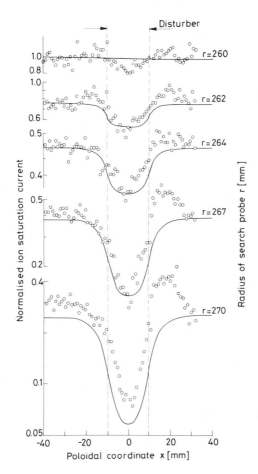

FIGURE 19.  Results from DITE two-probe experiment, Figures 17 and 18. Ion saturation current collected on search probe as it is swept through the "wake" of the disturber probe. Solid lines are solution to diffusion Eq. (59) for $D_\perp = 0.18$ m$^2$/s.

The SP was biased into ion collection and gave the results shown in Figure 19. Equation (59) was solved for different values of $D_\perp$ to find the best fit, and all of the solid curves in Figure 19 are for the same value, $D_\perp = 0.18$ m$^2$/s. One may note that the level of agreement between experiment and model is very good, spanning almost two orders of magnitude in $I_s^+$.

From this experiment one may draw a number of important conclusions:

(a) Cross-field transport to a probe in a very high $\vec{B}$ plasma is effectively diffusive (whatever the underlying cause may be), the density gradients are on the scale length of the probe size $d$, and the diffusion is isotropic (in directions perpendicular to $\vec{\mathbf{B}}$).

(b) The diffusion coefficient is well represented by the empirical Bohm value, Eq. (55), which was 0.15 $\mathrm{m}^2/\mathrm{s}$ for the conditions of Figure 19 (5 eV, $2T$).

On the basis of these experiments, then, it appears that the general sol model is indeed applicable to probes in very high magnetic fields. Such an important conclusion, of course, requires further testing in other plasma conditions and in different magnetic devices.

Other experiments with the DITE probe system are in progress to study the effect of biasing the disturber probe. One important result of the latter work is a test of whether applying a bias to the DP causes the plasma potential to change in the disturbed zone. It was found experimentally that biasing the DP from floating to $-100$ volts (into ion collection) causes less than $\sim 5$ volt change in the plasma potential measured by the SP located in the DP collection tube. This result is an important one regarding the measurement of $T_e$ using Langmuir probes. In Section IV.C it was concluded that the classical Langmuir probe I–V characteristic would hold for $V'_{\mathrm{pr}} \lesssim 0$. For this to be correct it is necessary that a change of bias of the probe to more negative values below $V'_{\mathrm{pr}} = 0$ should be entirely "soaked up" by the sheath. Since the cross-field transport mechanism is not understood it cannot be ruled out *a priori* that some of the voltage change will appear within the plasma, e.g., between the probe magnetic flux tube and adjacent ones leading to the reference electrode. It is therefore an important and reassuring result that the change in probe voltage from $V'_{\mathrm{pr}} = 0$ to $V'_{\mathrm{pr}} \ll 0$ is found to be, in fact, almost entirely taken up by the sheath, as assumed in the simplest picture, Eq. (19).

## VI. Effect of Fluctuations

The plasma edge region of fusion devices is characterized by very strong fluctuations, particularly of density and potential where $\tilde{n}/\bar{n}$ and $\tilde{V}_{\mathrm{pl}}/\bar{V}_{\mathrm{pl}}$ range up to 100% (Zweben and Gould, 1983; Ritz et al., 1984; Robinson and Rusbridge, 1969). These fluctuations are generally believed to be the cause of the observed, rapid, nonclassical cross-field transport. Langmuir probes have been used on a number of fusion devices (Zweben and Gould, 1983; Ritz et al., 1984; Robinson and Rusbridge, 1969) to measure $\langle \tilde{n}\tilde{E}_0 \rangle$, and the quantitative comparison with the observed transport is sometimes good.

These studies will not be considered further here. Rather, the focus of this section is to examine the distortion to the classical Langmuir probe

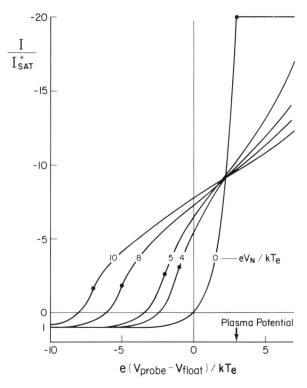

FIGURE 20. Examples of Langmuir characteristics observed in the presence of noise by a slow time response meter. Fluctuation amplitude $= V_N$. The electron current follows an exponential below solid point. Case shown approximates hydrogenic plasma ($V_{plasma} - V_{float}$ taken to be $3\ kT_e/e$ for $V_N = 0$).

I–V characteristic, Eq. (19), which occurs when fluctuations occur. Fluctuations in $V_{pl}$ can be particularly strong in plasmas which are wave-heated, e.g., at ion cyclotron resonance frequencies.

The behavior of Langmuir probes in the presence of fluctuations such as those induced by rf heating, has been widely studied experimentally and theoretically (Butler and Kino, 1963; Boschi and Magistrelli, 1963; Garscadden and Emeleus, 1962; Crawford, 1963; Kato, 1966; Stampa and Wulf, 1978). Consider a classical, single Langmuir probe characteristic, e.g., the curve marked "0" in Figure 20. Consider the effect of a periodic variation in the potential of the plasma $V_{p\ell}$ adjacent to the probe. When the plasma potential increases above its average value, the voltage drop across the sheath becomes greater, and so electron collection by the probe is reduced, possibly to negligible values compared to the ion saturation

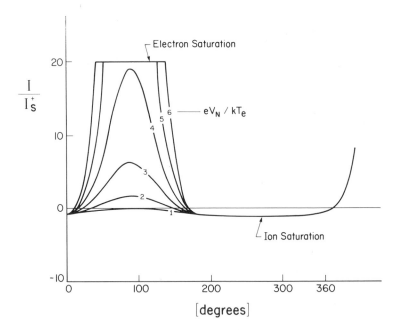

FIGURE 21. A single cycle of the probe characteristic with imposed DC and fluctuation voltage $V_N$. The DC bias is $eV_B/kT_e = -1$ here.

current $I_s^+$ (which continues to flow to the probe). This portion of the cycle corresponds to the 180°–360° section of Figure 21 In the other half of the cycle, the plasma potential falls below its average value and thus the potential drop across the sheath is reduced. This increases the electron current, while the ion current remains at $I_s^+$ (unless the oscillation potential amplitude is quite large). This part of the cycle corresponds to the 0°–180° portion of Figure 21. Note that the situation is equally well described by considering the plasma potential to be fixed while the probe has a DC bias of $V_B$ plus a fluctuation or noise potential of amplitude $V_N$. We adopt hereafter the latter picture. Thus $V_B \equiv V_{pr} - V_{p\ell}$ in Figure 20.

If the circuit and current monitoring components are sufficiently fast to respond at the fluctuation frequency, then a time-varying signal such as that of Figure 21 will be recorded. (Of course, even in fast circuits, some attenuation will occur, particularly of the higher harmonics, and since time profiles such as Figure 21 clearly contain harmonics, distortions occur in practice.)

A *very* large distortion occurs, however, if the current detection element and/or circuitry is too slow to follow the variations. If, for example, the

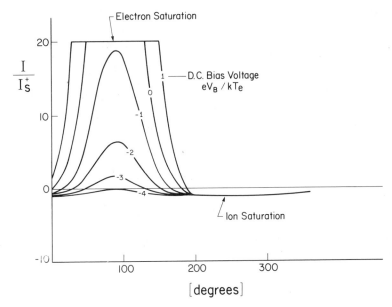

FIGURE 22. As figure 21 but for $V_N$ fixed at $eV_N/kT_e = 4$ and various DC bias levels.

probe is floating, then the total positive charge collected (180°–360°, Figure 21) must equal the total negative charge collected (0°–180°). This equality is achieved by the floating probe spontaneously readjusting its average or DC bias to more negative levels, see Figure 22 where as one can see a value of $eV_B/kT_e \approx -2.5$ will give about equal positive and negative charges in the two halves of each cycle for the case of $eV_N/kT_e = 4$. Thus, one can mark one point on the time-averaged Langmuir characteristic, Figure 20, namely the floating (or zero time-averaged current) potential. Note that the apparent floating potential registered in this way is more negative than if $V_N = 0$. Consider next what happens if one DC biases the probe by some amount above its new floating potential, e.g., consider the curve for $eV_B/kT_e = -2$ in Figure 22. Now a net, time-averaged electron current is collected by the probe. One can readily show that so long as the DC bias is not raised so far as to drive the probe into electron saturation current during part of the cycle, the time-averaged electron current to the probe will simply increase exponentially with $V_B$, just as in the case of $V_N = 0$. This break point (for nonexponential behavior) is shown as a solid dot on the curves of Figure 20.

Therefore, so long as one does not drive the DC bias too high above the measured (time-averaged) floating potential, the Langmuir characteristic is undistorted, except for a shift along the voltage axis. Thus measurements of $T_e$ and $n_e$ ($I_s^+$) will not be corrupted, provided one uses data only below, say, the apparent floating potential, a policy already recommended for other reasons, Section IV.C.

As is evident from Figure 20, the critical point moves down the characteristic as $V_N$ increases. For $V_N$ sufficiently large, the point falls below the $I = 0$ axis, and then measurements of $T_e$ and $n_e$ are compromised. It can be shown (Boschi and Magistrelli, 1963) that the time-averaged floating potential is

$$\frac{e\overline{V_F}}{kT_e} = -\ell n\left[I_0'\left(\frac{eV_N}{kT_e}\right)\right], \tag{60}$$

where $I_0'$ is the zeroth order hyperbolic Bessel function. The location of the critical point is simply calculated—it is such that the fluctuating potential will carry the probe into electron saturation for part of the cycle. Thus the critical potential $V_c$ is given by

$$V_c = \frac{3kT_e}{e} - V_N, \tag{61}$$

where the factor 3 applies to the particular case of hydrogen (taking the result from Eq. (18) to be approximately 3).

For sufficiently large noise amplitude, $V_N \geq V_N^*$, the critical point is on or below the $I = 0$ axis, i.e., $V_c = V_f$, thus

$$3 - \frac{eV_N^*}{kT_e} = -\ell n\left[I_0'\left(\frac{eV_N^*}{kT_e}\right)\right]. \tag{62}$$

This equation can be solved using the approximation for $x \gg 1$:

$$I_0'(x) \simeq 0.4e^x x^{-1/2}. \tag{63}$$

Thus

$$\frac{eV_N^*}{kT_e} \approx 64.5. \tag{64}$$

The value of $V_N^*$ is still higher for non-hydrogenic plasmas. Therefore, provided the $\tilde{V}_{pl}$ noise level is not extremely high, values of $n_e$ and $T_e$ can still be simply derived by adhering to the policy already recommended, viz., restrict data use to points below the apparent floating potential.

Fluctuations in density $\tilde{n}_e$ can be registered as fluctuations in $I_s^+$ with a sufficiently fast circuit. Otherwise, these fluctuations are averaged out and have no effect on the $I$–$V$ characteristic (Crawford, 1963).

Fluctuations in $T_e$ have also been considered, but experimentally it is found that $\tilde{T}_e/\overline{T}_e$ is generally small (Zweben and Gould, 1983).

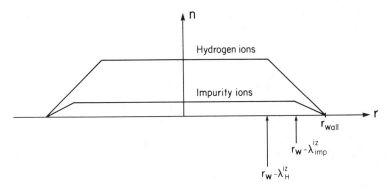

FIGURE 23. Radial profiles of hydrogen (plasma) ions and impurity ions show that impurity fraction can be highest near the edge. Assumes that each ion is transported only by diffusion and at flux rate proportional to its own density gradient with the same $D_\perp$; neutral atoms enter from wall and penetrate distance $\lambda^{iz}$ before ionizing, $\lambda_H^{iz} > \lambda_{imp}^{iz}$; density at wall = 0, i.e., strong edge sink.

## VII. Impurity Effects

Impurity ion fractions, $f = n_{imp}/n_e$, in the edge plasmas of fusion devices can exceed those in the main part of the plasma. It is often observed that impurity ions move radially in tokamaks subject to anomalous diffusion (Keilhacker et al., 1981; Gentle, 1984) sometimes plus a small inward pinch, and at a flux density rate $\Gamma_{imp}$ proportional to their own density gradient:

$$\Gamma_{imp} = -D_\perp \frac{dn_{imp}}{dr}. \tag{65}$$

A similar relation also often holds for the hydrogenic particles and with $D_\perp^H \simeq D_\perp^{imp}$. Assuming for simplicity that all neutral impurity atoms entering from the edge, at $r = a$, penetrate a distance $\lambda_{imp}^{iz}$ before ionizing, while the hydrogen penetrates $\lambda_H^{iz}$, then the solution of Eq. (65), together with the simplified boundary condition $n_H(a) = n_{imp}(a) = 0$, give the density profiles shown in Figure 23 (Engelhardt and Fenberg, 1978; Post and Lackner, 1986; McCracken and Stangeby, 1985). Since $\lambda_H^{iz} > \lambda_{imp}^{iz}$ generally, the impurity fraction is higher near the edge than in the center, e.g.,

$$f\left(a - \lambda_{imp}^{iz}\right)/f(0) = \lambda_H^{iz}/\lambda_{imp}^{iz}. \tag{66}$$

Thus, while central impurity fractions of oxygen and carbon are typically (spectroscopically) measured (Hawryluk et al., 1979) to be a few percent, edge fractions may be significantly higher.

Because of the critically important role which impurities play in fusion plasmas, potentially preventing energy break-even (Jensen et al., 1978) it is important to measure edge impurity densities, particularly since this region is clearly the impurity source. A theory is therefore required to interpret impurity measuring probes, e.g., deposition probes; this is discussed in Section VII.A.

The high impurity fraction near the edge raises the possibility that *all* probes—even ones not intended to measure impurities, e.g., Langmuir probes—may be showing effects due to impurities. This is discussed in Section VII.B.

## A. Impurity Measuring Probes

Insertion of a floating probe into the plasma causes a hydrogenic plasma flow to occur to the probe, as described in Section III.B. Impurity ions diffusing into the (hydrogenic) collection tube of the probe find themselves subject to two forces toward the probe: 1) friction with the plasma flow, 2) the ambipolar electrostatic field which causes the plasma flow (Post and Lackner, 1986). One wishes to know the relationship between the impurity flux reaching the probe $\Gamma_{pr}^{imp}$ and the impurity density far from the probe, $n_{imp}$.

A simplified treatment of the problem assumes:

(a) 1-D geometry with impurities diffusing into the probe flux tube uniformly over length $L_{col}$, Eq. (5), thus source strength

$$S_{imp} = \begin{cases} \text{constant}, & 0 \leq x \leq L_{col} \\ 0, & x > L_{col} \end{cases} \qquad (67)$$

(b) hydrogenic flow velocity along the flux tube varying as $v_H = c_s (x/L_{col})$ where $x = 0$ at the far end of the tube; electrostatic field $E(kT_e/eL_{col})(x/L_{col})$,

(c) impurity ion fraction $f = n_{imp}/n_e \ll 1$.

Thus, conservation of mass and momentum

$$n_{imp} v_{imp} = S_{imp} x \qquad (68)$$

$$n_{imp} v_{imp} \frac{dv_{imp}}{dx} = \frac{Z_{imp} e n_{imp} E}{m_{imp}} + \nu n_{imp}(v_H - v_{imp}) - v_{imp} S_{imp}, \qquad (69)$$

where $\nu$ is the momentum transfer frequency for collisions between impurity and hydrogenic ions.

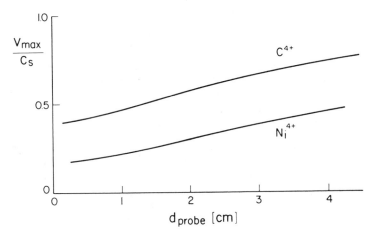

FIGURE 24. Examples of terminal velocity of impurity ions as they reach the sheath edge at the probe of collection area $d^2$. $n_e = 10^{18}$ m$^{-3}$, $T_e = T_{D^+} = 25$ eV, $D_\perp = 1$ m$^2$/s, D$^+$ plasma.

The solution of Eqs. (68) and (69) is:

$$n_{\text{imp}} = \text{constant} \tag{70}$$

and

$$\Gamma_{\text{imp}} = \beta n_{\text{imp}} c_s \tag{71}$$

where

$$\beta = \frac{\nu L_{\text{col}}}{4c_s} \left[ -1 + \left( \frac{1 + 8\alpha}{\nu^2} \right)^{1/2} \right] \tag{72}$$

$$\alpha = \frac{Z_{\text{imp}} k T_e}{m_{\text{imp}} L_{\text{col}}^2} + \frac{\nu c_s}{L_{\text{col}}} \tag{73}$$

$c_s$ = hydrogenic ion acoustic speed

Examples are given in Figure 24. One may note that $\Gamma_{\text{imp}}$ is dependent on probe size and that typically impurity ions reach the probe at a substantial fraction of the hydrogenic sound speed.

One consequence of the last fact is that sputtering caused by impurity ion impact can be significant since the impact energy is quite high. For example, a Ni$^{4+}$ ion in $D^+$ accelerated to speed $c_s$ has energy ~ 30 kT, which is larger than the sheath energy gain of ~ 3 ZkT = 12 kT. Partly because of this strong effect, it is necessary to take into account removal of already-deposited impurities by sputtering and the subsequent fate of such atoms if one is to reliably relate *net* deposition, i.e., that actually measured, to $n_{imp}$. Having been sputtered, the impurity atoms have a finite probability of being reionized within the probe's flux tube and thus redeposited. A complete treatment of this problem is not possible analytically, and Monte Carlo calculations are required. The simple assumptions concerning the shape of the hydrogenic flow and electrostatic fields near the probe should also be replaced by a more realistic model.

As indicated in Section D, the deduction of $n_e$ and $T_e$ from Langmuir probe characteristics is insensitive to uncertainties in $L_{col}$, i.e., the cross-field transport mechanism. Unfortunately, the same cannot be said for impurity flow to a probe which is influenced by the actual value of $L_{col}$, Eq. (73). For a sufficiently large probe, however, $\beta \to 1$ and this problem is avoided, at least for the simple model described above.

## B. The Interpretation of Langmuir and Other Probes in Impure Plasmas

Consider next the situation when the impurity fraction $f = n_{imp}/n_e$ is not vanishingly small. From Figure 24 it is evident that the impurity ions are not always perfectly coupled collisionally with the hydrogenic ions. For simplicity, however, consider the case where the coupling is perfect. Then the ions reach the sheath edge with sound speed $c_s$:

$$c_s^2 = \frac{kT_j \sum_j n_j + kT_e \sum_j Z_j n_j}{\sum_j m_j n_j}, \qquad (74)$$

where the sums are over all ion species present.

Fortunately, for typical impurities in the edge of fusion devices, Eq. (74) does not give a value of $c_s$ which differs much from the value for a pure hydrogenic plasma. For example, consider a rather extreme case of $n_{D^+} = n_{c^{4+}} = n_{0^{6+}}$ and $T_j = T_e$. Equation (74) gives $c_s = 0.68\,(kT/m_H)^{1/2}$ compared with $(kT/m_H)^{1/2}$ for a pure $D^+$ plasma.

For purposes of illustration it is sufficient to assume a pure non-hydrogenic plasma, comparing it with a pure hydrogenic one.

The ion *particle* flux density to a negative, including a floating surface, is

$$\Gamma_f^+ \simeq n_{imp} c_s, \qquad (75)$$

which, of course, is less than $n_e c_s$ since $n_e = Z_{imp} n_{imp}$. The ion *current* density, however, is

$$I_s^+ = Z_{imp} e \Gamma_f^+ \tag{76}$$

$$= n_e e c_s, \tag{77}$$

and since $c_s$ is almost unchanged from the pure hydrogenic case, we see that $I_s^+$ gives $n_e$ fairly accurately, even if the level of impurities is not well known. The measurement of $T_e$ is also not affected by unknown impurity levels.

It can be shown that the floating sheath potential drop is

$$\frac{eV_f}{kT_e} = \frac{1}{2} \ell n \left[ \left( \frac{2\pi m_e}{m_{imp}} \right) \left( Z_{imp} + \frac{T_{imp}}{T_e} \right) (1 - \gamma_{se})^{-2} \right], \tag{78}$$

which is also not much different for typical fusion plasma impurities compared with pure hydrogen. For example, for a pure $C^{4+}$ plasma, $eV_f/kT_e = -3.2$ compared with $-2.8$ for pure $D^+$. Thus the heat flux to a floating surface is not greatly altered, e.g., for pure $D^+$ and $T_e = T_i$, $\delta = 7.5$, while for pure $C^{4+}$ and $T_e = T_i$, $\delta = 6.5$.

Impurities have a somewhat stronger effect on the electron collection reduction parameter $r$, Eq. (42), primarily by lowering $\lambda_{ei}$. For example, a pure $C^{4+}$ plasma has a value of $r$ which is about one third of that for pure $D^+$.

## VIII.  Remaining Problems

The interpretation of plasma probes in very high magnetic fields is still a developing field. The major obstacle to the construction of a complete probe theory is the same one encountered in explaining magnetic confinement of plasmas, namely, the mechanism of cross-field transport is not understood. Fortunately, certain important aspects of probe interpretation appear to be insensitive to this uncertainty, but other aspects are not. Progress is dependent on work along two directions:

(a) further experiments in which measurements made by probes are compared with those made by non-probe diagnostic techniques,
(b) test *on* probes in which various predictions of probe theory are directly tested.

The latter type of work should be able to shed light on the mechanism of cross-field transport, ambipolar and nonambipolar, by measuring the lateral and longitudinal spatial extents and magnitudes of the probe disturbance zones. Plasma fluctuation studies carried out in conjunction with such probe systems may be able to answer the question of how nonambipolar cross-field transport is possible. These studies require that fully two dimensional plasma models be developed for probes, e.g., Figure 14.

Impurity probe modelling requires considerable development, including two-dimensional models for the background plasma and a treatment of the impurities which allows for sputter-removal, reionization and redeposition.

Flow to a probe, as to a limiter, is often not fully (self-) collisional, and so the use of a fluid model can be questioned (Post and Lackner, 1986). It is true that most of the predictions from collisional (fluid) and collisionless (kinetic) treatments give rather similar results, e.g., for the particle flux $\Gamma$ as a function of the undisturbed plasma parameters $n_e$ and $T$, Section IV.B. Other predictions, however, can differ substantially. The ambipolar electric field derived from simple 1-D fluid analysis (Woods, 1965; Stangeby, 1984e, 1986) has a singularity at the plasma-sheath interface, with very large values of $E$ predicted to exist near the probe or limiter. Some kinetic treatments (Bissel et al., 1986) also find this, others do not (Emmert et al., 1980). Further work is needed to resolve this matter, particularly for the interpretation of impurity behavior where a strong $E$-field near the probe would influence the net retention.

Probe and sol modelling share another unresolved problem, viz., the nature of the ion velocity distribution at the plasma-sheath interface (Stangeby, 1986). This question is not critical for low moments of the ion distribution, e.g., particle flux density $\Gamma$, as all treatments give nearly the same result, i.e., average ion velocity $\simeq c_s$. It is important for higher moments, however, such as the heat flux density where different treatments yield results differing by about a factor of two (Stangeby, 1986). Resolution of this problem will probably result from detailed studies of the transition region between the far-field, collisional part of the flow and the collisionless region extending about one mean free path from the probe.

The plasma temperatures are higher in the present generation of fusion devices than smaller machines and can be expected to increase further as break-even conditions are approached. As $T_e$ is increased, so is the secondary electron emission, with the result that the sheath potential drop diminishes affecting sputtering and heat transmission, Eq. (18). This does not greatly complicate probe interpretation for the most part, except if $\gamma_{se} \gtrsim 1$. In this case there is some uncertainty as to the existence of a sheath (Chen, 1985). A double sheath may exist, effectively clamping the value of $\gamma_{se}$ to $\sim 0.8$ (Hobbs and Wesson, 1967). Alternatively, the sheath may disappear and the probe float at $\simeq V_{pl}$. If the latter happens, the heat transmission will attain very high values with significant practical and interpretative implications for probes. An unstable, time-dependent solution is also possible.

## IX. Conclusions

From a fundamental point of view, the state of probe interpretation for very strong magnetic fields is unsatisfactory, a situation which will presumably continue until the mechanism of cross-field transport in highly ionized plasmas is fully understood.

From a practical point of view, the situation is less unsatisfactory, since it appears that the most important measurements of $n_e$ and $T_e$ can be made in such a way that the influence of uncertainties in the cross-field transport mechanism is not significant. Nevertheless, further experimentation of two types is required to substantiate this conclusion:

(a) careful and detailed comparisons of probe measurements with non-probe techniques carried out simultaneously in the same fusion plasma,
(b) experiments *on* probes to test various aspects of prediction, such as described in Section V.D.

In addition, a number of complications arise in the actual environment of fusion devices including limiter-shadows, fluctuations, and impurities, and the influence of these complications on probe interpretation is only partially assessed to date.

## References

Bissell, R. C. and Johnson, P. C. (1987). *Phys. Fluids*, **30**, 779

Bohm, D., Burhop, E. H., and Massey, H. S. (1949). In: *Characteristics of Electrical Discharges in Magnetic Fields* (A. Guthrie and R. K. Wakerling, eds.). McGraw-Hill, New York.

Boschi, A. and Magistrelli, F. (1963). *Il Nuovo Cimento* **29**, 487.

Brown, I. G., Compher, A. B., and Kunkel, W. B. (1971). *Phys. Fluids* **14**, 1377.

Budny, R. and Manos, D. (1984). *J. Nucl. Mater.* **121**, 41.

Butler, H. S. and Kino, G. S. (1963). *Phys. Fluids* **6**, 1346.

Callen, J. D. et al. (1982). *Proceedings of the Ninth International Conference on Plasma Physics and Contained Nuclear Fusion Research*, Baltimore, (IAEA, Vienna 1983), Vol. I, p. 297.

Chen, F. F. (1965). In: *Plasma Diagnostic Techniques* (R. H. Huddlestone and S. L. Leonard, eds.). Academic Press, London.

Chen, F. F. (1984). *Introduction to Plasma Physics and Controlled Fusion*, 2nd edition. Plenum Press, New York.

Chen, F. F. (1985). *Modern Uses of Langmuir Probes*. IPP J-750, Nov. 1985, Nagoya.

Chodura, R. (1986). In: *Physics of Plasma-Wall Interactions in Controlled Fusion* (D. E. Post and R. Behrisch, eds.). Plenum Press, New York.

Chung, P. M., Talbot, L., and Touryan, K. J. (1975). *Electric Probes in Stationary and Flowing Plasmas*. Springer-Verlag, Berlin.

Clements, R. M. (1978). *J. Vac. Sci. Technol.* **15**, 193.

Cohen, S. A. (1978). *J. Nucl. Mater.* **76 & 77**, 68.

Crawford, F. W. (1963). *J. Appl. Phys.* **34**, 1897.

de Chambrier, A. et al. (1984). *J. Nucl. Mater.* **128 & 129**, 310.

Ditte, U. (1983). *Elektrostatische Sonden in Starken Magnetfeld*, Ph.D. Thesis, University of Essen.

Ditte, U. and Grave, T. (1985). IPP III/102, December 1985.

Ditte, U. and Muller, K. G. (1983). *Proceedings of the XVI International Conference on Phenomena in Ionized Gases*, Dusseldorf, 29.8.

Emmert, G. A., Wieland, R. M., Mense, A. T., and Davidson, J. N. (1980). *Phys. Fluids* **23**, 803.

Engelhardt, W. and Feneberg, W. (1978). *J. Nucl. Mater.* **76 & 77**, 518.

Erents, S. K., DITE, Culham, private communication.

Erents, S. K., Tagle, J. A., McCracken, G. M., Stangeby, P. C., and de Kock, L. (1986). *Nucl. Fusion*, **26**, 1591.

Ertl, K. (1984). *J. Nucl. Mater.* **128 & 129**, 163.

Franklin, R. N. (1976). *Plasma Phenomena in Gas Discharges.* Oxford, Chapter 2.

Garscadden, A. and Emeleus, K. G. (1962). *Proc. Phys. Soc.* **79**, 535.

Gentle, K. W. et al. (1984). *Plasma Phys. and Controlled Fusion* **26**, 1407.

Harbour, P. J. and Proudfoot, G. (1981). *Proceedings at the IAEA Technical Committee Meeting on Divertors and Impurity Control, Garching, July 6–10*, 1981, 45.

Harbour, P. J. and Proudfoot, G. (1984). *J. Nucl. Mater.* **121**, 222.

Harrison, M. F. (1979). In: *Physics of Plasma-Wall Interactions in Controlled Fusion* (D.E. Post and R. Behrisch, eds.). Plenum Press.

Hawryluk, E. J., Suckewer, S., and Hirshman, S. P. (1979). *Nucl. Fusion* **19**, 607.

Heifetz, D. B. (1986). In: *Physics of Plasma-Wall Interactions in Controlled Fusion* (D. E. Post and R. Behrisch, eds.). Plenum Press.

Hershkowitz, N., Nelson, B., Pew, J., and Gates, D. (1983). *Rev. Sci. Instrum.* **53**, 29.

Hobbs, G. D. and Wesson, J. A. (1967). *Plasma Phys.* **9**, 85.

Hugill, J. S. (1983). *Nucl. Fusion* **23**, 331.

Ivanov, Yu. A. et al. (1976). *Sov. Phys. Tech. Phys.* **21**, 830

Jensen, R. V. et al. (1978). *Nucl. Sci. and Eng.* **65**, 282.

Johnson, D., Grek, B., Dimock, D., Paladino, R., and Tolans, E. (1986). *Rev. Sci. Instrum.* **57**, 1810.

Kato, K. et al. (1966). *J. Phys. Soc. of Japan* **21**, 2036.

Katsumata, I. and Okazaki, M. (1967). *J. Appl. Phys. Japan* **6**, 123.

Keilhacker, M. et al. (1981). *Proceedings of IAEA Technical Committee Meeting on Divertors and Impurity Control, Garching, July 6–10*, 1981, 23.

Kilpatrick, S. J., Manos, D. M., Budny, R. B., Stangeby, P. C., Ritter, R. S., and Young, K. M. (1986). *J. Vac. Sci. Technol.* **A4**, 1817.

Kimura, H. et al. (1978). *Nucl. Fusion* **18**, 1195.

LaBombard, B. and Lipschultz, B., Alcator C, MIT, private communication.

Laframboise, J. G. (1966). University of Toronto, Institute for Aerospace Studies, Report No. 100.

Langmuir, I. (1929). *Phys. Rev.* **33**, 954; **34**, 876.

Lipschultz, B., Hutchinson, I., LaBombard, B., and Wan, A. (1986). *J. Vac. Sci. Technol.* **A4**, 1810.

Manos, D. M. (1985). *J. Vac. Sci. Technol.* **A3**, 1059.

Manos, D. M., Budny, R. V., Kilpatrick, S., Stangeby, P., and Zweben, S. (1986). *Rev. Sci. Instrum.* **57**, 2107.

Manos, D. M., Budny, R., Satake, T., and Cohen, S. A. (1982). *J. Nucl. Mater.* **111 & 112**, 130.

Manos, D. and McCracken, G. M. (1986). In: *Physics of Plasma-Wall Interaction in Controlled Fusion* (D. E. Post and R. Behrisch, eds.). Plenum Press.

Matthews, G. F. (1984). *J. Phys.* D, 17, 2243.

Matthews, G. F., Stangeby, P. C., and Sewell, P. (1987). *J. Nucl. Mater.*, 145–147, 220.

McCormick, K. et al. (1985). *Rev. Sci. Instrum.* 56, 1063.

McCracken, G. M. and Stangeby, P. C. (1985). *Plasma Phys. and Control Fusion* 27, 1411.

Poschenrieder, W., Venus, G. et al. (1982). *J. Nucl. Mater.* 111 & 112, 29.

Post, D. E. and Lackner, K. (1986). In: *Physics of Plasma-Wall Interactions in Controlled Fusion* (D. E. Post and R. Behrisch, eds.). Plenum Press.

Proudfoot, G., to be published.

Ritz, C. P., Bengston, R. D., Levinson, S. J., and Powers, E. J. (1984). *Phys. Fluids* 27, 2956.

Robinson, D. C. and Rusbridge, M. G. (1969). *Plasma Phys.* 11, 73.

Rubinstein, J. and Laframboise, J. G. (1982). *Phys. Fluids* 25, 1174.

Sanmartin, J. R. (1970). *Phys. Fluids* 13, 103.

Shmayda, W., Winter, J., Waelbroeck, F., and Wienhold, P. (1987). *J. Nucl. Mater.*, 145–147, 201.

Simpson, J. A. (1986). *Rev. Sci. Instrum.* 32, 1283.

Smy, P. R. (1976). *Adv. Phys.* 25, 517.

Staib, P. (1980). *J. Nucl. Mater.* 93 & 94, 351.

Staib, P. (1982). *J. Nucl. Mater.* 111 & 112, 109.

Stampa, A. and Wulf, H. O. (1978). *J. Phys.* D, 11, 1119.

Stangeby, P. C. (1982a). *J. Phys.* D, 15, 1007.

Stangeby, P. C. (1982b). *J. Nucl. Mater.* 111 & 112, 84.

Stangeby, P. C. (1984a). *J. Nucl. Mater.* 121, 36.

Stangeby, P. C. (1984b). *J. Nucl. Mater.* 121, 55.

Stangeby, P. C. (1984c). *J. Nucl. Mater.* 128 & 129, 969.

Stangeby, P. C. (1984d). *Phys. Fluids* 27, 682.

Stangeby, P. C. (1984e). *Phys. Fluids* 27, 2699.

Stangeby, P. C. (1985a). *J. Phys.* D, 18, 1547.

Stangeby, P. C. (1985b). *Phys. Fluids* 28, 644.

Stangeby, P. C. (1986). In: *Physics of Plasma-Wall Interactions in Controlled Fusion* (D. E. Post and R. Behrisch, eds.). Plenum Press.

Stangeby, P. C. (1987). *J. Nucl. Mater.*, 145–147, 105.

Stangeby, P. C., McCracken, G. M., Erents, S. K., and Matthews, G. (1984). *J. Vac. Sci. Technol.* A2, 702.

Stangeby, P. C., McCracken, G. M., and Vince, J. E. (1982). *J. Nucl. Mater.* 111 & 112, 81.

Swift, J. D. and Schwar, M. J. (1969). *Electric Probes for Plasma Diagnostics.* American Elsevier, London.

Tagle, J. A., Stangeby, P. C., and Erents, S. K. (1987). *Plasma Phys. and Controlled Fusion* 29, 297.

Wampler, W. (1982). *Appl. Phys. Lett.* 41, 335.

Woods, L. C. (1965). *J. Fluid Mech.* 23, 315.

Zuhr, R. A., Roberto, J. B., and Appleton, B. R. (1984). *Nucl. Sci. Applic.* 1, 617.

Zweben, S. J. and Gould, R. W. (1983). *Nucl. Fusion* 23, 1625.

# 6 Analysis of Surfaces Exposed to Plasmas by Nondestructive Photoacoustic and Photothermal Techniques

B. K. Bein and J. Pelzl

*Ruhr-Universitat Bochum, Institut für Experimentalphysik VI*
*D-4630 Bochum, P. O. Box 102148, Federal Republic of Germany*

**211**

## I.  Introduction

Thermal waves arise in a body consisting of gas, liquid, or solid matter if this body contains heat sources which are periodic in time. The heat diffusing from the sources into the surrounding matter produces a temperature distribution which is oscillatory in time and space. A characteristic feature of these *heat waves* or *thermal waves* is the strong attenuation of the temperature amplitude with distance from the heat source. Thermal waves are well known phenomena in nature and in engineering. Periodic heating by the sun generates low frequent thermal waves in the earth's crust. Rapidly oscillating heat waves are produced in the walls of combustion engines, causing materials problems with thermal stresses and fatigue.

In basic and applied research, thermal waves have been a field of scientific investigation for more than a century. By the mid-nineteenth century Angström used periodical heating of copper bars to measure the thermal diffusivity of this metal (Angström, 1863), and in 1880 A. G. Bell discovered the so-called *photoacoustic effect* (PAE) of solid materials (Bell, 1880 and 1881). In order to observe the photoacoustic effect, a sample is placed inside an airtight cell and then illuminated with light which is intensity-modulated by a mechanical chopper, for example. The heat waves generated inside the solid by absorption of the optical radiation penetrate into the surrounding body of gas, giving rise to an acoustic pressure. Bell detected the pressure fluctuation of the gas with a hearing tube; nowadays, sensitive microphones are employed.

The first theoretical interpretation of the photoacoustic effect of solids on the basis of thermal waves was given by Rosencwaig and Gersho (1976). Allan Rosencwaig also established photoacoustic spectroscopy (PAS) as a tool for optical studies of solids (Rosencwaig, 1973). Since then, thermal wave studies have experienced a renaissance, and new techniques for the generation and detection of thermal waves have dramatically developed. In addition to optical spectroscopy and calorimetry, nondestructive evaluation of materials has become an important new field of application of thermal waves (Birnbaum and White, 1984).

Presently, thermal wave studies include excitation by electromagnetic radiation extending from microwaves (Evora et al., 1980, Netzelmann et al., 1984) and infrared (Low and Parodi, 1980; Vidrine, 1980) to x-rays (Mascarenhas et al., 1984), as well as heat generation by particle beams such as electrons (Brandis and Rosencwaig, 1980) or argon ions (Satkiewitz et al., 1985). In order to measure the time-dependent thermal response of a solid due to the absorption of a modulated radiation input, several alternative detection schemes have been developed. In addition to the audioacoustic method used in photoacoustic experiments, other important new techniques are photothermal laser beam deflection (Boccara et al., 1980) and observation of the ultrasonic response (White, 1963; Hutchins, 1986). Measurements of the infrared radiation generated by thermal waves which have been used already in earlier studies of the thermal diffusivity (Cowan, 1961) again have received increasing attention, particularly in nondestructive evaluation (NDE) applications (Kanstad and Nordal, 1986).

In the context of this review, we shall limit the discussion to those excitation and detection techniques which have already been used or which are of potential interest for the characterization of solid-state materials treated by plasma processes. For a more complete and extensive description of the great variety of sensing techniques for thermal waves in solid as well as in gas and liquid matter, we refer to a recent review by Tam (1986). Earlier reviews covering this subject or parts of it were published, e.g., by Pao (1977), Rosencwaig (1978), Patel and Tam (1981), West et al. (1983), and Lai and Chan (1985).

Finally, the nomenclature to be used in this article will be reviewed. Thermal wave methods developed recently are commonly denoted as *photothermal*, with an appendix specifying the detection technique, such as *photothermal beam deflection* or *photothermal radiometry*. Other denotations have historical origins. The term *photoacoustic* is currently used if the photothermal response of a solid sample is detected audioacoustically with a gas-contact microphone (Rosencwaig, 1980). The term *optoacoustic* is frequently applied when heat diffusion plays a minor role and the acoustic response to the absorption of radiation is detected by a direct coupled ultrasonic transducer or a microphone. This applies to gases at audioacoustic frequencies (Pao, 1977) and to solids in the ultrasonic frequency or time regime below a microsecond (Patel and Tam, 1981).

This article is divided into five sections. In Section II the theoretical background and some basic experimental concepts of thermal wave physics are discussed. The theoretical description is elaborated in a one-dimensional model; supplementary excitation processes associated with the generation of heat waves, such as the generation of elastic deformations or photocarriers, are also taken into consideration.

In Section III, the experimental aspects of excitation and detection of thermal waves are presented. Among the various kinds of detection schemes available, only those which are relevant to plasma-solid interaction studies will be discussed. Some important applications of thermal wave methods in solid state research and materials characterization are briefly reviewed. This section is closed by a comparison of the thermal wave concept with heat pulse experiments.

Section IV is devoted to the applications of thermal wave methods to plasma-treated solid surfaces. Section IV.A deals with semiconductor materials, plasma-deposited films, and plasma-etched semiconductor devices. Lately, this field is developing very rapidly and seems to be promising for thermal wave techniques, though the quantitative understanding of the measured phenomena is not yet complete. In semiconductor materials, light-induced excitations of the electron-hole plasma, carrier diffusion, and recombination lead to heat sources delayed in time and space, which may give additional information but complicate the understanding as well. In Section IV.B the thermal wave analysis of plasma-sprayed coatings is reviewed in brief. In this technological field, thermal waves have a longer history of application.

Section V provides a survey on the state of art on thermal wave assisted research of plasma-solid interactions in nuclear fusion research. Most of the existing research work deals with diagnostics and characterization of the thermal properties of graphite limiters (Section V.B) and metallic divertor target plates (Section V.C) which already had been exposed to the plasma in fusion experiments. Finally, in Section V.D, the relationship between thermal wave analysis of solid state materials and plasma diagnostics in fusion devices is discussed.

## II. Basic Theory of Thermal Waves

### A. HEAT DIFFUSION EQUATION AND WAVE SOLUTIONS FOR A SOLID IMMERSED IN A GAS ATMOSPHERE

Thermal pulses and waves can be excited in a body of gas, fluid, or solid state matter by pulsed or intensity-modulated heating processes. The heating mechanism due to absorption of electromagnetic radiation will be discussed in Section III.A. Here basic concepts on the heat diffusion in solids and thermal waves are briefly reviewed. More detailed theoretical treatments can be found in recent work and reviews by Aamodt et al. (1977), Mandelis and Royce (1979), Pelzl and Bein (1983), and Rosencwaig (1980). In solid-state materials, heat generation and propagation are governed by the heat diffusion equation

$$\rho c \frac{\partial T(\mathbf{x}, t)}{\partial t} = -\operatorname{div} \mathbf{F}(\mathbf{x}, t) + Q(\mathbf{x}, t), \tag{1}$$

by the appropriate boundary conditions, and by the strength and localization of the heat sources $Q(\mathbf{x}, t)$. The heat flow $\mathbf{F}(\mathbf{x}, t)$ in Eq. (1) is related to the temperature distribution $T(\mathbf{x}, t)$ by

$$\mathbf{F}(\mathbf{x}, t) = -k \operatorname{grad} T(\mathbf{x}, t). \tag{2}$$

Here, $\rho$, $c$, and $k$ are the mass density, specific heat capacity, and thermal conductivity, respectively, of the solid, which in general can vary with the space-coordinates $\mathbf{x}$ and the time $t$. $Q(\mathbf{x}, t)$ represents a time-dependent volume or surface heat source.

If an homogeneous solid of constant thermophysical parameters $k_s$, $\rho_s$, and $c_s$ is considered, Eq. (1) can be simplified in one-dimensional geometry to

$$\frac{\partial T(x, t)}{\partial t} = \alpha_s \frac{\partial^2 T(x, t)}{\partial x^2} + \frac{Q(x, t)}{(\rho_s c_s)}, \qquad (3)$$

where $\alpha_s = k_s/(\rho_s c_s)$ is the thermal diffusivity of the solid. If, furthermore, prompt heating by electromagnetic radiation is assumed, the heat source is given by

$$Q(x, t) = -\eta_s \frac{dI(x, t)}{dx} = \eta_s \beta_s I(x = 0, t) \exp(-\beta_s x). \qquad (4)$$

Here the intensity of the incident radiation may be described by

$$I(x = 0, t) = \frac{I_0}{2} \operatorname{Re}[1 + \exp(i\omega t)], \qquad (5)$$

where $\omega$ is related to the modulation frequency $f$ of the intensity by $\omega = 2\pi f$. In Eq. (4), $\beta_s$ is the optical absorption constant of the solid, and $\eta_s$ is the ratio of the intensity of the unreflected part of the radiation to the total incident intensity. The heat source in the solid is directly related to the absorbed intensity if intermediate excitations and nonthermal deactivation channels are excluded from consideration. The optical parameters $\beta_s$ and $\eta_s$, in general, are functions of the wave length $\lambda$ of the incident radiation. For simplification of the following derivations, we only consider radiation of a definite wave length $\lambda$. The partial differential equation resulting from the insertion of Eqs. (4) and (5) into Eq. (3) can be solved by an ansatz $T(x, t) = \overline{T}(x) + T'(x, t)$, which allows separation of the stationary from the time-dependent problem

$$\frac{\partial T'(x, t)}{\partial t} = \alpha_s \frac{\partial^2 T'(x, t)}{\partial x^2} + \frac{I_0 \eta_s \beta_s}{(2\rho_s c_s)} \exp(-\beta_s x) \operatorname{Re}[\exp(i\omega t)]. \qquad (6)$$

Subsequently, only the time-dependent problem is treated; for the solid its complex solution can be written as

$$T_s'(x, t) = [A_s \exp(\sigma_s x) + B_s \exp(-\sigma_s x) + C_s \exp(-\beta_s x)] \exp(i\omega t). \qquad (7)$$

Neglecting convective motions and heat sources in the gas region in front of

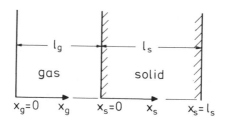

FIGURE 1.    Schematic of a solid immersed in a gas atmosphere.

the solid (Figure 1), a first-order solution is given there by

$$T_g'(x, t) = \left[ A_g \exp(\sigma_g x) + B_g \exp(-\sigma_g x) \right] \exp(i\omega t). \qquad (8)$$

By inserting the solutions (7) and (8) into the homogeneous equations

$$\frac{\partial T'(x, t)}{\partial t} = \alpha_{s, g} \frac{\partial^2 T'(x, t)}{\partial x^2},$$

the complex quantities $\sigma_s$ and $\sigma_g$ are determined,

$$\sigma_{s, g} = (1 + i)\sqrt{\frac{\pi f}{\alpha_{s, g}}}, \qquad i = \sqrt{-1}, \qquad (9)$$

where $\alpha_g$ is the thermal diffusivity of the gas. The quantity $C_s$ is determined from the inhomogeneous equation (6) of the solid region,

$$C_s = - \frac{\eta_s I_0}{\left[ 2k_s \beta_s \left(1 - \frac{\sigma_s^2}{\beta_s^2}\right) \right]}. \qquad (10)$$

If volume heat sources can be neglected for the gas region, such term $C_g$ does not exist.

The integration constants $A_s$, $B_s$, $A_g$ and $B_g$ can be determined from the boundary conditions of negligible temperature variations at the outer boundaries $x_g = 0$ and $x_s = l_s$ of the system gas/solid (compare Figure 1)

$$T_g'(x_g = 0, t) = 0, \qquad (11)$$

$$T_s'(x_s = l_s, t) = 0, \qquad (12)$$

and from the conditions of temperature and heat flow continuity at the solid/gas interface,

$$T_g'(x_g = l_g, t) = T_s'(x_s = 0, t), \qquad (13)$$

$$F_g'(x_g = l_g, t) = F_s'(x_s = 0, t). \qquad (14)$$

For a rather extended gas region and a thick solid, in the limit of a semi-infinite gas and solid sample region,

$$\exp\left(-2|\sigma_g|l_g\right) \ll 1, \tag{15}$$

$$\exp\left(-2|\sigma_s|l_s\right) \ll 1, \qquad \exp\left[-\left(\beta_s + |\sigma_s|\right)l_s\right] \ll 1, \tag{16}$$

the integration constants for the solid region are calculated as

$$A_s = 0, \qquad B_s = -C_s \frac{\beta_s}{\sigma_s} \frac{\left(1 + g_{gs}\dfrac{\sigma_s}{\beta_s}\right)}{\left(1 + g_{gs}\right)}, \tag{17}$$

and the complex temperature distribution for the solid of a finite optical absorption constant then results as

$$T_s'(x, t) = \frac{\eta_s I_0}{2k_s\sigma_s} \frac{\exp\left(i\omega t\right)}{\left(1 - \dfrac{\sigma_s^2}{\beta_s^2}\right)} \left[ \frac{\left(1 + g_{gs}\dfrac{\sigma_s}{\beta_s}\right)}{\left(1 + g_{gs}\right)} \exp\left(-\sigma_s x\right) - \frac{\sigma_s}{\beta_s} \exp\left(-\beta_s x\right) \right]. \tag{18}$$

**Table 1.** Thermophysical Parameters of Some Solids and Gases at Room Temperature

| | Thermal conductivity W/(m°K) | Specific heat capacity J/(kg°K) | Mass density kg/m$^3$ | Thermal diffusivity $10^{-7}$ m$^2$/s | Effusivity Ws$^{1/2}$/(Km$^2$) |
|---|---|---|---|---|---|
| Aluminum | 237 | 890 | 2700 | 980 | 24000 |
| Beryllium | 170 | 1840 | 1850 | 500 | 24000 |
| Copper | 398 | 380 | 8933 | 1163 | 36900 |
| Molybdenum | 140 | 240 | 10240 | 550 | 18500 |
| Nickel | 75 | 460 | 8900 | 190 | 17500 |
| Titanium | 21 | 520 | 4500 | 90 | 7000 |
| Tungsten | 160 | 134 | 19300 | 660 | 20300 |
| Inconel | 14.5 | 455 | 8415 | 38 | 7460 |
| TiAl$_6$V$_4$ | 7.1 | 565 | 4430 | 29 | 4210 |
| Steel V2A | 15.1 | 480 | 7860 | 40 | 7530 |
| Graphite | 5–150 | 660–750 | 1700–2200 | 30–1300 | 2800–13000 |
| Silicon | 148 | 710 | 2330 | 880 | 15600 |
| Quartz glass | 1.36 | 710 | 2200 | 8.71 | 1460 |
| Teflon | 0.16–0.23 | 1050 | 2200 | 0.7–1.0 | 600–750 |
| Air (atmospheric pressure) | 0.026 | 717 | 1.184 | 308 | 4.71 |
| Helium | 0.15 | 3210 | 0.164 | 2847 | 8.88 |
| Argon | 0.018 | 317 | 1.634 | 342 | 3.03 |

In Eqs. (17) and (18) the quantity $g_{gs} = \sqrt{(k\rho c)_g} / \sqrt{(k\rho c)_s}$ is the ratio of the effusivity $\sqrt{(k\rho c)}$ of the gas and the solid material, respectively. For most gas/solid interfaces the quantity $g_{gs}$ is relatively small (Table 1).

## B. MAIN PROPERTIES OF THERMAL WAVES

Equation (18) can considerably be simplified if the condition $g_{gs} \ll 1$ is taken into account and if, additionally, the limit of a surface-heated opaque solid, $|\sigma_s|/\beta_s \ll 1$, is considered. From the resulting complex solution,

$$T'(x, t) = \frac{\eta_s I_0}{2k_s\sigma_s} \exp(-\sigma_s x + i\omega t), \tag{19}$$

the real part solution can be easily derived,

$$T'(x, t) = \frac{\eta_s I_0}{2\sqrt{(k\rho c)_s}} \frac{1}{\sqrt{2\pi f}} \exp\left[-\sqrt{\frac{\pi f}{\alpha_s}}\, x\right] \cos\left(2\pi ft - \sqrt{\frac{\pi f}{\alpha_s}}\, x - \frac{\pi}{4}\right), \tag{20}$$

which reveals that in a periodically heated solid, space- and time-dependent temperature fluctuations are induced which can be interpreted as thermal waves. The principal features of such thermal waves are:

a) From the wave number $|k|_{\text{th.w.}} = \sqrt{\pi f/\alpha_s}$ the thermal wave length and the propagation velocity of the temperature maxima or minima can be determined,

$$\lambda_{\text{th.w.}} = \frac{2\pi}{|k|_{\text{th.w.}}} = \sqrt{4\pi\alpha_s/f} \tag{21}$$

$$v_{\text{th.w.}} = \lambda_{\text{th.w.}} f = \sqrt{4\pi\alpha_s f}. \tag{22}$$

b) Between the periodic heating process according to Eq. (5), $I'(x = 0, t) = (I_0/2)\cos(2\pi ft)$, and the thermal response (20), there is a phase lag,

$$\Delta\varphi = \sqrt{\frac{\pi f}{\alpha_s}}\, x + \frac{\pi}{4}, \tag{23}$$

which increases with the propagation distance $x$ of the thermal wave.

c) It can be seen that the wave amplitude is strongly damped; at a distance of

$$x = \mu = \sqrt{\frac{\alpha_s}{(\pi f)}} \tag{24}$$

it decays to $1/e = 0.368$ of its initial value, and at a propagation distance of a wavelength it is damped by a factor of $\exp(-2\pi) = 0.0019$, which means the solution Eq. (20) for the opaque semi-infinite solid can even be applied to geometrically relatively thin samples as long as their thickness is comparable to the thermal wave length.

d) Since the attenuation length $\mu$ of the amplitude, the so-called thermal diffusion length, and the phase shift vary with the modulation frequency of the heating process, a systematic variation of the modulation frequency $f$ can be used for subsurface depth inspections of solid samples, whereby the amplitude damping and the phase shift are the measurable quantities. There is, however, a natural limitation at larger penetration depths, due to the exponential damping factor of the amplitude.

As can be seen from Eq. (20), low values of the quantity $\sqrt{(k\rho c)}$, the effusivity, lead to high oscillation amplitudes of the surface temperature, and low values of the thermal diffusivity lead to a very rapid attenuation of the amplitude below the surface. High values of the effusivity, on the other hand, lead to low surface temperature oscillations, and high diffusivity values contribute to a relatively deeper penetration of the thermal wave.

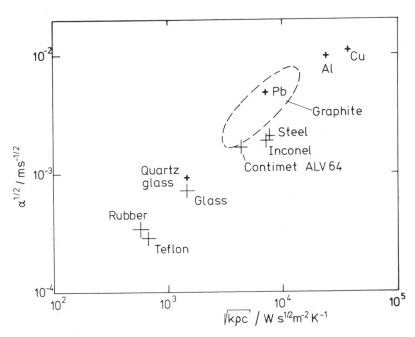

FIGURE 2.    Thermal diffusivity/effusivity diagram of solid state materials.

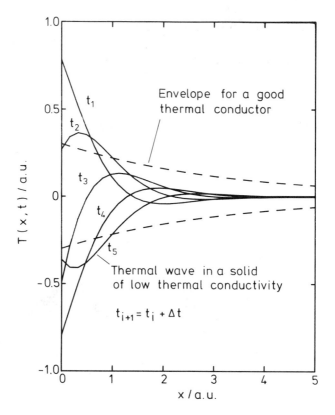

FIGURE 3.  Thermal waves, excitation and propagation.

High or low values of the effusivity of solids normally combine with high or
low values of the thermal diffusivity (Figure 2), which means that both with
good thermal conductors and thermal insulators the measurement of thermal waves has to face serious problems. For insulators it is easy to measure
the relatively large surface temperature oscillations, but the information
they can give comes from a region just at the surface (Figure 3); for good
thermal conductors the information can come from deeper below the
surface, but it is difficult to measure the smaller temperature fluctuations.

Deviations from the general behavior (Figure 2) of high or low effusivity
values, combined with a high or low thermal diffusivity, are possible in
rough and porous solids and in plasma-deposited materials. According to a
theoretical model on the influence of the surface roughness on the effective
thermal properties, open pores or cracks perpendicular to the surface imply
an increasing effusivity, while the thermal diffusivity may decrease (Bein

and Pelzl, 1986, Section V.B.2). On the other hand, spherical pores, according to Maxwell's model, give rise to an effusivity decreasing linearly with the volume porosity, while the thermal diffusivity is less affected (Bein et al., 1986a). Due to the stratified structure of plasma-sprayed coatings, the diffusivity might be as low as 25% of the value of the corresponding compact material (McPherson, 1984), whereas the effusivity is less reduced. On the other hand, it was observed that the effusivity of plasma-sprayed surfaces was strongly affected by thermal treatment, whereas the thermal diffusivity remained nearly unchanged (Cielo et al., 1986).

## C. PHYSICAL RELEVANCE OF THE PARAMETERS GOVERNING THE BEHAVIOR OF THERMAL WAVES

Among the parameters which affect the excitation and propagation of thermal waves according to Eqs. (18) or (20), the thermal diffusivity is the relevant parameter for time-dependent diffusion processes within homogeneous isotropic materials (Carslaw and Jaeger, 1959), whereas the combined quantities $\sqrt{(k\rho c)}$, the effusivity, or the thermal inertia $(k\rho c)$ are the relevant parameters for time-varying heating or cooling processes of surfaces and heat transport across composite layered bodies.

The physical importance of the effusivity easily can be understood if we study time-dependent surface heating processes. If the net heat flux absorbed at the surface of an homogeneous isotropic semi-infinite solid is given by $F_s(t)$, the evolution of the temperature distribution (Carslaw and Jaeger, 1959) can be calculated by

$$T(x, t) = \frac{1}{\sqrt{(k\rho c)_s \pi}} \int_0^t F_s(t - t') \exp\left[-\frac{x^2}{(\alpha_s t')}\right] \frac{dt'}{\sqrt{t'}}. \quad (25)$$

Eq. (25) shows that the combined quantity $\sqrt{(k\rho c)_s}$ is the relevant thermophysical parameter, rather than the thermal conductivity, mass density, and specific heat capacity separately. This quantity $\sqrt{(k\rho c)_s}$ alone determines the surface temperature $T_s(x = 0, t)$, and it is a measure for the heat energy stored in a solid per degree of temperature rise after the beginning of a surface heating process (Grigull and Sandner, 1979).

At first glance the effusivity, which has the dimension W $s^{1/2}/(m^2 K)$, might be a rather abstract quantity, but in reality we are very familiar with it. If we touch bodies of equal temperature but of different effusivities, they do not seem to be equally hot or cold, instead we feel that one body is "hotter" or "colder." According to

$$T_m = \frac{\sqrt{(k\rho c)_1}\, T_1 + \sqrt{(k\rho c)_2}\, T_2}{\sqrt{(k\rho c)_1} + \sqrt{(k\rho c)_2}}, \tag{26}$$

the time-independent contact temperature $T_m$ between two semi-infinite bodies of different effusivity and initially different temperatures $T_1$ and $T_2$ is nearer to the temperature of the body of the higher effusivity value such that the contact temperature we feel is a function of the effusivity of the body we touch.

At ideal surfaces of compact homogeneous solids the value calculated for the effusivity from the bulk parameters $k$, $\rho$, and $c$ and the value which effectively governs a surface heating process may be identical; under realistic conditions, however, changes of the surface due to machining, heat treatment, roughness, and porosity can contribute to a changed effective effusivity at the surface, which can affect cooling or heating processes across surfaces.

Though the effusivity is the relevant thermophysical parameter for surface heating or cooling processes, as observed in the re-entry of spacecraft (Hartunian and Varwig, 1962), in quenching and ignition processes as anode spot formation (Rich et al., 1971), it is a rather unknown quantity in physics. The relevance of the effusivity in plasma-surface interactions and nuclear fusion research has been recognized both for the diagnostics in fusion devices (McCracken and Stott, 1979) and for the erosion of first-wall materials (Behrisch, 1980).

The optical parameters $\eta_s$ and $\beta_s$ describe the strength and localization of the heat source, respectively. If opaque or optically thick samples are considered where the inverse of the optical absorption constant $\beta_s$ is smaller than the thickness of the sample, and if only nonradiative de-excitation processes are considered, the parameter $\eta_s$ can be understood as photothermal conversion efficiency. In general, the value, $0 < \eta_s(\lambda) \leq 1$, is unknown for most materials and individual surfaces. Following the definition, $\eta_s = 1 - R$, the quantity $\eta_s$ can be determined from measurements of the reflectivity $R$, where for diffusively reflecting surfaces the local intensity distribution of the reflected radiation has to be taken into account. In some cases, weakly absorbing materials like glasses and for orthogonal incidence on smooth surfaces, the reflectivity can be predicted theoretically from Beer's formula

$$R = \frac{(n-1)^2}{(n+1)^2}, \tag{27}$$

where $n$ is the refractive index.

## D. REFLECTION AND SCATTERING OF THERMAL WAVES

Since the amplitude and the phase of thermal waves depend on the effusivity and thermal diffusivity, their propagation (damping and phase retardation) will change when regions of different thermal properties are reached. In the terms of wave propagation, these phenomena can be interpreted as reflection and scattering of thermal waves. Subsequently, the one-dimensional solution for a composite layered solid will be derived to study reflection phenomena which will be used later to analyze plasma-sprayed coatings, limiters, and divertor plates. Scattering can be observed at vertical boundaries of samples (Aithal et al., 1984; McDonald et al., 1986).

If we consider a system consisting of a thick substrate ($b$), a surface layer ($s$) of thickness $l_s$ where optical absorption takes place and a gas region ($g$) in front of the solid (Figure 4), the complex wave solutions for the three regions are given by

$$T_b'(x, t) = [A_b \exp(\sigma_b x) + B_b \exp(-\sigma_b x)] \exp(i\omega t) \qquad (28)$$

for the substrate region and by Eqs. (7) and (8) for the surface layer and the gas region, respectively. In the quantity

$$\sigma_b = (1 + i)\sqrt{\frac{\pi f}{\alpha_b}}, \qquad (29)$$

$\alpha_b$ is the thermal diffusivity of the substrate. The integration constants $A_b$, $B_b$, $A_s$, $B_s$, $A_g$, and $B_g$ of the complete system are now determined from the boundary conditions of negligible temperature oscillations at the outer boundaries of the system at $x_g = 0$, Eq. (11), and at $x_b = l_b$,

$$T_b'(x_b = l_b, t) = 0 \qquad (30)$$

and from the conditions of temperature and heat flux continuity at both the

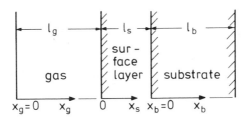

FIGURE 4. Schematic of a two-layer solid.

gas/surface layer interface, Eqs. (13) and (14), and the surface layer/substrate interface,

$$T_s'(x_s = l_s, t) = T_b'(x_b = 0, t), \tag{31}$$

$$F_s'(x_s = l_s, t) = F_b'(x_b = 0, t). \tag{32}$$

If according to relation (15), once again a rather extended gas region and a thick substrate are considered,

$$\exp(-2|\sigma_b|l_b) \ll 1, \tag{33}$$

and if, additionally, an optically thick surface layer is assumed,

$$\exp(-\beta_s l_s) \ll 1, \tag{34}$$

the integration constants are calculated to be

$$A_s = C_s \frac{\beta_s}{\sigma_s} \frac{(1 + g_{gs}\sigma_s/\beta_s)(1 - g_{sb})\exp(-2\sigma_s l_s)}{(1 + g_{gs})(1 + g_{sb}) + (1 - g_{gs})(1 - g_{sb})\exp(-2\sigma_s l_s)} \tag{35}$$

and

$$B_s = -C_s \frac{\beta_s}{\sigma_s} \frac{(1 + g_{gs}\sigma_s/\beta_s)(1 + g_{sb})}{(1 + g_{gs})(1 + g_{sb}) + (1 - g_{gs})(1 - g_{sb})\exp(-2\sigma_s l_s)}. \tag{36}$$

The constant $C_s$, Eq. (10), remains unchanged. The resulting expression for the temperature distribution in the surface layer can be simplified considerably if, according to Table 1, the relation $g_{gs} \ll 1$ holds and, additionally, the relation $g_{gs}|\sigma_s|/\beta_s \ll 1$ is also fulfilled,

$$T_s'(x, t) = \frac{\eta_s I_0}{2k_s\sigma_s} \frac{\exp(i\omega t)}{\left(1 - \dfrac{\sigma_s^2}{\beta_s^2}\right)} \left[\exp(-\sigma_s x)\frac{[1 + R_{sb}\exp(-2\sigma_s(l_s - x))]}{[1 - R_{sb}\exp(-2\sigma_s l_s)]}\right.$$

$$\left. - \frac{\sigma_s}{\beta_s}\exp(-\beta_s x)\right]. \tag{37}$$

In the limit of surface heating, $|\sigma_s|/\beta_s \ll 1$, this solution for the composite layered solid simplifies to

$$T_s'(x, t) = \frac{\eta_s I_0}{2k_s\sigma_s} \frac{[1 + R_{sb}\exp(-2\sigma_s(l_s - x))]}{[1 - R_{sb}\exp(-2\sigma_s l_s)]} \exp(-\sigma_s x + i\omega t) \tag{38}$$

and can be compared with the surface-heated semi-infinite solid, Eq. (19). The quantity

$$R_{sb} = -\frac{(1 - g_{sb})}{(1 + g_{sb})} \tag{39}$$

in Eqs. (37) and (38) can be understood as the *thermal reflection coefficient*, corresponding to the transition from the surface layer to the substrate (Bennett and Patty, 1982). If the surface layer and the substrate have equal effusivities, $g_{sb} = \sqrt{(k\rho c)_s} / \sqrt{(k\rho c)_b} = 1$, the thermal reflection coefficient is $R_{sb} = 0$ and solution (38) will be identical to solution (19) for the homogeneous semi-infinite solid. If the surface layer has a lower effusivity value than the substrate, $0 < g_{sb} < 1$, the thermal reflectivity will have negative values, $-1 < R_{sb} < 0$. By comparing Eq. (38) with Eq. (19), the temperature of the surface layer will be lower in general. If the substrate has a lower effusivity value than the surface layer, $g_{sb} = \sqrt{(k\rho c)_s} / \sqrt{(k\rho c)_b} > 1$, the reflectivity will be positive, $0 < R_{sb} < +1$. In this latter case, the temperature within the surface layer will be higher than in the corresponding homogeneous semi-infinite solid, Eq. (19), whereby it is justified to call $R_{sb}$ the *thermal reflection coefficient*.

E. THERMOELASTIC GENERATION OF STRESS AND STRAIN

The excitation and propagation of thermal waves in solids can be accompanied by secondary effects, e.g., generation of thermoelastic stresses and strains, which may be used for the detection of thermal waves (Section III.B.3) and which, on the other hand, also may contribute to problems of pulse- or wave-heated solid bodies (thermal stresses and fatigue, Section V.A).

General relations between the stresses and time-dependent heating of a solid body, independent of special design geometries and boundary conditions, can be derived starting from Hooke's law of thermoelasticity (Chadwick, 1964),

$$\sigma_{ij} = 2\mu_L \epsilon_{ij} + \lambda_L \epsilon_{kk} \delta_{ij} - (3\lambda_L + 2\mu_L)\beta_e T(\mathbf{x}, t)\, \delta_{ij} \qquad (40)$$

and from the momentum balance applied to a body free of external forces,

$$\rho \frac{\partial^2 u_i}{\partial t^2} = \frac{\partial \sigma_{ij}}{\partial x_j} = 2\mu_L \frac{\partial \epsilon_{ij}}{\partial x_j} + \lambda_L \frac{\partial \epsilon_{jj}}{\partial x_i} - (3\lambda_L + 2\mu_L)\beta_e \frac{\partial T(\mathbf{x}, t)}{\partial x_i}. \qquad (41)$$

From the balance of the internal energy of the thermoelastic solid, follows the equation

$$\rho c \frac{\partial T(\mathbf{x}, t)}{\partial t} = -\frac{\partial F_j}{\partial x_j} + Q(\mathbf{x}, t) + T(\mathbf{x}, t)\frac{\partial \sigma_{ij}}{\partial T}\frac{\partial \epsilon_{ij}}{\partial t}, \qquad (42)$$

which is a generalization of the heat diffusion equation and in which the last term on the right-hand side describes the thermal contributions of the deformation field. Here, $\mu_L$ and $\lambda_L$ are Lamé's constants of elasticity, $\beta_e$ the linear thermal expansion coefficient, $\sigma_{ij}$ and $\epsilon_{ij}$ the components of the stress and strain tensor, respectively, and $u_i$ the components of the deformation vector.

If pulsed heating on the time-scale of a millisecond or frequencies up to several 10 KHz are considered, order of magnitude considerations based on the typical values of the density $\rho$ and elastic constants $\mu_L$, $\lambda_L$ and on dimensions of the solid body in the millimeter or centimeter range show that the inertia term $\rho\, \partial^2 u_i/\partial t^2$ is small and can be neglected in comparison to the elastic restoring forces on the right-hand side of Eq. (41). From the resulting quasistationary momentum balance, a quite general differential can be derived by further differentiation,

$$\Delta\left[\operatorname{div}\mathbf{u} - \frac{(3\lambda_L + 2\mu_L)}{(\lambda_L + 2\mu_L)}\beta_e T(\mathbf{x}, t)\right] = 0 \qquad (43)$$

and a general solution of Eq. (43) is given by

$$\operatorname{div}\mathbf{u} = \frac{(3\lambda_L + 2\mu_L)}{(\lambda_L + 2\mu_L)}\beta_e T(\mathbf{x}, t) + S(\mathbf{x}), \qquad (44)$$

where $S(\mathbf{x})$ is an arbitrary solution of the Laplacian $\Delta S(\mathbf{x}) = 0$ (Chadwick, 1964). With the help of Eq. (44), which relates the deformation dynamics to the time-dependent temperature field, the contribution of the deformations to the energy balance in Eq. (42) can be evaluated and transformed to appear as a correction term of the specific heat capacity $c$,

$$\rho c\left[1 - \frac{(3\lambda_L + 2\mu_L)^2\beta_e^2}{(2\mu_L + \lambda_L)\rho c}T(\mathbf{x}, t)\right]\frac{\partial T(\mathbf{x}, t)}{\partial t} = -\operatorname{div}\mathbf{F}(\mathbf{x}, t) + Q(\mathbf{x}, t). \qquad (45)$$

For typical solid-state parameters and moderate temperatures, this correction term is small in comparison to one (Pelzl and Bein, 1983), and thus the temperature distribution can be treated according to the usual linear heat diffusion Eq. (1), decoupled from the stress and strain fields.

In order to calculate the thermoelastic response of a heated sample, the heat diffusion equation and the momentum balance have to be solved simultaneously. For the momentum balance the appropriate boundary conditions (clamped or free surfaces) and the geometry of the sample have to be considered (Nowacki, 1966). Apart from principal analytical solutions

(White, 1963; Opsal and Rosencwaig, 1982), numerical solutions are required to interpret the measured signals.

## F. Delayed Heat Generation

The preceding equations in Section II have been derived under the assumption of prompt heat generation. This means that the time delay between the absorption of radiation and the heat available in the sample is small compared to the time resolution of the experiment. In nonradiative decay of electronic states by electron-phonon coupling, for example, the heat is delivered within a few picoseconds, a time interval much smaller than the accessible time resolution of microseconds to nanoseconds when ultrasonic or optical detection probes are used (Section III).

Longer time delays may occur when the de-excitation process involves transient states. Retardation effects have already been observed with photothermal methods in optically excited organic molecules and semiconductors. Singlet states of organic molecules at energies corresponding to the UV wave length are normally transferred with a high quantum efficiency into triplet states, e.g., by intersystem crossing. The lifetime of these intermediate states is typically on the order of microseconds, which can be resolved with the ultrasonic probe (Heihoff, 1986). In semiconductors, interband absorption simultaneously provides a prompt and a delayed heat source. Instantaneous heating results from the electrons relaxing inside the conduction band; photoexcited electron-hole pairs at the band gap can diffuse into the solid before recombining after a characteristic lifetime $t_r$ (Sablikov and Sandomirskii, 1983).

Taking into consideration the delayed heat production, the source term in Eq. (3) has to be replaced by

$$Q(x, t) = Q_0(x, t) + \text{const.} \frac{N(x, t)}{t_r}, \qquad (46)$$

where $Q_0(x, t)$ describes the prompt heating, and the delayed heating is proportional to the density of transient states $N(x, t)$ and to the decay or recombination rate $1/t_r$, respectively. In the case of semiconductors, the space- and time-dependent density $N(x, t)$ of the photoexcited electron-hole pairs is governed by the carrier diffusion equation. For a quantitative interpretation of the thermal response, the two coupled diffusion equations have to be solved.

The photocarriers, which before recombining modify the optical properties of the semiconductor, can be monitored directly with optically based photothermal methods such as the internal beam deflection (Fournier et al., 1986) and the modulated reflectance (Opsal and Rosencwaig, 1985).

## III. Excitation and Detection Techniques for Thermal Waves

### A. GENERATION OF THERMAL WAVES

Photothermal methods are based on generation and detection of a time-dependent thermal excitation due to the absorption of either electromagnetic radiation or of particle beams. The photothermal experiments fall into two major categories, which are the time-domain measurements with a pulsed excitation and the frequency-domain measurements proceeding from a periodic excitation in time, giving rise to thermal waves. The subsequent discussion is restricted to the latter category. A comparison with time-domain photothermal experiments will be given in Section III.C.

A simplified block diagram of a frequency domain photothermal experiment is shown in Figure 5. The main components are the radiation source; the modulator, which interrupts the beam flux periodically; the detector transforming the thermal wave response into an electrical signal; and,

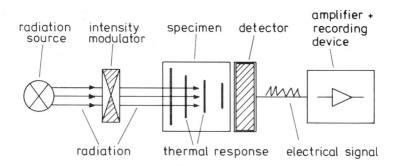

PHOTOTHERMAL   EXPERIMENT

FIGURE 5.   Schematic of a frequency-domain photothermal experiment based on the generation and detection of thermal waves.

finally, the devices for the treatment of the electrical signal, comprised of a lock-in detector as the central unit.

Radiation sources suitable for thermal wave studies generally require high luminosities because of the fundamental limitation imposed by the detector noise and the efficiency of the radiation input-to-heat and the heat-to-electrical signal conversion. To get a rough estimate of the minimum power density necessary, e.g., for a photoacoustic measurement or a laser beam deflection experiment, we consider the case of heat wave excitation in a surface-heated semi-infinite solid. According to Eq. (20), the amplitude of the oscillating surface temperature $T_0'$ (in K) and the radiation flux $I_0$ (W/m$^2$) converted into heat in the surface layer are related by the simple formula,

$$I_0 = T_0'\sqrt{\omega}\,\sqrt{(k\rho c)_s}\,,\qquad (47)$$

where $\omega$ is the angular modulation frequency, $\sqrt{(k\rho c)_s}$ the effusivity of the surface material, and where the photothermal conversion efficiency is assumed to approximate $\eta \approx 1$. The detection limit $T_0'$ of the probes considered here is about $10^{-4}$ K. The effusivity values may be of the order of 5000 Ws$^{1/2}$/Km$^2$ (Table 1) at room temperature, and a convenient modulation frequency may be 1000 Hz. Inserting these values, the minimum radiation power required for a sample of 1 cm$^2$ surface area is then of the order of a few mW. Thus, in more favorable cases, i.e., for lower frequencies and effusivity values, a few tenth of mW at the sample site could be sufficient.

Radiation powers of this magnitude are easily provided by monochromatic sources such as lasers in the optical regime or klystrons in microwave spectrometers. For wavelength dependent spectroscopy, however, incoherent light sources such as arc discharges and incandescent lamps are normally used in combination with spectrometers. Commercially available 1000 W lamps such as xenon, mercury arc, or tungsten filament lamps have a spectral radiance in the visible range at 550 nm of about 50 mW/(sr nm). High luminosity grating monochromators typically have a numerical aperture of 4 and a diffraction efficiency of 60% in the neighborhood of the blaze wavelength. The detection limit therefore requires a band pass of about 1 nm. The derived numbers demonstrate that the high wavelength resolution ($10^{-3}$ nm) required for spectroscopic photothermal experiments can only be achieved with laser excitation. A more extensive discussion of light sources suitable particularly for photoacoustic spectroscopy was published recently (Bechthold, 1984).

For x-rays and particle beams, a similar treatment does not provide a useful estimate, because the thermal wave generation is obscured by supplementary effects. Employing high energetic radiation or particle beams, a portion of the radiation energy is lost in the process of defect creation. Particularly when charged particles are used, mechanisms other than thermal wave excitation may generate an ultrasonic response, which is the most currently used detection probe for this type of experiments. Electron currents needed in studies with an electron microscope are about $10^{-7}$A at an electron energy of 20 keV, corresponding to a power deposition of mW on the sample at typical scanning velocities (Ermert et al., 1984).

Thermal waves require a periodic heating process which is achieved by modulating the incident radiation intensity at the frequency $f$. External modulation devices which can be inserted into the beam emitted from a DC source are mechanical choppers, electro-optic, or acousto-optic modulators. Mechanical choppers offer the advantage of being able to be employed for all types of radiations, but the frequency range is limited from a few hertz to about 5 kHz. For most purposes the shape function of the modulated intensity is of minor importance. Normally, the thermal wave response is analyzed using lock-in techniques, thereby probing selectively the amplitude and the phase of the signal. The shape function is only of immediate importance when nonlinear effects, e.g., the response of the second harmonic, are studied (Balk, 1986).

## B.   DETECTION TECHNIQUES FOR THERMAL WAVES

Thermal waves, temporally and spatially oscillating temperature distributions, are associated with changes of other physical quantities. Consequently, they can be detected either directly by a measurement of the local temperatures or heat fluxes or indirectly by monitoring temperature induced modifications of optical, mechanical, or electrical properties. For both fields, a great variety of more or less sophisticated detection techniques have been developed in the past. The classification of the various methods in Figure 6 proceeds from the physical aspects involved in the detection process. The most currently used detection techniques are divided into three categories depending on the contact between the sensor probe and the sample. The other scale takes into consideration the frequency scale covered by each method. Full lines indicate the time resolutions or frequency ranges which have already been attained in experiments reported in literature. Generally, the theoretical limitations imposed, e.g., by the time resolution of the detector would admit much larger dynamical ranges. Subsequently, a few characteristic features of those techniques which have already been used for the inspection of plasma treated solid materials will be briefly discussed.

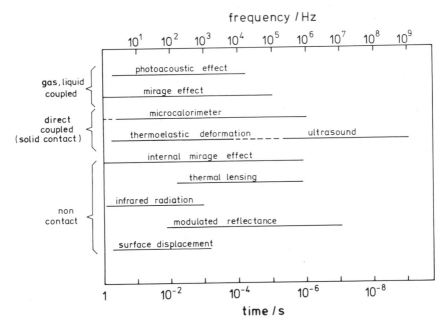

FIGURE 6. Most currently used photothermal detection techniques. The full lines indicate the time resolution or frequency range already attained in experiments reported in literature.

## 1. The Photoacoustic Effect

In the photoacoustic sensing technique the periodic thermal response of the sample surface is transferred to the surrounding gas in a gas-tight cell. The pressure fluctuations induced in the gas volume by the heat flux across the solid/gas interface are detected by a microphone mounted inside the cell. The first theoretical description of the signal generation process associated with a solid in a photoacoustic cell was given by Rosencwaig and Gersho (1976). It was based on the assumption of a one-dimensional thermal boundary layer expanding and contracting at the solid/gas interface, thereby acting as a piston on the major part of the gas volume, which is governed by the adiabatic gas law. Subsequent models (McDonald and Wetsel, 1978; Korpiun and Buechner, 1983) treated refinements and extensions of the piston model. In a more general derivation based on the principles of continuum mechanics, the pressure signal was directly related to the modulated heat flux at the solid/gas interface, independent of special model assumption (Bein and Pelzl, 1983).

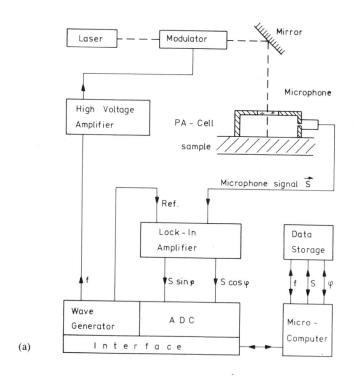

(a)

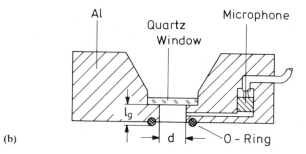

(b)

FIGURE 7. Schematic of the experimental setup used for the photoacoustic studies of limiter graphite (a) and cross section of the photoacoustic cell (b) (after Bein et al., 1984).

The conventional experimental photoacoustic setup consists of a periodically modulated light source and the photoacoustic cell containing the sample and the microphone. Figure 7a shows a schematic of the arrangement used for the thermal wave studies of graphite limiter plates (Section V.B). The light source is an argon ion laser. The laser beam is intensity-modulated by an electro-optical modulater. The body of the photoacoustic cell (Figure 7b) is made of aluminum to minimize the signal contributions from the cell walls. At the top, the cylindrical gas cavity is closed by a quartz window and at the bottom by the sample itself. A rubber ring between the cell body and the sample is used for acoustic sealing. To prevent the microphone from direct illumination of light, it is separated from the main cavity by a narrow channel. As a consequence of this construction, the acoustic transfer function of the photoacoustic cell shows a Helmholtz type resonance in the KHz region (Nordhaus and Pelzl, 1981). The increased signal-to-noise ratio of the resonance allows extension of the measurements to higher frequencies.

The in-phase and out-off-phase components of the microphone signal with respect to the modulator reference are analyzed with a lock-in amplifier. The amplitude and the phase angle of the photoacoustic signal are recorded and interpreted as a function of the modulation frequency $f$ at a fixed wavelength when thermal properties of the sample are to be analyzed. For the spectroscopic analysis of solids (PAS), the photoacoustic signal is analyzed as a function of the wavelength $\lambda$ of the incident radiation at a fixed modulation frequency.

Based on the one-dimensional piston model (Rosencwaig and Gersho, 1976), an estimate of the acoustic pressure amplitude is given by

$$\delta p = \gamma \bar{P} \mu_g \frac{T_0'}{\left( l_g \bar{T} \right)} \tag{48}$$

where $\gamma$ is the adiabatic coefficient, $\bar{P}$ and $\bar{T}$ are the steady state pressure and temperature and $l_g$ the length of the gas cavity, respectively. According to Eq. (24), the thermal diffusion length of the gas is given by $\mu_g = \sqrt{2\alpha_g/\omega}$ and the amplitude $T_0'$ of the thermal wave at the transition from the opaque solid to the gas can be inserted from Eq. (47) to obtain

$$\delta p = \gamma \bar{P} \frac{I_0 \sqrt{\alpha_g}}{\bar{T} l_g \sqrt{(k\rho c)_s} \, \omega} . \tag{49}$$

Although approximate, this relation can lead to two important conclusions:

a) The signal decreases with the inverse power of the modulation frequency. Considering only white noise, the signal-to-noise ratio remains indepen-·

dent of the modulation frequency. Thus, the noise figure is mainly determined by the peculiar noise sources of the experimental arrangement. In the case of depth-dependent thermal inhomogenities or of thermally thin samples, deviations from the $1/\omega$ dependence of the photoacoustic amplitude are expected (Pelzl, 1984).

b) If relation (48) is applied to the cell of Figure 7, which has a gas length between the sample and the window of $l_g = 1$ cm and is filled with air at ambient pressure and temperature, one obtains an acoustic pressure amplitude of about $\delta p = 10 \, (T_0/K)$ (in Pa) for a modulation frequency of 1000 Hz. Typical detection limits of commercially available electret or condensor microphones are in the mPa range. Photoacoustic sensing therefore requires a minimum amplitude $T_0'$ of the oscillating surface temperature of about $10^{-4}$ K, as stated in Section III.A.

To improve the sensitivity of the photoacoustic cell, the average gas pressure $\bar{P}$ can be increased or the gas volume (considered in the one-dimensional treatment by the gas length $l_g$) can be decreased. This reduction, however, is limited to sizes larger than the thermal diffusion length in the gas, otherwise the heat losses through the top window would falsify the interpretation. The most efficient enhancement of the photoacoustic response occurs at low temperatures of both the sample and the gas. Due to the combined effect of the solid and the transmitting gas, the signal increases with an inverse power law of the temperature of $T^{-2.75}$ for He as pressure transmitting gas. Thus, the signal amplitude can be increased by a factor of 500 by lowering the temperature from 300 K to 15 K (Pelzl et al., 1982). Deviations from the semi-infinite solid towards thin layers or small particles can increase the sensitivity already at room temperature (Netzelmann et al., 1988).

The influence of the cell, cell dimensions and materials, on the photoacoustic signal, have been analyzed systematically in the past (Parker, 1973; Aamodt et al., 1977; Quimby and Yen, 1979). To reduce signal contributions from the cell walls due to light reflected at the sample surface, the effusivity value of the cell materials should be high or the cell walls should be transparent (Kordecki et al., 1986). In depth profiling studies with graphite and amorphous metal foils, stray light corrections procedures were applied in order to widely eliminate the signal contributions from both opaque and transparent cells (Krueger et al., 1987; Kordecki et al., 1988).

## 2.   The Photothermal Beam Deflection (Mirage Effect)

The photothermal deflection of a probe laser beam, the so-called *mirage* effect (Boccara et al., 1980), provides an alternative detection technique for

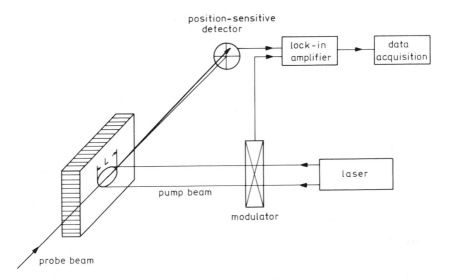

FIGURE 8. Schematic of a photothermal experiment using the laser beam deflection (mirage effect) technique.

thermal waves, which can offer some fundamental advantages when compared to photoacoustic detection. In the typical mirage effect experiment (Figure 8), the solid sample is periodically heated by an intensity-modulated pump laser beam. The thermal waves propagate from the solid into the surrounding gas or liquid medium, where space and time-dependent variations of the refractive index are produced. The gradient of the refractive index perpendicular and parallel to the sample surface is probed by a second laser beam, e.g., from a He-Ne laser. The deflection angle of this probe beam is determined by position sensitive detectors such as quadrant or lateral diodes. Angular variations down to $10^{-8}$ radians can be measured with commercially available position sensors.

The vertical deflection angle $\theta_z$ experienced by the probe beam interacting along a distance $L$ of its path with the thermal waves is roughly given by (Fournier and Boccara, 1984)

$$\theta_z = \left(\frac{L}{n}\right)\left(\frac{dn}{dT}\right)\left(\frac{\partial T}{\partial z}\right) = \left(\frac{L}{n}\right)\left(\frac{dn}{dT}\right)\left(\frac{T_0'}{\mu_g}\right). \tag{50}$$

Here, $T_0'$ is the temperature oscillation amplitude at the sample surface; $n$ and $dn/dT$ are the refractive index and its temperature coefficient, and $\mu_g$ the thermal diffusion length of the surrounding gas or liquid. In detail the mirage effect was treated both in experiment and theory by Jackson et al.

(1981), Murphy and Aamodt (1980), Aamodt and Murphy (1981 and 1983), and Mandelis (1983).

Applying relation (50) to a sample immersed in air ($n = 1$, $dn/dT = 10^{-6}$/K) at a modulation frequency of 1000 Hz ($\mu_g = 0.1$ mm) and with an interaction length of $L = 10$ mm, the detection limit is about $T_0' = 10^{-4}$ K, which is comparable to the temperature resolution of the photoacoustic technique. Much higher sensitivities are achievable by immersing the sample in an organic liquid such as $CCl_4$ ($n = 1.45$, $dn/dT = 6.10^{-4}$/K, $\mu_g = 0.045$ mm), where at similar conditions the detection limit becomes $T_0' = 10^{-7}$ K. The noise level is determined by two main sources, namely the pointing noise of the He-Ne probe laser beam and mechanical vibrations of the detection arrangement consisting of sample, probe laser, and diode.

In comparison with photoacoustic sensing the mirage detection provides three advantages: 1) The laser beam deflection does not require a closed cell, thus avoiding parasitic signals generated in cell walls or windows. 2) The frequency range is not limited by a detector response function. 3) The mirage effect is a local detection probe admitting measurements of lateral thermal anisotropics, which, however, require a three-dimensional treatment of the thermal problem (Lepoutre et al., 1986).

As the mirage effect sensitively responds to small changes of the probe beam distance from the surface, this technique is less favorable in the case of powders or samples with rugged surfaces; the calibration with respect to a reference sample (Section V.B) also faces much more experimental difficulties than in the case of photoacoustic detection.

### 3. The Thermoelastic Response (Ultrasonic Waves)

The thermoelastic sensing techniques are based on the detection of strains or stresses associated with locally and temporally varying temperature fields in solids. Above all, the detection of thermoelastically generated ultrasonic signals offers a very high time resolution up to frequencies of GHz. To illustrate some fundamental features of the thermoelastic method, we consider a semi-infinite solid periodically heated at the surface (Figure 1). Due to the thermoelastic coupling the heat waves induce oscillating longitudinal displacements, which can be determined by solving the equation of motion (41). The thermal expansion coefficient, $\beta_e$, is the coupling constant relating the time-dependent temperature distribution to the strain and stresses, Eq. (40). The thermoelastic response can be divided into two regimes:

a) The thermoelastic deformation regime (see Figure 6) represents the quasistatic limit at low modulation frequencies, where the inertia term in the equation of motion can be neglected (Section II.E; Pelzl and Bein, 1983; Junge et al., 1983) and where the signal amplitude has approximately a $1/\omega$ dependence (Jackson and Amer, 1980), like the photoacoustic signal, Eq. (49).

b) The ultrasonic regime at elevated modulation frequencies is characterized by the propagation of the thermoelastically induced displacement at sound velocities. For a mechanically free surface of the solid half space, the efficiency of conversion $\eta$ of the absorbed radiation intensity $I_0$ into elastic wave energy is (White 1963):

$$\eta = \frac{k \, \beta_e^2 \, \omega \, I_0}{2 \, v \, \rho^2 \, c^3},\tag{51}$$

where $k$, $\beta_e$, $\rho$, and $c$ are the thermal conductivity, the thermal expansion coefficient, mass density, and specific heat capacity, respectively, of the solid; $v$ is the sound velocity and $\omega$ the angular modulation frequency. As Eq. (51) shows, the conversion efficiency increases with the frequency and the ultrasonic signal itself is a nonlinear function of the incident radiation intensity. Inserting data for Al, the conversion coefficient is about $10^{-15} * I_0 / (\text{W cm}^{-2})$ at 10 MHz. Experiments with transducers of a typical detection limit of 0.1 Pa therefore require radiation intensities of roughly 1 W/cm$^2$, which could be achieved by focusing, for example, the beam of a low power laser. This estimate shows that the ultrasonic sensing requires higher power deposition rates, but, with respect to the temperature amplitude, the sensitivity compares with that of the photoacoustic or mirage effect.

Analyzing the ultrasonic response at elevated frequencies in terms of the characteristic lengths of optical absorption, thermal diffusion, and geometrical size, two distinct regimes can be recognized: (1) In the limit of very strong optical absorption, the sound generation is confined to a surface layer determined by the thermal diffusion length. Detection of the ultrasonic signal as a function of the modulation frequency provides a means of a high resolution nondestructive evaluation of subsurface properties (Rosencwaig and Busse, 1980). (2) For very low optical absorptivity, the ultrasonic signal is produced in a region that is determined by the dimensions of the radiation beam rather than by the thermal diffusion length. In this low absorptivity limit, ultrasonic detection is most suited for spectroscopic studies (Patel and Tam, 1981). The use of the pulsed laser for optoacoustic investigation of low absorbing media provides two major advantages: High peak powers are available and de-excitation processes can directly be observed in a time domain analysis (Section III.C).

The ultrasonic response is most currently detected by means of resonant or nonresonant piezo-electric transducers (e.g., $PVF_2$ foils) attached to the sample by glue or grease. The ultrasonic detection provides a real complement to the photoacoustic and the mirage effect. Ultrasonic measurements are compatible with vacuum conditions and offer time resolutions not accessible to the other methods (Figure 6).

## 4. Photothermal Radiometry (Infrared Radiation)

Photothermal radiometry relies on the observation of changes in the thermal radiation as a result of the excitation of thermal waves. The incremental radiant emittance dW due to a thermal wave amplitude $T_0'$ is determined by the Stefan-Boltzmann law in differential form.

$$dW = 4 \, \epsilon \, \sigma \, T^3 \, T_0'. \tag{52}$$

Here, $\sigma$ is the Stefan-Boltzmann constant, and $\epsilon$ the emissivity of the material. For opaque and thermally thick samples, the amplitude $T_0'$ of the surface temperature oscillation follows from Eq. (47). The standard experimental setup consists of two optical arrangements, one for the pump radiation path and the other collecting the thermal radiation (Nordal and Kanstad, 1981). Sensitivity aspects of this method have been treated by Kanstad and Nordal, 1986, so we can refer to the values obtained. Using commercial IR detectors, the detection limit at 300 K is about $T_0 = 10^{-3}$ K, which is an order magnitude smaller than the sensitivity of the techniques discussed before. Due to the lower efficiency, thermal wave radiometry is limited to modulation frequencies in the KHz range, although the time constants of IR detectors permit time resolutions of nanoseconds. Equation (52) indicates that measurements at high temperature are more advantageous. In case where the thermal radiation originates from regions inside the sample, reabsorption effects can complicate the analysis of depth dependent properties.

Photothermal radiometry provides a true remote technique, particularly attractive for optical and thermal studies of samples in vacuum at high pressure and temperatures and in inaccessible environments. Typical experimental arrangements use liquid nitrogen cooled semiconductor detectors to measure the local IR emission at the front or rear surface of a sample (e.g., Busse et al., 1986). More recently, video-compatible infrared field imagers are applied that directly provide the ac-temperature distribution image of a surface; these images can be analyzed with respect to the material properties (Reynolds, 1986) or with respect to the power deposition process as used in nuclear fusion research as a diagnostic tool (Section V.D). Pulsed excitation offers some advantages when the signals from a modulated cw source are too low, e.g., at high frequencies, since in the pulse experiment all deposited energy contributes to the detected temperature amplitude (Leung and Tam, 1984).

## 5. Other Detection Techniques

Until now, the remaining techniques drawn into Figure 6 are less frequently used in the field covered by this review.

The *microcalorimeter* methodologically is a promotion of conventional calorimeter with a time resolution improved due to a reduced thermal inertia. The quantity accessible to the measurement is the heat flux from the sample into the calorimeter. Microcalorimeters currently consist of a pyroelectric $PVF_2$ foil, with the sample directly deposited on it (Melcher and Arbach, 1982). Most promising applications are the investigation of thin film properties (Coufal, 1984, 1986) and the thermal characterization of desorption and adsorption processes (Träger et al., 1982).

The *internal mirage effect* and the *thermal lensing* sense internal temperature fields by a probe laser beam and therefore require transparent samples. The collinear or transverse deflections of a laser beam inside the sample is related to the gradient of the index of refraction generated by a thermal wave (Jackson et al., 1981). Thermal lensing, on the other hand, probes the thermally induced curvature of the refractive index (Swofford and Morell, 1978). With respect to sensitivity and resolution, the characteristics of both techniques are very similar (Jackson et al., 1981). A most interesting result is obtained with the transverse mirage effect in semiconductors. Photocarriers excited simultaneously with heat wave produce an additional deflection, which can be used to deduce intrinsic electronic parameters (Fournier et al., 1986).

The *modulated reflectance* has a very similar physical basis as the methods mentioned before. A laser beam reflected on the sample surface probes the change of the refractive index induced by an intensity modulated pump laser beam (Rosencwaig et al., 1985). Applied to semiconductors, the modulated reflectance signal has two sources: Optically excited thermal waves and plasma waves produced by photocarriers (Opsal and Rosencwaig, 1985). The plasma waves are of practical significance, as they probe the electronic properties. Some applications to plasma etched materials will be discussed in Section IV.A.

*Photothermal displacement* spectroscopy relies on the surface distortion produced by optical heating with a pump laser beam. The small local displacement of the illuminated surface area can be detected either by interferometric methods or by the deflection of a laser probe beam (Amer, 1983). The displacement technique offers the ability to differentiate between surface and bulk absorption and is particularly suitable for investigations requiring vacuum conditions (Olmstead et al., 1983).

## 6.   Applications of Photothermal Methods in Solid-State Research

With a few exceptions (Murphy and Aamodt, 1977; Pichon et al., 1979; Bechthold et al., 1981a; Bechthold and Campagna, 1981b; Pelzl et al., 1982), the application of the photothermal methods has been mainly restricted to the range around ambient temperature. Most of the photothermal studies are devoted to the determination of optical or thermal properties which intervene in the generation process and propagation of heat waves. Primary quantities are the absorptivity, the efficiency for nonradiative de-excitation, and the thermal properties such as the thermal diffusivity and effusivity (Section II.C). In addition—and this is the most unique feature of the thermal wave methods—the photothermal signal depends on the modulation frequency of heating (Section II.B). This effect can be used to control the variation of the mentioned primary quantities as a function of depth.

Proceeding from this point of view, three main areas of applications of the photothermal (PT) methods are apparent: photothermal spectroscopy, photothermal calorimetry, and depth profiling and subsurface imaging of optical and thermal properties. A complete discussion of all noteworthy experiments would deserve an entire article. Here, a few of the typical results are mentioned in order to demonstrate the potentialities and particularities of the photothermal method in view of their application to plasma-treated solid materials.

**a. Photothermal Spectroscopy.** Direct information on the optical properties can be obtained if the thermal wave response is measured as a function of the wavelength of the exciting radiation. With regard to other conventional spectroscopic methods, PT detection offers a substantial advantage when the sample absorbs either very strongly (Parke et al., 1982) or very weakly. In the latter case, pulsed excitations combined with piezodetectors are more favorable (Patel and Tam, 1981). In the last years photoacoustic and mirage effect detection have become very useful techniques for Fourier transform infrared spectroscopy of surface species and powders (Vidrine, 1980; Low et al., 1982; Varlashkin et al., 1986). The thermal wave signal is proportional only to that part of the absorbed radiation which is dissipated as heat. In systems with different de-excitation paths such as fluorescence decay, photochemical reactions, and phonon induced de-excitation, the PT signal thus can be used for discrimination measurements. In this context the photoacoustic effect has been utilized to study surface plasmons (Inagaki et al., 1981) and to determine fluorescence efficiencies (Adams et al., 1980) and photovoltaic yields (Cahen and Halle, 1985).

**b. Photothermal Calorimetry.** At an intermediate stage the PT signal is a heat wave, subjected to heat sources and sinks, to damping and retardation. Thus, the PT signal provides information about thermal features of the sample. The first PT calorimetric studies had been devoted to the investigation of phase transitions (Florian et al., 1978; Korpiun and Tilgner, 1983). Other applications comprise measurements of the local thermal diffusivity of polymer films (Merte et al., 1983) and of metallic foils (Kordecki et al., 1986) and the inspection of thermal coupling of particulates deposited on substrates (Utterback et al., 1985).

A new method based on the transverse mirage effect has been used to measure local thermal diffusivities of solid materials (Kuo et al., 1986a, 1986b). A still developing field of thermal wave assisted measurements comprises studies of adsorption and desorption processes on surfaces (Träger et al., 1982; Korpiun et al., 1986, Hussla et al., 1986).

**c. Depth Profiling and Imaging.** Photothermal methods provide the unique ability of depth sensitive exploration of physical properties as a consequence of the strong attenuation and retardation of the heat waves (Section II.B). Figure 9 reviews the principle features of depth profiling studies with thermal waves. The characteristic attenuation length of a thermal wave Eq. (24), the thermal diffusion length, decreases with increasing modulation frequency of heating. Reflected heat waves (Section II.D) and waves originating from absorbing centers from below the surfaces by a distance larger than the diffusion length, cannot contribute significantly to the PT signal probes at the surface. Thus, by varying the diffusion length via the modulation frequency, absorbing centers or thermal inhomogeneities at different depths can be thermally connected or disconnected from the surface where the gas-acoustic signal is generated, beam deflection occurs, or from which IR radiation is emitted. This depth sensitivity is used for two

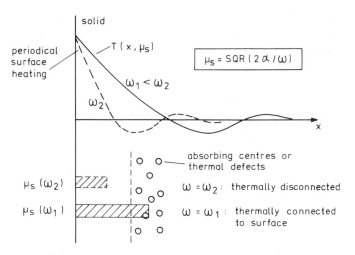

FIGURE 9. Thermal wave depth profiling illustrated on a semi-infinite solid which is subjected to periodical surface heating at the modulation frequencies $\omega_1$ and $\omega_2$, respectively.

principal purposes:

a) Subsurface thermal and mechanical defects are visualized by heat waves generated in the surface layer of the solid material, e.g., by a laser or electron beam focused on a metal surface. By scanning the beam across the sample, subsurface images can be obtained for the various depths appropriate to the selected modulation frequencies (Wong et al., 1978; Busse, 1979; Rosencwaig, 1982; Balk, 1986).

b) Depth-dependent distribution of the active absorbing centers can be explored by monitoring the amplitude and phase lag of the heat waves reaching the surface of the sample. "Optical" depth profiling has been applied to study the color pigment layers in color reversal films (Helander et al., 1981; Görtz and Perkampus, 1982; Uejima et al., 1985) and to investigate magnetization as a function of distance from the surface in layered magnetic tapes (Netzelmann and Pelzl, 1984; Netzelmann et al., 1986).

C. THERMAL PULSE METHODS
   (TIME DOMAIN EXPERIMENTS)

Thermal waves are the response to a cyclic heating process, and their propagation is governed by the diffusion equation. The amplitude and phase angle of the heat waves depend on the modulation frequency. In order to derive a complete solution of the thermal problem, the thermal response has to be investigated as a function of the modulation frequency.

Thermal pulses, on the other hand, are the response to a flash-like heating process. The time domain procedure involves recording the magnitude and time delay of the heat arriving at the detector related to the radiation absorbed in the sample. Formally, results of time domain and frequency domain measurements are convertible and related to each other by a Fourier transformation (Balageas et al., 1986; Pelzl, 1984). In practice, however, data recorded either as a function of time or of frequency do not provide the same information or do not have the same accuracy with regard to the same quantity. This is due to limitations imposed by the excitation and detection systems. Considering general attributes, such as noise sources and dynamical ranges accessible to the particular experimental setups, the following crude appointment is possible: Pulse experiments tend to be more favorable when processes on time scales of microseconds and smaller are studied, whereas frequency domain measurements are more suitable for the investigation of longer time behavior.

Pulsed excitation offers a better yield, because the deposited heat contributes completely to the thermal signal, whereas the modulation depth of the thermal wave decreases with increasing modulation frequency. On the other hand, frequency domain experiments provide a better duty cycle, particularly at low frequency where they overbalance the yield difference. Noise considerations lead to a similar conclusion. Pulse measurements require broad band amplifiers and the $1/\omega$ noise becomes most disturbing when long time scales are emphasized. Conducting a measurement at a constant frequency, the white noise can be suppressed considerably by use of the narrow band lock-in technique. Among the conditions that may favor a pulse experiment, one has to mention in the first place nonlinear excitation processes such as the thermoelastic generation of ultrasound (Section III.B.3) or heat production by two photon absorption. Systems subjected to a kinematic process may require very rapid recording of the complete thermal response, as it is provided by a time domain experiment. Apart from these cases, thermal waves may provide the optimum approach for the majority of the detection techniques listed in Figure 6 and most of the applications considered in the scope of this survey.

## IV. Thermal Wave Analysis of Plasma Treated Materials

The application of thermal wave methods to plasma-treated solid surfaces is discussed in this section. Section IV.A refers to semiconductor materials, where the physical basis of thermal wave analysis can be more complex due to the diffusion of charge carriers coupled to the heat diffusion equation by the recombination process (Section II.F). Plasma-sprayed coatings, a classical field of application of thermal wave techniques in nondestructive testing and evaluation of materials are treated in Section IV.B. Both of these subsections deal with materials and components exposed to plasmas of relatively low temperatures, with the intention of influencing the physical properties of the solid in a controlled sense. This differs from the materials exposed to plasmas in fusion devices, analyzed in Section V, where plasma-surface interactions normally take place in an uncontrolled form.

## A. Plasma Exposure of Semiconductor Materials

Semiconductor material is normally plasma-exposed and treated for two mainly technological processes: 1) plasma assisted deposition of amorphous or crystalline semiconductor films on various substrates and 2) surface alteration of semiconductor devices by plasma etching. A characteristic feature of plasma chemical reactions are the high specific reactivities due to high local temperatures, whereas the bulk heat content is maintained at a

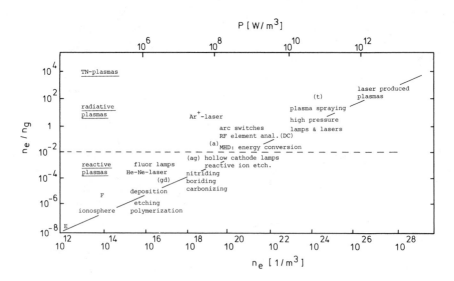

FIGURE 10. Characterization of process plasmas in a diagram where $n_e/n_g$ is the ratio of the electron to neutral gas density and $n_e$ the electron density. Abbreviations: gd: glow discharge; ag: anomalous glow; a: arc; t: thermal plasma (after Schram, 1987).

very low level. If the plasmas are classified in a $n_e/n_g$ versus $n_e$ diagram (Figure 10), where $n_e$ is the electron density and $n_g$ the neutral gas density, respectively, the nonmagnetically confined plasmas roughly group along the line $n_e/n_g = \text{const}(n_e)^{1/2}$ (Schram, 1987). For plasma deposition and plasma etching, normally rf-heated plasmas with pressures of about 10 Pa are used. In the scope of this contribution, we will not discuss the plasma conditions and the plasma-solid reaction processes studied elsewhere in this series (Fauchais et al., Donnelly, Chapman et al.). The subsequent discussion is limited to the characterization of the final product by photothermal methods.

## 1. Optical Properties of Plasma-Deposited Semiconductor Films

In recent years much effort has been devoted to improving the plasma-assisted preparation of semiconducting films. Amorphous silicon ($a$-Si) and microcrystalline silicon ($\mu c$-Si) have attracted most attention because of their protruding technical importance for microelectronic devices and solar cells. For these applications the device performance strongly depends on the electronic levels and defect states in the forbidden gap, which can be explored by optical spectroscopy.

Conventional optical studies of band gap states in thin film amorphous semiconductors include transmission and reflection experiments (Iqbal et al., 1983), luminescence studies (Street et al., 1981), and photoconductivity measurements (Wronski et al., 1982). Currently, these optical experiments are complemented by electron spin resonance (ESR) and light-induced electron spin resonnance (LESR) studies, which permit determination of the spin defect concentrations (Jousse et al., 1986). Lately, photothermal beam deflection and photoacoustic spectroscopy have received access as alternative sensitive tools for the determination of optical absorption coefficients in thin film semiconductors. Absorption coefficients down to 10 cm$^{-1}$ can be easily measured by the photothermal methods, which, in addition, are less sensitive to optical inhomogeneities of the samples and offer the possibility for in situ studies. So far, the photothermal techniques were mainly applied to the investigation of optical defects in doped and undoped hydrogenated $a$-Si ($a$-Si : H), and thus we limit the following discussion to this material.

Two principle procedures are published in the literature to prepare ($a$-Si : H): sputtering (SP) and plasma-induced chemical vapor deposition in a glow discharge (GD). Plasma-induced deposition is performed in plasma reactors, where the radio frequency power is coupled into the plasma capacitively (Wagner et al., 1983) or inductively (Beyer et al., 1981). Chemical transport in a direct current hydrogen glow discharge is also used (Veprek, 1985). Silane $SiH_4$. is most commonly utilized as starting material in glow discharges to produce $a$-Si : H. In the discharge, silane is rapidly decomposed, but instead of decreasing to the thermodynamic equilibrium value of $3 \times 10^{-8}$ mol%, it reaches a partial chemical equilibrium at 0.26 mol% (Veprek, 1984). Films of hydrogenated silicon doped with acceptor or donator atoms such as B or P can be produced by plasma decomposition of silane doped with, for example, $PH_3$ or $B_2H_6$.

The electronic properties of the deposited films sensitively depend on the preparation conditions. Therefore, research activities in this field attempt to control the variation of the electronic properties as a function of experimental parameters such as the substrate temperature, plasma conditions, or doping concentrations and to correlate these quantities with structural properties. In this context information on the band gap and on localized gap states which are directly provided by the measurement of the optical absorption coefficient are indispensable.

Since photothermal spectroscopy has been recognized as a sensitive tool for optical spectroscopy in the visible and infrared wavelength range, this technique frequently gets used in absorptivity measurements and luminescence quantum efficiency studies of $a$-Si : H films. Conventional optical methods require reliance upon experimental assumptions such as optical homogeneity, which can hardly be verified, whereas photoacoustic and photothermal deflection spectroscopy are much less sensitive to scattering and directly provide the optical absorption constant. To enhance the sensitivity, photothermal beam deflection measurements are most commonly conducted using a liquid in contact with the film, such as $CCl_4$ at ambient temperatures (Boulitrop et al., 1985) or liquid nitrogen in the vicinity of 77°K (Jackson and Nemanich, 1983). Recent results obtained with photothermal methods from plasma-produced silicon films, complemented by ESR and LESR studies, are now discussed.

The sub-band-gap states of hydrogenated amorphous silicon films ($a$-Si : H) deposited by glow discharge have been investigated at room temperature by Jackson and Amer (1982) using photothermal deflection spectroscopy. These authors measured the absorption coefficient in the range from 0.4 eV to 2.4 eV of undoped samples at various rf powers and of films doped with $PH_3$— and $B_2H_6$—. The main features displayed by both the undoped and doped films are a broad absorption shoulder in the forbidden below 1.5 eV and an exponential tail at higher energies (Figure 11).

In the undoped samples the optical defect density agrees with the number of dangling bonds determined by electron spin resonance over a range

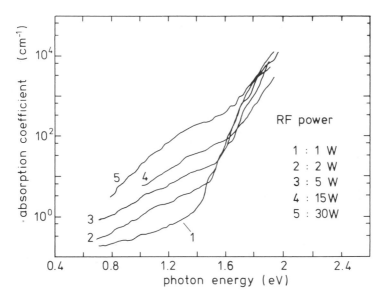

FIGURE 11.   Absorption coefficient as a function of photon energy of undoped $a$-Si : H at various rf powers measured at room temperature using photothermal deflection spectroscopy (after Jackson and Amer, 1982).

covering three orders of magnitude (Figure 12). Therefore, the authors attribute the source of the absorption shoulder in the band gap to optical transitions from singly occupied silicon dangling bond states. Doping with either $PH_3$ or $B_2H_6$ introduces defects at about the same energy level as the dangling bonds, but doping with both centers simultaneously removes the defects.

*Sputtered Films of $a$-Si : H*, which have been investigated by Bustarret et al. (1985) and Jousse et al. (1985), do not show the proportionality between optical and spin defect density. Undoped $a$-Si : H films of 1 $\mu$ thickness had been deposited by reactive sputtering on silicon substrates with the substrate temperatures varied between 300 K and 725 K. For samples prepared at elevated substrate temperatures, the number of dangling bond states deduced by EPR is an order of magnitude smaller than the number of gap states obtained from photothermal deflection spectroscopy. The origin of the spinless defects in sputtered $a$-Si : H is attributed to Si—H bonds in $SiH_n$ configurations associated with large structural defects.

*Doping studies* of hydrogenated amorphous silicon prepared by sputtering techniques have been performed by Jousse et al. (1986). Again, simulta-

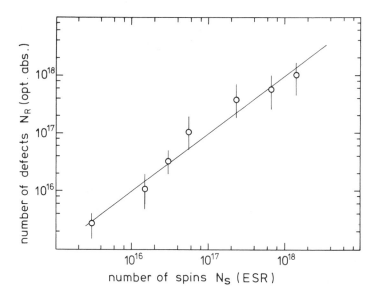

FIGURE 12.   Number of defects calculated from the excess optical absorption versus the number of dangling bonds as determined by electron spin resonance (ESR) (after Jackson and Amer, 1982).

neous PDS and ESR measurements have been applied to investigate changes of the optical gap related to the hydrogen content, the local order, and the B doping effects. With increasing boron content, a shoulder emerges in the sub-band-gap absorption spectra and progressively smears out the exponential Urbach tail. The low-energy shoulder is attributed to an optical transition from the singly occupied dangling-bond state $T_3^0$ to extended states of the conduction band. Analyzing the optical defect density $N_R$ as a function of the gas phase dopant concentration, the authors found two distinct regimes: a region of decreasing defect density at low doping levels and an increase of $N_R$ above a nominal doping of $6 \times 10^{18}$ cm$^{-3}$ (Figure 13).

The spin concentration $N_S$ from ESR shows a similar behavior in elevated concentration regimes, but $N_S$ remains constant at low doping levels. As in the undoped $a$-Si : H films, doped samples prepared by either glow discharge or sputtering behave remarkably different. An increase of the optical defect densities as a square root of the gas phase doping concentration has been found for doping of $a$-Si : H with $B$, as well as with P and As, by glow discharge (Street et al., 1981; Wronski et al., 1982; Street, 1985). The difference between the samples from both preparation

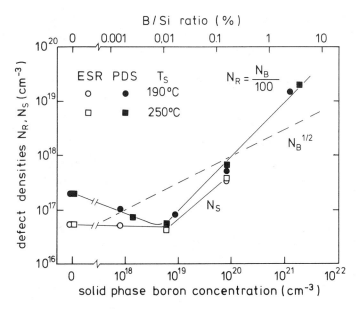

FIGURE 13. Defect densities from optical absorption $N_R$ and electron spin resonance (ESR) $N_S$ as a function of solid phase dopant concentration in sputtered $a$-Si : H. The dashed line represents $N_R$ values obtained from glow discharge deposited films (after Jousse et al., 1986).

techniques is explained by the authors by a different growth process which, in the case of sputtering, is much farther from equilibrium than in the glow discharge method.

Recent photothermal and ESR measurements on *chlorinated amorphous silicon films* prepared by glow discharge indicate that the correlation between spin and optical defects can already be modified by changing the composition of the reactive gas added to the plasma (Mostefaoui et al., 1985). Two series of $a$-Si : H,Cl films have been produced using the same glow discharge apparatus, one starting with the addition of small amounts of $SiCl_4$ in a $SiH_4 + H_2$ gas source and the other using a $SiCl_4 + H_2$ mixture with no silane. Samples from the first series show a good correlation between the spin concentration and the number of defects deduced from optical absorption measurements, in agreement with the results of Jackson and Amer (1982). However, in films originating from the second procedure, the number of optical defects was two orders of magnitude higher than that determined by ESR. This behavior compares with that observed by the other groups in sputtered films.

Comparative studies of the effect of the rf power and of gas composition have been undertaken by Hata et al. (1985), who investigated amorphous silicon nitride films (a-*SiN*: *H*) using photothermal deflection spectroscopy. Samples were prepared in an inductively coupled plasma reactor at 300°C by an rf glow discharge of a $SiH_4 - N_2 - H_2$ mixture at either constant rf power and different $N_2SiH_4$ ratios or at varying rf powers, with the gas mixture maintained constant. Increasing rf power and increasing nitrogen content have the same effect on the Urbach-like exponential tail, both shifting the absorption edge to larger wavelengths. At the lowest photon energy achieved in these experiments, 1.4 eV, no evidence has been found for an extra absorption in the sub-band-gap, as it was observed by Jackson and Amer in *a*-Si : H (Figure 11).

Dramatic effects of passivation on the deep gap absorption of amorphous silicon films prepared by glow discharge of silane have been observed by PDS measurements (Frye et al., 1987). Plasma, as well as chemical, passivation has been found to decrease the number of deep bulk states by a factor of 3 to 10, as deduced from the absorption coefficient. The authors conclude that a very high density of surface states in the unpassivated film could be one of the reasons for this strong reduction during passivation.

In those cases where optical absorption spectra of compact samples have to be investigated, the use of a *piezoelectric transducer* may provide a sensitivity as good as the photothermal deflection technique. Although the former is not contactless, it is applicable in a vacuum and most suitable for the high frequency modulation regime (Section III.B.3). Mikoshiba and Tsubouchi (1981) used ZnO films coated on the rear surface of the sample to investigate the influence of defects on the optical spectra of Si, *a*-Si, GaAs, and InP. Inspecting the thermoelastic signal in the optical wavelength region, a strong enhancement of the nonradiative decay was found if a thin *a*-Si layer was sputtered on Si crystal or if dark line defects had been generated in the semiconductor crystal surface by laser irradiation. Mikoshiba and Tsubouchi (1981) also made use of the modulation frequency dependence of the signal (Section III.B.6.c) to deduce information on the location of the defects.

Absorption spectra in the *far-infrared region* provide information on vibrational excitations which are very sensitive to bonding configurations. With photothermal methods, spectra of good spectral contrast with flat baselines and good spectral intensity can also be obtained on those samples which would require sophisticated preparation if conventional detection schemes are applied. Photoacoustic detection has been used to measure the infrared absorption spectrum of reactively sputtered *a*-Si : H in the range from 130 to 1000 cm$^{-1}$ in conjunction with inelastic neutron and Raman scattering experiments (Kamitakahara et al., 1984). The IR measurements were performed with a Fourier transform IR spectrometer, which was operated with a photoacoustic cell (Shanks et al., 1983). In comparison to the partial phonon density of states for the H-vibration delivered by the inelastic neutron scattering results on $a$-SiH$_{0.12}$, the IR spectrum from the same sample shows two additional prominent peaks at about 27 meV and 110 meV. The former is interpreted to be a local acoustic mode in which a Si—H atomic pair vibrates together against the Si atoms, whereas the latter feature is associated with a vibration of the SiH$_2$ configuration. Although these studies give evidence of the great potentiality of vibrational spectroscopy in characterization of local configurations in hydrogenated semiconductors, these tools, including photothermally detected IR absorption, are still scarcely used in this field.

## 2. Photothermal Evaluation of Plasma Etched Semiconductors

For the production of integrated circuits in silicon technology and micronscaled patterning of thin semiconductor films, plasma etch processes have found wide applications. Plasma etching of a solid material occurs when the solid and an active species produced in a low pressure glow discharge react to form a volatile compound. Particularly when good uniformity over a large substrate and high selectivity are required, plasma etching (PE) and reactive ion etching (RIE) offer essential advantages over wet chemical etching (Timmermanns, 1985). A common drawback of the plasma assisted techniques is the damage of the materials in the form of structural defects, deep level traps, and surface states.

Etching induced modifications of the substrate are currently analyzed using in situ XPS and Auger electron spectroscopy. Optical and UV emission spectroscopy are frequently applied to monitor the etching process in the plasma (Timmermanns, 1985). Nowadays, thermal wave methods are also becoming progressively recognized as an adequate diagnostic tool for the evaluation of microscale defects in semiconductor material devices (Rosencwaig, 1982). The unique feature of the so-called *thermal wave microscopy* is the depth sensitivity providing nondestructively a three-dimensional thermal or optical image of the subsurface region. The resolution in thermal wave microscopy is determined by the thermal wavelength or diffusion-length and the spot size of the incident radiation beam generating the thermal waves. In order to obtain an image resolution on micron or submicron scale, coherent laser light or particle beams, such as electrons in an electron microscope, are suitable as exciting radiation. To meet the resolution requirement for the thermal diffusion length, modulation frequencies in the MHz range are necessary. Thermal waves at these elevated frequencies are most suitably detected via ultrasonic or optical responses.

The first high resolution thermal wave image of a semiconductor device has been observed by Rosencwaig and Busse (1980) using an argon-ion laser beam, intensity modulated at 185 kHz and focused to about 5 $\mu$m on the surface of an electronic integrated circuit. The ultrasonic response was detected with a lead-zirconate-titanate transducer glued beneath the electronic circuit.

Thermal wave microscopy with an electron beam has been developed and is applied in conjunction with conventional scanning electron microscopes (Cargill, 1980; Brandis and Rosencwaig, 1980). As the thermal wave response is detected by ultrasonic probe, this technique is also called scanning electron acoustic microscopy (SEAM). Nowadays, SEAM has become a powerful tool for imaging of subsurface features in semiconductors and metals (Balk, 1986).

A further recently developed method relies on the modulated optical reflectance (Rosencwaig et al., 1985). The optical reflectivity is changed locally by a focused intensity modulated laser beam, and the alterations are detected via the reflection of a second probe laser beam. A most intriguing feature of this technique appears when it is applied to semiconductor materials. The modulated reflectance signal in a semiconductor is produced by the laser induced thermal waves as well as by the photoexcited plasma waves (Pelzl et al., 1988). These plasma waves are very sensitive to defects and lattice disorder, thus providing an ideal tool for monitoring etching-induced damage in semiconductors (Opsal and Rosencwaig, 1985).

In the scope of this section, some applications of the scanning electron acoustic microscopy and of the modulated optical reflectance on semiconductors with particular emphasis to plasma treated materials are demonstrated. These two methods are considered as most promising techniques for the morphological evaluation of microscale damage in semiconductor devices.

In the SEAM technique, the intensity modulation of the primary electron beam is used to generate thermal waves at the entry point of the beam. The elastic deformations associated with the strongly damped thermal waves give rise to ultrasound at the same modulation frequency or at even harmonics of it. The ultrasonic waves easily propagate to the bottom of the sample, where they are detected by a PZT. However, the signal observed in SEAM is not yet completely understood. It might be possible that ultrasound is also generated directly, e.g., by electrostrictive effects or piezoelectric coupling (Kultscher and Balk, 1986). Both the amplitude and phase of the ultrasound signal as a function of the position of the primary electron beam can be used for producing electron acoustic micrographs (Balk and

Kultschner, 1983). The intensity and the contrast of the micrograph is determined by the elastic and thermal properties and, particularly in semiconductors, the signal also strongly depends on the concentration and lifetime of the major free carriers. The spatial resolution achievable in semiconductors at high modulation frequencies (up to 50 MHz) is of the order of a few tenth of microns.

Applied to semiconductors, SEAM provides information on dislocations, dopant inhomogeneities, and process-induced damage such as that caused by plasma etching or by ion implantation. An example demonstrating the potentiality of SEAM for semiconductors has been reported by Rosencwaig and White (1981), who made a SEAM image of a wafer that had been

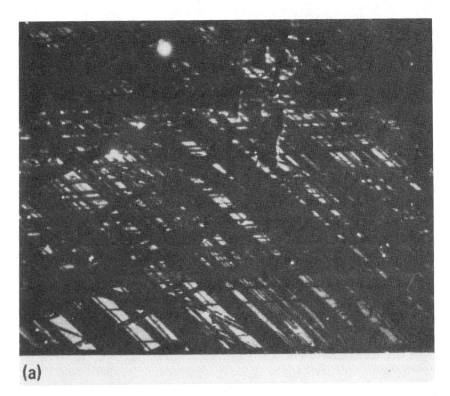

FIGURE 14. Image of an oxided and P-doped silicon wafer with the oxide layer removed subsequently in the region shown in the lower part of the micrographs. The geometrical patterns had been etched into the oxide layer before doping. First photograph (a): backscatter electron micrograph showing the geometrical pattern only in the oxided region at the top. Second photograph (b): SEAM image showing the pattern in the bare silicon region near the bottom of the two micrographs (after Rosencwaig and White, 1981).

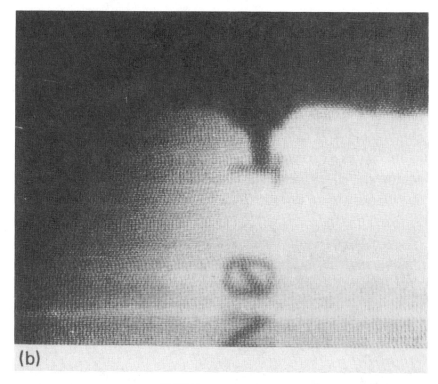

(b)

FIGURE 14. Continued.

subjected to ion-implantation and etching procedures. Figure 14 shows two electron micrographs, the first one obtained by conventional backscattering techniques and the second one by using acoustic detection, with the electron beam modulated at 640 kHz. The region of the wafer, which is shown at 100 × magnification, had been treated as follows: After coating with a $SiO_2$ layer, a geometrical resolution pattern was etched into the oxide layer and subsequently phosphorous atoms were diffused into this pattern. Thereafter, the remaining oxide layer was completely removed from one half of the wafer. The area around the boundary is shown in the micrographs with the bare Si at the bottom and the oxided region in the upper part of the two photographs. In the backscattering image, the geometrical pattern can be seen only in the oxided region, whereas with SEAM the pattern is also visible on the bar silicon. This is due to the fact that the generation of the acoustic signal sensitively depends on modifications of the local electronic properties, which in the present case are influenced by the diffusion of phosphor into the silicon matrix.

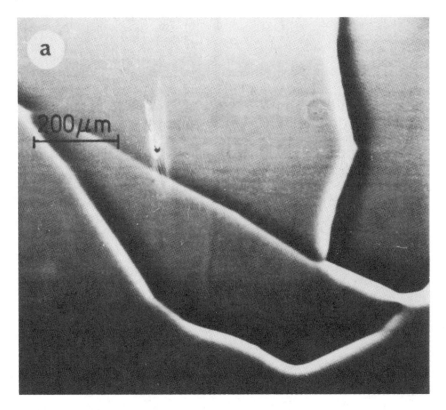

FIGURE 15. Time resolved SEAM image of a polycrystalline Si sample showing grain boundaries. Delay times with respect to the primary beam pulse (width = 20 ns) are 2.6 $\mu$s and 3.8 $\mu$s for the first (a) and second (b) micrograph, respectively (after Balk, 1986).

A particular advantage provided by the thermal wave methods is the ability for imaging of subsurface regions. Depth dependent micrographs are obtained by varying the thermal wavelength via the modulation frequency. At high frequencies the thermal wave image reproduces the regions very close to the surface, and with decreasing modulation frequency the penetration depth of the thermal waves increases. In the time domain, high and low frequencies correspond to short and long delay times between excitation and detection of the thermal waves, respectively. Figure 15 shows a time-resolved SEAM image of a polycrystalline Si sample. The lines are due to grain boundaries. The middle boundary seems to be located nearest to the surface; its response disappears first with increasing time delays.

An intriguing, still unexplained, feature of SEAM is the observation of strong nonlinear effects in most materials. For harmonic excitation using a

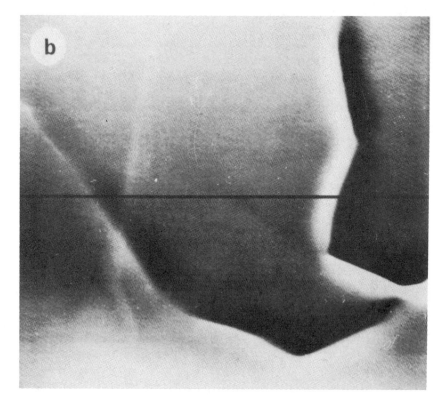

FIGURE 15.  Continued.

primary electron beam modulated at the frequency $f$, the second harmonic
signal at $2f$ can reach amplitudes of the order of 10% of the fundamental.
Applying nonlinear SEAM to the polycrystalline sample of Figure 15;
denuded zones of lower oxygen and carbon concentration around the grain
boundaries could be visualized (Balk, 1986).

Although, until the present, SEAM has not been used extensively for the
evaluation and characterization of plasma treated semiconductors, the given
examples demonstrate the potential of this technique. An experimental, but
not severe, drawback might be the necessity of preparing the sample to fit
into the electron microscope.

The modulated optical reflectance, the second method to be discussed
here, provides a noncontact technique suitable for in situ studies of semiconductor devices in the manufacturing process. The method is based on
the dependence of the complex optical refractive index upon the absorption

of optical radiation which in a semiconductor is associated to the production of heat and of photocarriers. At the focus of an intensity modulated laser beam, the primary excitations are heat waves and electron-hole plasma waves, giving rise to a periodically oscillating refractive index (Opsal and Rosencwaig, 1985; Skumanich et al., 1985). Experimentally the amplitude and the phase of the modulated optical reflectance is detected by a probe laser beam reflected at the entry point of the primary exciting beam (Rosencwaig et al., 1985). Besides the reflected beam, also the transmitted intensity or the scattered photons can be utilized to yield information on

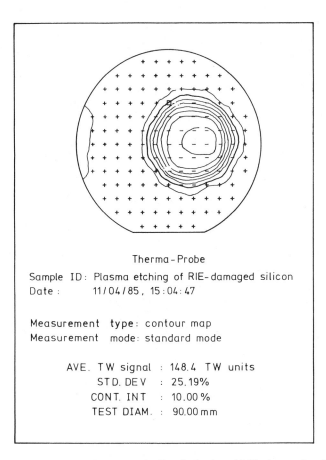

FIGURE 16.    Thermal wave (TW) map of a Si wafer having a highly damaged surface region from plasma etching. The bold line represents the average TW value, the values of the light contour lines differ from the average by intervals of 10% (provided by Therma-Wave, Inc., Freemont, CA).

the light induced changes of the optical properties (Rosencwaig et al., 1986).

The modulated optical reflectance has been proved to be a most suitable tool to detect the radiation damage in silicon wafers due to reactive ion (RIE) and plasma etch (PE) with good sensitivity (Geraghty and Smith, 1986; Smith and Geraghty, 1986). As the method permits mapping of the whole sample, etch uniformity or presence of polymer material on the patterned wafer can be measured. Figure 16 (Therma-Wave Inc., Application Note 200.02, March 1986) shows a thermal wave map of a Si wafer, illustrating the lateral distribution of plasma etch-induced damage. The thermal wave contour maps have been recorded with a Therma-Probe system (Therma-Wave Inc.) working with an Ar-ion pump laser modulated at 1 MHz and focused to a spot diameter of 1 $\mu$m. The mappings of the modulated reflectance signals are composed of 137 measured points. The + and − symbols represent sites showing a thermal wave amplitude (TW) greater than or less than the average signal represented by the square symbol. The dark and the lighter contour lines trace the average reading and those above and below the average, respectively. The wafer under

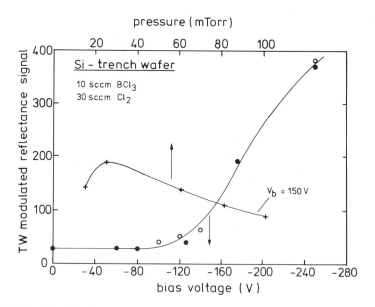

FIGURE 17.   Modulated reflectance signal of a trench wafer as a function of bias voltage (lower scale) and of the gas pressure (upper scale) in a cleanup etch process. The light and dark dot symbols represent data from two different measurements which were two weeks apart. (after Geraghty and Smith, 1986).

inspection had been damaged by a prior RIE-oxide etch step before it was subjected to a plasma etching process in a planar-electrode etcher in order to remove the RIE induced layer. The gross uniformity now appearing after the plasma etching on the wafer is caused by a hot plasma spot. Along a horizontal line, the thermal wave signal amplitude varies from about 170 TW units in the undamaged region to about 60 TW units in the center of the damaged pattern.

A parametric study of the RIE damage process using modulated reflectance has been reported by Geraghty and Smith (1986). The authors investigated the RIE damage of a trench wafer sample as a function of the DC bias and of the pressure in the silicon cleanup process performed with a commercial etcher (AME 8130). The results are shown in Figure 17. The cleanup etch step follows the main etch process with which most of the trench is etched. Its purpose is to remove the damage of the former step while etching is completed. Varying the DC self-bias voltage in the cleanup step between $-60$ V and $-250$ V, the RIE damage increases markedly above a certain threshold value of about $-120$ V. The measured signals at lower bias voltage are close to the values obtained in undamaged silicon. The effect of varying gas pressure in the cleanup process is shown in the same figure. With increasing pressure the damage to the crystal lattice becomes smaller. This observation is explained by an increasing chemical nature of the etch process as the mean free path of the reactive ions becomes shortened at higher pressures.

## B. THERMAL WAVE TESTING OF PLASMA SPRAYED COATINGS

Plasma spraying is used in technological applications to protect surfaces and to adapt the product surface to specific requirements. Plasma sprayed coatings of tungsten or chromium carbide are used to increase the abrasion resistance and sprayed coatings of aluminum are applied to protect against environmental corrosion (Almond and Reiter, 1980). Rough aluminum coatings are used for their antiskid properties and ceramic coatings (alumina) for their electrical or thermal insulation properties. Heat resistant thermal barrier coatings of zirconium dioxide sprayed onto an intermediate bond layer of, e.g., NiCrAlY, are applied in jet engines and gas turbines to protect metal surfaces from the effects of high heat fluxes whereby thermal expansion effects are reduced. Thus, the lifetime of components is extended, the cooling requirements may be lower, or the engine efficiency can increase with higher combustion temperatures (Liebert and Miller, 1984; Brandt, 1981). For application in nuclear fusion devices, plasma sprayed coatings of low atomic number $Z$, e.g., SiC/Al and TiC, are analyzed with respect to their thermal properties and behavior under H and He irradiation (Roth and Smith, 1986; Fournier et al., 1987). Coatings, with a ratio of refractory material to metal component gradually changing from the surface to the bulk might give the combination of materials properties required for the first wall, namely, low $Z$ surface for low plasma radiation losses, accommodation between the thermal expansion coefficients of the ceramic surface and metal bulk for thermal shock resistance, and good thermal contact to the metal heat sink.

## 1. Plasma Spray Technique and Structure and Thermal Properties of Sprayed Coatings

In plasma spraying the coating material is injected in powder form into a plasma arc flame, made molten there, and deposited by a high-velocity plasma gas stream onto the relatively cool substrate surface. Plasma sprayed coatings exhibit a lamellar-stratified structure owing to the superposition of impact-flattened droplets solidificated at high cooling rates during successive passages of the plasma jet over the substrate (Patel and Almond, 1985; Pawlowski et al., 1984). The properties of coatings depend on the spraying conditions, where a large variety of parameters, process data, and material properties can be identified: 1) temperature, flow rate, spray distance, composition (e.g., Ar or Ar-H mixtures), and electrical power level of the plasma jet; 2) powder flow rate and injection velocity; 3) thermal properties and size of powder particles; and 4) the properties, surface preparation, and cooling conditions of the substrate (Vardelle et al., 1985). Thermal treatment of the coating after spraying and use of the sprayed component in high-temperature environments can additionally affect the properties.

Owing to the stratified structure of sprayed coatings, a porosity of only 10% can be accompanied by a thermal conductivity reduced to values between 20% and 90% of the respective solid phase. Such an extreme variation exceeds by far the previsions of a model of approximately spherical pores (Brailsford and Major, 1964) but easily can be explained by a model of thin planar pores and limited regions of thermal contact between adjacent lamellae (McPherson, 1984). Based on a planar-pore model, the irreversible increase of the thermal conductivity with heat treatment, which is observed with nearly unchanged porosity (Vardelle et al., 1985), also can be understood: With a slight change of the planar pores to a more spherical form, the contact region can increase considerably, while the total pore volume remains constant.

Apart from eventually different thermal properties of the substrate and coating material, the porosity of the sprayed coating can increase the conductivity differences and thus thermal wave methods are especially sensitive and suited to analyze plasma sprayed coatings.

## 2. Inspection Techniques for Plasma Sprayed Coatings

The optimization of the coating structure to meet the required mechanical or thermophysical properties is the task of the plasma spraying process. This implies an appropriate substrate preparation and the control of the spraying parameters. For ceramic zirconia coatings, for example, it has been shown that the porosity depends on the particle temperature and impact velocity, which in turn are governed by the plasma process parameters (Pawlowski et al., 1984; Vardelle et al., 1985). For application under extreme conditions, an inspection of the sprayed product is necessary, additionally, to guarantee the coating thickness and the desired thermophysical or mechanical properties.

Most important in the inspection of coatings, however, is the detection of flaws, adhesion failures, excessive porosity, or lack of coherence within the coating. These types of defects are inherent to the spray process and can be caused by different thermal expansion behavior between the substrate and coating, by too large powder particles of the coating material, or by embrittlement due to residual stresses in thicker coating layers (Steffens and Crostack, 1980).

Such a reliable quality control should preferentially be nondestructive, noncontact, single-sided, and should be directly applicable to the coated product without the necessity of sample preparation.

When applied to plasma sprayed coatings, various techniques of nondestructive testing (NDT) have natural limitations, giving incomplete information:

a) Optical interferometry cannot be used for depth inspections, owing to the opaqueness and thickness (up to about 100 $\mu$m) of sprayed coatings.
b) Electrical and magnetic test methods (eddy current, magnetic flux, or potential probes) are not sensitive to insufficient coating-substrate bonding or delaminations parallel to the substrate surface (Steffens and Crostack, 1980).
c) The application of ultrasonic interferometry is limited due to the strong attenuation of ultrasound in sprayed coatings, typically 250 dB/cm at 5 MHz for sprayed nickel aluminide compared to about 0.5 dB/cm for low carbon steel (Cox et al., 1981; Patel and Almond, 1985).

d) Acoustic emission analysis under stress load is not completely nondestructive.

e) Holographic interferometry, which is sensitive to geometrical features
(Steffens and Crostack, 1981), may be used to detect delaminations when
surface deformations are first induced by a heating process. The signal
generation process, however, is rather indirect (heat accumulation and
subsequent thermoelastic deformation) and may also be affected by local
variations of the reflectivity and absorptivity for radiation heating.

Thermal methods of testing materials have been applied to plasma
sprayed coatings for more than 20 years (Green, 1966), and photoacoustics
and photothermal wave methods are increasingly discussed and used in
connection with plasma sprayed coatings during recent years (Luukkala,
1980). Thermal wave methods may be especially suited when we are directly
interested in the thermophysical performance of a coating, such as in the
case of thermal barrier coatings, where it may be of interest to know and
eventually adjust the *thermal reflection coefficient* (Section II.D) of the
coating or of the system coating/substrate to the specific heat transfer
conditions.

Among the various schemes of thermal testing of solids, e.g., the modulated heating-beam technique following Cowan (1961) or the laser-flash
method (Parker et al., 1961), the single-sided photoacoustic or photothermal
detection techniques offer a principal advantage when applied to plasma
sprayed coatings: The single-sided technique can be applied to thick
samples, e.g., a coating layer on a substrate without removal of the coating
from the substrate.

During use, plasma sprayed surfaces may suffer changes due to wear and
heat treatment (morphological changes, annealing, etc.), whereby the porosity and required insulating properties may be affected. A repeated control
of the thermal properties of components would be desirable under such
circumstances and one-sided noncontact thermal wave techniques which are
applicable without special sample preparation on the original equipment
parts would be especially advantageous for such inspections.

A disadvantage of double-sided techniques (excitation of the thermal
wave or pulse on the front surface and detection of the thermal response on
the rear surface) on porous coatings also may be that the retarded and
strongly damped rear surface signal is relatively more affected by even
smaller heat sources in the interior of the coating, i.e., at the bottom of
open pores and cracks or due to partial optical transmittance, than the
front surface signal. Thus the rear surface signal shows a smaller phase shift
and subsequently contributes to a systematic error, resulting in higher
values for the thermal diffusivity (Brandt, 1981; Pawlowski, 1984).

### 3. *Application Examples of Photoacoustic and Photothermal Measurements to Plasma-Sprayed Coatings*

Subsequently, a few examples of successful application of thermal wave analysis to plasma sprayed coatings are presented. In a first approximation, the system plasma sprayed coating/substrate can be represented thermally by a two-layer semi-infinite solid (Section II.D), where the effusivity of the first layer, the sprayed layer, in general is smaller than the effusivity of the substrate below. In a thermal wave experiment, the heat sink effect of the substrate will be visible in a relatively low surface temperature. If, on the other hand, an adhesion defect is considered, the heat sink effect

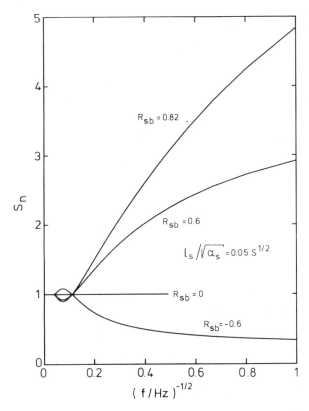

FIGURE 18. Signal amplitudes for thermally thin plasma sprayed coatings normalized against thermally thick plasma sprayed material. [Theoretical curves based on the two-layer model, delaminated ($R_{sb} > 0$) and bonded ($R_{sb} > 0$) coatings.]

of the substrate is reduced and the surface temperature is higher. Theoretical curves of the signal amplitude of a two-layer solid divided (normalization of signals, Section V.B) by the signal amplitude of a thermally thick homogeneous solid are plotted in Figure 18 as functions of the inverse square root of the modulation frequency. The two-layer solid may represent the system plasma sprayed coating/substrate, and the homogeneous reference solid may be a plasma sprayed thick sample. For low frequencies,

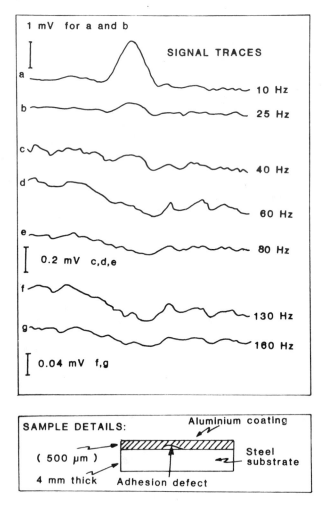

FIGURE 19.   Line scan of the photothermal amplitude across a delaminated plasma sprayed coating (after Patel and Almond, 1985).

where the thermal diffusion length is comparable to the coating thickness, low normalized surface temperatures are obtained owing to negative thermal reflection coefficients $R_{sb}$ (Eq. 39). High normalized surface temperatures are obtained for positive thermal reflection coefficients $R_{sb}$ corresponding to relatively higher effusivity values $\sqrt{(k\rho c)_s}$ in the coating layer than in the second layer below. This is the case for a delamination, which may be represented by a gas layer introduced between the substrate and the coating.

Such a delamination effect observed experimentally is shown in Figure 19. The photothermal line scans along the surface of a plasma sprayed aluminum coating on a steel substrate were monitored with the help of an infrared detector at different modulation frequencies $f$ (Patel and Almond,

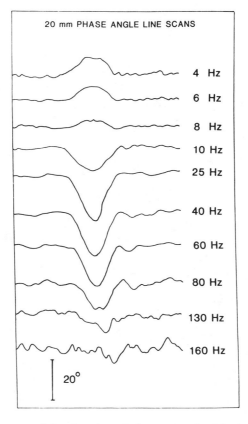

FIGURE 20. Line scan of the photothermal phase across the delaminated coating (after Patel and Almond, 1985).

1985). At low frequencies (25 Hz and 10 Hz), the delamination is manifest in the higher signal amplitude. At higher frequencies, however, above 60 Hz the thermal diffusion length is too short to give an image of the delamination. In Figure 20 are presented the corresponding phase measurements which at low frequencies undergo a change of sign. Such an inversion of sign at low frequencies is also shown by a theoretical curve (Figure 21) obtained for a delamination, $R_{sb} = 0.82 > 0$, in the two-layer model.

The influence of different thicknesses of coatings which are well bonded to the substrate ($R_{sb} < 0$) is shown in Figure 21 for a fixed frequency of $f = 16$ Hz. In a line scan of the phase angle along the surface of a step coated sample (plasma sprayed nickel chrome carbide on steel) this effect is verified experimentally (Figure 22). The change of the phase angle with the steps of coating thickness between 60 $\mu$m and 240 $\mu$m is clearly evident.

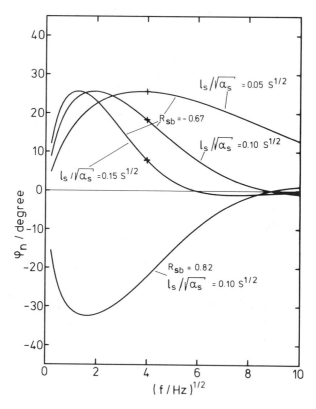

FIGURE 21. Phase shift for thermally thin bonded and delaminated plasma sprayed coatings. (Theoretical curves based on the two-layer model).

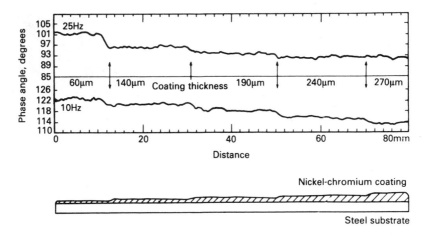

FIGURE 22. Changes of the phase angle with the steps of coating thickness (after Almond et al., 1985).

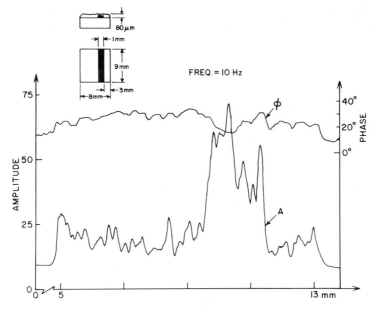

FIGURE 23. Photoacoustic line scan (signal amplitude and phase) along the surface of a plasma sprayed sample of limited width, showing an adhesion defect (at the position 10 mm) and the lateral boundaries of the sample at 5 mm and 13 mm (after Aithal et al., 1984).

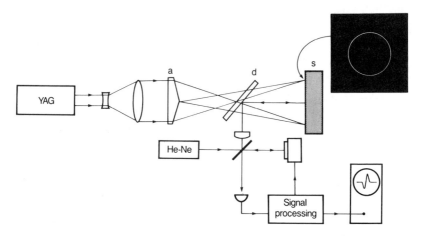

FIGURE 24.   Schematic of an experimental arrangement for noncontact thermoelastic testing. (Converging wave technique after Cielo, 1985).

Apart from inhomogeneities below the surface, lateral discontinuities can
also be localized thermally. Figure 23 shows a photoacoustic line scan taken
at a constant modulation frequency of 10 Hz along the surface of a plasma
sprayed nickel chrome alloy on a copper substrate. Between 10 mm and 11
mm an alumina defect between substrate and coating can be detected
(Aithal et al., 1984). Additionally, at the edges of the sample, 5 mm and 13
mm, peaks of the signal amplitude and rather abrupt changes of the phase
are observed as well. These lateral inhomogeneities show the limitation of
the one-dimensional theoretical description of thermal waves (Section II)
and suggest, on the other hand, that apart from volume porosity, surface
textures, or roughness may also affect the effective thermal properties
(Aamodt and Murphy, 1982; Bein and Pelzl, 1986; Bein, 1988).

As an example of a qualitative noncontact thermoelastic testing technique (Section III.B.3) of plasma sprayed surfaces, a converging wave
technique (Cielo, 1985) is presented here. The schematic of the experimental arrangement is shown in Figure 24. In principle, the YAG laser is
focused to an annular shape by a combination of a converging telescope
and a lens of conical cross section. An ultrasonic surface wave is generated
by the pulsed (10 ns) annular surface heat source, which converges into the
center of the heated annulus. There, the resulting elastic surface displacement is monitored interferometrically with a He-Ne laser. An unbonded
coating layer can be detected by abnormally high deformations of longer
duration, Figure 25a, in comparison to a well bonded area in Figure 25b.

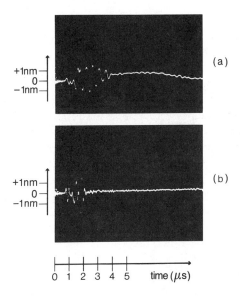

FIGURE 25.   Response of a 250 μm thick plasma-sprayed coating (aluminum on steel) to the converging wave technique: (a) Delaminated area and (b) well-bonded area (after Cielo, 1985).

The signal obtained for the well bonded area is much more irregular than signals obtained with the same excitation method from homogeneous samples. The difference may be due to the inhomogeneity of the plasma-sprayed coating (250 μm aluminum coating on steel).

## V.   Thermal Wave Analysis of Plasma-Surface Interactions in Controlled Fusion

### A.   The Role of the Thermophysical Properties of Materials in Plasma-Surface Interactions

Mainly, three areas of plasma-surface interactions in magnetic-confinement based nuclear fusion have been studied in the past: 1) erosion of the solid materials exposed to the plasma, 2) contamination of the plasma by impurities released at the walls, and 3) implantation, retention, and release of hydrogen in and from the wall materials. Another aspect of plasma surface interactions, namely, the energy deposition by the plasma on the surrounding material surfaces, has not been studied so extensively until now. This may be due to various reasons:

a) First of all, in the confinement experiments in the past, the discharge durations were short and the heat loads onto the limiters and other first-wall components were small. With further progress in fusion research, however, and in fusion reactors, thermal effects at the limiters, armors, and divertor target plates will become more important and the interdependence between power deposition, the wall erosion mechanisms, and plasma contamination will become more visible.

b) The second reason may be that not all energy loss channels from the plasma (electromagnetic radiation and charge exchange neutrals, conduction/convection losses) are easy to measure and even after the measurement of the power deposition on the divertor target plates or limiters by time-resolved infrared thermography, the tokamak energy balance could not be closed completely (Mueller, et al., 1983). Thus a general insecurity about the energy transfer from the plasma to the material surfaces still persists. A systematic effort to look for all energy losses after hard disruptions even has not yet been undertaken.

c) The third reason is that quantitative investigations and evaluations with respect to the thermal effects are difficult, owing to the rather uncertain thermophysical parameters of the solid surfaces and materials in the large temperature interval involved.

## 1. Power Deposition on Material Surfaces in Fusion Devices—Relevant Thermophysical Parameters

Thermal effects are concentrated on armor plates, limiters, and divertor plates, which protect the liner and the vacuum vessel from the direct contact with the hot plasma. Here the concentrated conduction/convection power losses of the plasma or the beam shine-through have to be absorbed, in addition to the radiational or charge-exchange losses. Average limiter heat loads up to 10 MW/m$^2$, with pulse durations up to a few seconds, are possible and, after disruptions of the confinement, the limiters have to withstand heat pulses of 100–1000 MW/m$^2$, with pulse lengths of several milliseconds. The temperature rise of a limiter due to the absorption of a heat pulse of arbitrary time dependence can be treated in the approximation of a semi-infinite solid (Section II.C). If a homogeneous solid with constant initial temperature distribution inside the volume and with spatially uniform surface heating, independent of the surface coordinates, is considered, the surface temperature rise is given according to Eq. (25) by

$$\Delta T_{\rm s}(t) = \frac{1}{\sqrt{k\rho c \pi}} \int_0^t dt' \frac{{\rm F}_{\rm s}(t - t')}{\sqrt{t'}}. \tag{53}$$

Here, ${\rm F}_{\rm s}(t)$ is the absorbed net power deposition. For the purpose of the present discussion, which aims at the thermal properties of first-wall materials, such effects as vapor shielding of the incident power (Hassanein et al., 1982) may be negligible. Eq. (53) is valid as long as the thermal diffusion time $\tau_{\rm s} = l_{\rm s}^2/\alpha_{\rm s}$ of, e.g., a limiter plate, is small when compared to the pulse duration. Here $l_{\rm s}$ is the thickness of the plate and $\alpha_{\rm s}$ its thermal diffusivity. For a limited plate thickness, actively cooled limiters or temperature-dependent thermal properties (Mast and Vernickel, 1982), the solution for the surface temperature takes a more complicated form than Eq. (53), but, in principle, in first approximation, the combined quantity $\sqrt{k\rho c}$ is the decisive thermophysical parameter for the temperature of solid surfaces in contact with the plasma. The importance of this quantity for plasma-surface interactions was recognized very early (Behrisch, 1972; McCracken and Stott, 1979). Research about the effective thermal properties of first-wall materials and their changes under plasma surface interactions in fusion devices only started recently (Bein et al., 1984).

The difficulty in applying Eq. (53) to calculate the surface temperature of wall components arises from the fact that the corresponding thermal properties are rather uncertain:

First, these properties here are needed for technical materials, alloys, e.g., produced by a certain fabrication process, and cannot be considered as physical parameters of chemically pure materials. Due to machining or mechanical or thermal treatment, the surface value $\sqrt{k\rho c}$ may be different from the quantity, which can be calculated from the bulk parameters $k$, $\rho$, and $c$ given in literature or supplied by the manufacturer.

Second, the thermal properties can vary over the wide temperature range that the wall components must withstand in fusion devices. The measurement of the thermophysical properties over this temperature interval is difficult.

Third, the surface value $\sqrt{k\rho c}$ may change locally and with time, due to plasma exposure (cyclic heat treatment, erosion, and redeposition of material).

## 2. Thermal Effects and Failures of the Material Surfaces in Contact with the Plasma

Failures of the wall components that can directly be related to repeated high pulses and subsequent high surface temperatures are erosion due to melting and/or sublimation and fracture, deformation, cracking, and fatigue induced by thermal stresses. The melt limit can be easily derived from Eq. (53) for a pulse of constant power deposition $F_0$ and duration $\Delta t$,

$$T_s(\Delta t) = T_s(t = 0) + 2F_0\sqrt{\Delta t}/\sqrt{k\rho c\pi} \leq T_m, \qquad (54)$$

where $T_m$ is the temperature of melting.

A general relation between the thermal stresses and the time-dependent temperature distribution in a solid, independent of special design geometries, can be derived starting from the basic laws of thermoelasticity (Section II.E). By inserting the integral Eq. (44), obtained in the quasistationary limit which perfectly applies to the time scales of heat pulses in fusion devices, into the contracted form of Hooke's law,

$$\sigma_{jj} = (3\lambda_L + 2\mu_L)\left[\epsilon_{jj} - 3\beta_e T(\mathbf{x}, t)\right], \qquad (55)$$

an order of magnitude equation is obtained that relates the averaged normal stresses to the temperature distribution,

$$\bar{\sigma} = \frac{2E}{3(1 - \nu)}\beta_e T(x, t). \qquad (56)$$

Here, Lamé's constants have been substituted by Young's modulus $E$ and Poisson's number $\nu$. Combining Eq. (54) with Eq. (56), it can immediately be seen that the thermal stresses due to surface heat pulses also directly depend on the combined quantity $\sqrt{k\rho c}$, the effusivity,

$$\bar{\sigma} = \frac{4E}{3(1 - \nu)}\beta_e\frac{F_0}{\sqrt{\pi}}\frac{\sqrt{\Delta t}}{\sqrt{k\rho c}}. \qquad (57)$$

For the selection of materials, electron beam heating tests have been done in the past to classify candidate limiter and first-wall materials with respect to increasing resistance against thermal shock and fatigue (Ulrickson, 1979). As criteria to define the shock resistance, the melt limit or excessive sublimation (graphite) and the fracture limit were used. In the thermal fatigue tests, cracking of the surface under repeated heat cycles is used as qualitative criteria. Such qualitative tests help to optimize the various components, and, in general, the classification sequence of materials found in the heating simulation experiments agrees with theoretical estimates, though the absolute values of the tests widely deviate from theoretical estimates (DeConinck and Snykers, 1978). This may be partially due to the fact that the effective values of the physical quantities involved are not known with sufficient accuracy and to the fact that it is not yet recognized what parameters are effectively involved. In the case of graphite, e.g., it was recently possible to correlate an improved erosion and cracking behavior qualitatively to smaller open porosities and smaller electrical resistivities (Delle et al., 1986). Here the characterization of the transport properties of graphite by its electrical resistivity replaced the thermal characterization.

Excessive surface temperatures also influence the plasma performance by evaporation/sublimation of solid material, contributing to an increased impurity content of the plasma. As the evaporation rate from a solid depends on the surface temperature $T_s(t)$, it also indirectly depends on the effective value of the effusivity at the surface, $\dot{N}(T_s) \sim (k\rho c)_s^{1/4} \exp[-C\sqrt{(k\rho c)_s}]$, and, with an effusivity decreased due to irradiation damage, the impurity production will change. Thus the integral evaporation yields related to the absorption of heat pulses might grow after a longer time of operation of a device and affect the plasma performance more than predicted by the estimates based on literature values of the thermal properties (Behrisch, 1980).

## 3. Temperature Dependence of Erosion Effects, Desorption, and Hydrogen Release

Apart from the direct failures due to the heat load on the material surfaces, other erosion mechanisms are also temperature dependent and thus are affected by the effective thermophysical parameters of the wall materials.

Strong temperature variations were observed for the erosion yields from graphite (Roth et al., 1982; Philipps et al., 1982; Bohdansky and Roth, 1985), and, under the simultaneous bombardment of both atomic hydrogen and energetic electrons, hydrogen, or argon ions, pronounced temperature-dependent maxima have been found (Auciello et al., 1983; Haasz et al., 1984; Vietzke et al., 1982) for the methane yield of graphite.

The impurity production in fusion devices due to unipolar arcs also may depend on the wall temperatures, as experiments with various metals showed an increase of the crater sizes with the sample temperatures (Zhao et al., 1984).

The desorption of impurities from solid surfaces is temperature dependent, and the diffusion and recombination coefficients of hydrogen in solids vary with temperature (Doyle and Brice, 1985). Consequently, the interdependent effects of hydrogen recombination, diffusion, and trapping in the solid walls are functions of temperature (Wienhold et al., 1982; Langley, 1984), whereby the recycling process, finally, also may be affected by the effective thermal properties of the wall materials.

## 4. Changes of the Thermophysical Properties Due to Plasma-Surface Interactions

Energetic particle bombardment, in general, affects the transport properties of solid-state matter. The thermal and electrical conductivity normally decrease. The related phenomena have widely been studied in connection with neutron irradiation and fission reactor applications. Irradiation damage in graphite, e.g., was reviewed by Kelly (1981). In limiters and first-wall fusion reactor materials, collision processes with charge-exchange neutrals, and charged particles also have to be considered. The reduction of transport properties is caused on different levels:

a) On the microscopic level, lattice damage due to atomic displacement and radiation-induced changes associated with the crystallite boundaries in polycrystalline materials have to be considered.

b) On larger length scales, swelling and bubble formation (due to helium, for example) below the surface, blistering, changes of the surface morphology, roughness, and macroporosity have to be considered. Additionally, in graphite, the void volume and the microcracks which resulted from cooling down from the graphitization temperature may be modified by pulsed heating during incorporation in fusion devices.

Changes of the thermal properties are also expected due to deposition of material by the plasma, where thermal contact resistances may be built up between the deposition layer and the solid surface. The thermal cycling in fusion experiments may act as annealing, heat treatment, also leading to changes of the thermal properties.

For graphite, the changes of thermal properties with irradiation damage have been studied for fission applications where, under steady-state heat flux conditions, the thermal conductivity $k$ is the relevant thermal parameter. A lack of information exists here for higher temperatures, which can be reached on limiters or armors in fusion applications. Under time-dependent surface heating, i.e. for thermal cycling and the absorption of heat pulses after hard disruptions, the relevant thermal parameter is the effusivity $\sqrt{k\rho c}$ . No systematic information about the changes of this quantity under radiation damage is available for nuclear graphite.

## 5.  Thermal Wave Techniques as a Diagnostic Tool for Solid Surfaces in Controlled Fusion

To study plasma surface interactions in controlled fusion, there are two opposite approaches. The first, most commonly followed, is to isolate and identify an effect for the study of its mechanism and consequences. Then, eventually, it is analyzed in correlation to other effects. An example of such a procedure in connection with the heat deposition on limiters and first walls are the simulation experiments, where the melt limit, sublimation behavior, and thermal stresses are analyzed. Here, the heat load is generally applied with the help of energetic electron beams; the erosion effects due to hydrogen or heavier particles which accompany the power deposition in a fusion device under realistic conditions are excluded, and thus the purely thermal effects can be studied separately.

In the opposite approach to plasma surface interactions, one tries to measure and quantify a physical parameter of a probe, e.g., of a limiter, exposed to the boundary plasma. Here, the different plasma-solid interaction mechanisms leave their traces. Eventually one can be successful in identifying dominant effects, but in general one is confronted with the complicate interdependence and superposition of effects. An example of such a procedure are the measurements of the electrical resistivity in carbon samples exposed to the plasma (Wampler, 1982).

If we are interested in measuring and quantifying the thermal properties on limiters or other first-wall components, these properties change locally and with time of incorporation in the fusion experiment due to radiation damage, thermal cycling, erosion, and redeposition of material. On the other hand, the changing thermal properties can also affect the maximum surface temperatures which may trigger the erosion effects. The wall erosion, in turn, affects the discharge behavior and the power deposition as an impurity source of the plasma (schematic in Figure 26). Thus in the synergistic interaction between irradiation and erosion damage, discharge performance and power deposition, the effective thermal properties of the material surfaces also play a certain role.

The latter approach to plasma surface interactions, the measurement of samples exposed to the plasma in the fusion device, will become increasingly more important and necessary with the progress in fusion research. Laboratory simulation tests can give first hints for future research activities, but ultimately the behavior of the material surfaces has to be controlled under plasma exposure in the fusion environment. For such tests and measurements, nondestructive techniques have to be developed and thermal wave testing may be such a technique among others such as optical inspection, acoustic transmission/reflection, stress wave generation, and propagation (Mattox and Davis, 1982). Thermal wave techniques can provide

a) experimental values of useful quantities, namely of the effective thermophysical parameters, which are needed for quantitative predictions for the fusion device in general or for a specific component, e.g., the system of active cooling of heat shields, and

b) these methods also potentially offer the means of noncontact measurements in the hostile inaccessible environment of a fusion experiment or reactor. Thus it might be possible to develop diagnostics and control

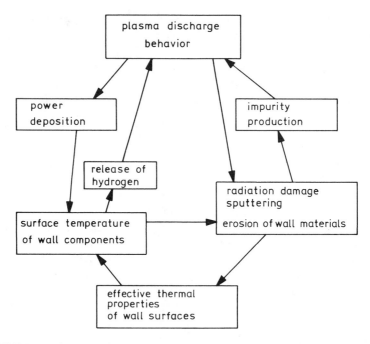

FIGURE 26. Schematic of interdependences between the thermal properties of first-wall components and plasma-surface interactions in controlled fusion.

devices based on thermal waves, if IR radiometry (Section III.B.4 and V.D) is applied as a detection technique.

In the photoacoustic measurements presented subsequently, limiters and divertor target plates exposed in tokamak devices were analyzed. These measurements were done after the samples had been withdrawn from the tokamak devices, and an acoustic technique was used to detect the thermal waves. In principle, infrared radiometry, as used in the thermal wave analysis of plasma sprayed coatings (Sections III.B.4 and IV.B), gives the same results, and infrared thermography to monitor surface temperatures (Section V.D) is available at the same time and tested in many fusion devices (Mueller et al., 1983). Other techniques for measurements in an hostile inaccessible environment, such as the detection of thermally induced stresses (Section III.B.3) also might be available.

Thermal wave techniques are sensitive to the physical quantities governing heat production and propagation (Section II), but as the thermal properties also depend on irradiation damage or grain boundaries, porosity, and surface roughness, the general mechanical state of limiters or first-wall components might also be monitored by such techniques.

### B. PHOTOACOUSTIC ANALYSIS OF GRAPHITE AND GRAPHITE LIMITER PLATES

Graphite or carbon limiters have successfully been used in recent tokamak experiments (TFR, JET, TFTR) and low-$Z$ limiter tokamaks seem to be one of the possible alternatives for future fusion reactors (Lackner et al., 1984). With respect to the fusion plasma, relatively large low-$Z$ impurity concentrations can be tolerated, since the radiation losses from the core plasma are low and the temperature-dependence of radiation losses additionally favors the thermal stability. Low-$Z$ materials, such as boron carbide, in general have low values of thermal conductivity or effusivity $\sqrt{k\rho c}$ and thus contribute to higher surface temperatures. Fortunately, graphite has, apart from the wide variation of thermal properties due to the individual fabrication process, relative high effusivity values, comparable to those of metal alloys (Table 1), and thus graphite probably will also be applied in the next research steps towards a commercial fusion reactor.

Frequency-dependent photoacoustics were applied until now to two graphite qualities (Bein et al., 1984 and 1985) which were used as limiter material in the tokamak experiments ASDEX and JET. In the toroidal limiter experiment in ASDEX (Vernickel et al., 1982a) the quality FP159I (Schunk & Ebe) was used, and in JET (Dietz et al., 1984) the quality 5890PT (Carbon Lorraine).

Measurements have been conducted with the experimental device and the one-side open-ended cell of Figures 7a and 7b, described in Section III.B.1. The amplitude and phase angle are recorded at distinct modulation frequencies of heating, mainly in the range between 4 Hz and 7.8 kHz. The corresponding thermal diffusion length of graphite, 50 $\mu$m $< \mu = \sqrt{\alpha/\pi f} <$ 1500 $\mu$m, should cover the range of macroporosity. In frequency-dependent photoacoustic measurements, the signals of the sample of unknown thermal properties are compared with the signals of a reference material of known parameters by a "normalization" procedure, consisting of a division of the measured complex signals (Bein et al., 1984). The frequency response function of the experimental device (including photoacoustic cell, microphone, and detection electronics) is eliminated by this procedure, and the properties of the sample are directly related to the parameters of the reference material such that the "normalized" signal amplitude and phase shift are only governed by the parameters of the sample and reference material, respectively.

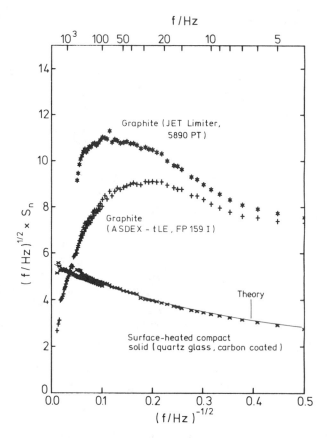

FIGURE 27. Normalized signal amplitudes for graphite of different mass density and a surface-heated compact sample.

## 1. Thermal Depth Profile of Graphite Before Plasma Exposure

In Figure 27 are presented the normalized amplitudes obtained from the two graphite qualities before exposure to the plasma and a compact homogeneous surface-heated test sample. The amplitudes are presented in the form $\sqrt{f}\,S_n = \sqrt{f}\,S_r/S_s$ versus $1/\sqrt{f}$. The index $s$ refers to the sample of unknown thermal properties, graphite for example, and the index $r$ to the reference material, e.g., neutral density glass (Schott NG10) used here for the normalization procedure. The surface-heated test sample is quartz glass, which, with a coating of 9 $\mu g/cm^2$ carbon, simulates the heating behavior of a homogeneous absolutely opaque solid with a surface heat source. The normalized amplitude of the surface-heated test sample in Figure 27 is in agreement with the theory of signal generation in compact solid samples (Bein et al., 1984 and 1985),

$$\sqrt{f}\, S_n\!\left(f^{-1/2}\right) = \frac{\eta_r}{\eta_s}\sqrt{(k\rho c)_s}\,\frac{\beta_r}{\sqrt{2\pi}\,\rho_r c_r}\left[1 - \sqrt{\frac{\alpha_r}{4\pi}}\,\beta_r f^{-1/2} + \frac{\alpha_r}{8\pi}\beta_r^2 f^{-1}\right],$$

(58)

whereas the amplitudes of the two graphite qualities significantly deviate from the behavior of the compact opaque model. In Eq. (58), $\sqrt{(k\rho c)_s}$ and $\eta_s$ are the effusivity and the photothermal conversion efficiency of the opaque solid, and $\alpha_r$, $\beta_r$, $\eta_r$, $\rho_r$, and $c_r$ are the thermal diffusivity, optical absorption constant, photothermal conversion efficiency, mass density, and specific heat capacity, respectively, of the reference material.

The normalized phase of the surface-heated test sample plotted in Figure 28 in the form $\sqrt{f}\,[\cot(\varphi_r - \varphi_s) - 1]$ versus $\sqrt{f}$ also confirms the theory of signal generation in the limit of compact opaque solids, $\sqrt{\alpha_s}\,\beta_s \to \infty$, whereas the phases of the two graphite qualities deviate from the behavior of compact solid samples in principle: They are beyond the limit of absolute opacity and cannot be approximated by the theoretical solution (Bein et al., 1986a)

$$\sqrt{f}\,[\cot(\varphi_r - \varphi_s) - 1] = \sqrt{\frac{\alpha_r}{\pi}}\,\beta_r \frac{\left[1 + \frac{2}{\beta_s}\sqrt{\frac{\pi f}{\alpha_s}}\left(1 + \frac{1}{\beta_r}\sqrt{\frac{\pi f}{\alpha_r}}\right)\right]}{\left[1 - \sqrt{\frac{\alpha_r}{\alpha_s}}\,\frac{\beta_r}{\beta_s}\right]} \quad (59)$$

for compact homogeneous more or less opaque materials by simply adjusting the optical or thermal parameters $\beta_s$ and $\alpha_s$. This is demonstrated by the broken curves in Figure 28 with their arbitrary finite values.

From the normalized amplitudes in Figure 27 follows that graphite, even before exposure to plasmas, shows a thermally effective two-layer or multi-layer structure: For a short thermal diffusion length, $\mu \sim f^{-1/2} \to 0$, the quantity $\sqrt{f}\,S_n$, which, according to Eq. (58), is proportional to the effusivity $\sqrt{(k\rho c)_s}$ at the surface, has a relatively low value. At intermediate frequencies, 25 Hz $< f <$ 200 Hz, corresponding to a deeper penetration of the thermal wave, the value $\sqrt{f}\,S_n \sim \sqrt{(k\rho c)}$ increases. Comparing the phases of the two graphites, that of the JET limiter is closer to the theoretical limit of the opaque compact solid than the other. The different mass densities of the two materials, 1.83 g/cm$^3$ for the JET and 1.7 g/cm$^3$ for the ASDEX limiter, respectively, suggest that the different volume

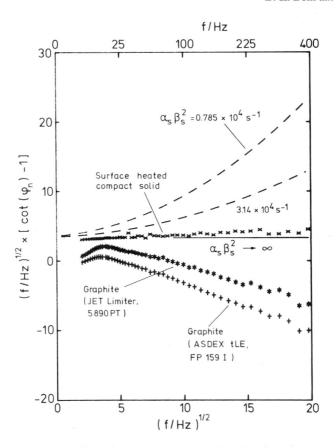

FIGURE 28.   Normalized phases for graphite and the surface-heated compact solid.

porosity and the related surface roughness might be responsible for the differences between the normalized phases and their location beyond the opacity limit $\sqrt{\alpha_s}\,\beta_s \to \infty$.

The normalized amplitudes obtained from graphite can be compared with theoretical solutions of the two-layer model (Bein et al., 1986b)

$$
\sqrt{f}\,S_n = \frac{\sqrt{(k\rho c)_s}}{\eta_s}\left[\frac{1 - R_{sb}e^{-2\sqrt{\pi f}\,\Lambda_s}\left[2\cos\left(2\sqrt{\pi f}\,\Lambda_s\right) - R_{sb}e^{-2\sqrt{\pi f}\,\Lambda_s}\right]}{1 + R_{sb}e^{-2\sqrt{\pi f}\,\Lambda_s}\left[2\cos\left(2\sqrt{\pi f}\,\Lambda_s\right) + R_{sb}e^{-2\sqrt{\pi f}\,\Lambda_s}\right]}\right]^{1/2}
$$

$$
\times \frac{\eta_r\beta_r}{\sqrt{2\pi}\,\rho_r c_r}\left[1 + \sqrt{\frac{\alpha_r}{\pi}}\,\beta_r f^{-1/2} + \frac{\alpha_r}{2\pi}\,\beta_r^2 f^{-1}\right]^{-1/2}. \qquad (60)
$$

Here, $\sqrt{(k\rho c)_s}$ and $\eta_s$ are the effusivity and the photothermal conversion efficiency of the surface layer. The square root of the thermal diffusion time $\Lambda_s = l_s / \sqrt{\alpha_s}$ is a measure for the thickness $l_s$ of the surface layer characterized by the thermal diffusivity $\alpha_s$. $R_{sb} = (g_{sb} - 1)/(g_{sb} + 1)$ is the thermal reflection coefficient of the two-layer model as defined in Section II.D and $g_{sb} = \sqrt{(k\rho c)_s} / \sqrt{(k\rho c)_b}$ is the ratio of the effusivity of the surface layer to that of the material below. In Figure 29 the measured amplitudes are approximated by curves with a fixed ratio $g_{sb} = 0.235$ and a variation of the parameter $l_s / \sqrt{\alpha_s}$. With a deeper penetration of the thermal wave, below $f < 10$ Hz, slight deviations between the measured values and the theoretical two-layer solution can be seen in Figure 29. The

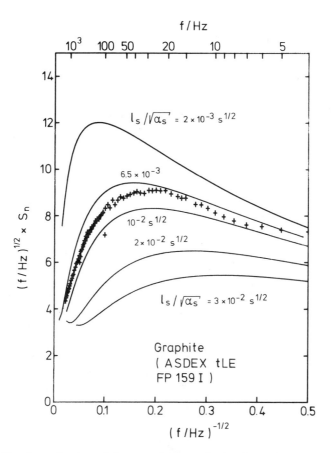

FIGURE 29. Approximation of the normalized amplitude of graphite by the two-layer model.

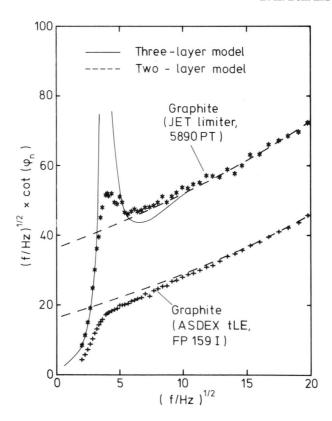

FIGURE 30. Quantitative interpretation of the normalized phase based on the two-layer and three-layer models. (Reference sample for normalization is a surface-heated compact solid).

phases normalized with a surface-heated compact reference material and presented in Figure 30 in the form $\sqrt{f}\cot(\varphi_r - \varphi_s)$ versus $\sqrt{f}$, are approximated by theoretical solutions based on the two-layer model (broken curve) and the three-layer model (continuous curve). At higher frequencies, $f > 100$ Hz, the first approximation based on the two-layer model

$$\sqrt{f}\cot(\varphi_r - \varphi_s) = \frac{0.5\sqrt{f}}{\sin(2\sqrt{\pi f}\,\Lambda_s)}\left[\frac{e^{2\sqrt{\pi f}\,\Lambda_s}}{(-R_{sb})} + R_{sb}e^{-2\sqrt{\pi f}\,\Lambda_s}\right] \quad (61)$$

confirms the approximation of the signal amplitudes (Figure 29). If lower frequencies are also considered, a transition to a third region, presumably the bulk material is reached with the deeper penetration of the thermal wave near $f = 25$ Hz. In the solution for the three-layer model, the position of the transition peak and the abrupt change of the slope of the curve on the left of the peak are governed by the ratio of the effusivity of the second layer to that of the bulk material, $g_{bp} = \sqrt{(k\rho c)_b} / \sqrt{(k\rho c)_p}$ and the thermal diffusion time $l_b^2/\alpha_b$ of the second layer. The values $g_{sb}$ and $g_{bp}$ found for the graphite used in the JET limiter (Bein et al., 1986b) indicate that the effusivity profile of graphite can in principle be represented by a three-layer model consisting of

a) a thin surface layer with a thermal diffusion time $l_s^2/\alpha_s$ in the range of several 10 $\mu$s and with an effusivity value $\sqrt{(k\rho c)_s}$ characteristic for the material just at the surface,
b) a second thicker layer with a diffusion time $l_b^2/\alpha_b$ in the range of some 10 milliseconds and an increasing effusivity, $\sqrt{(k\rho c)_b} > \sqrt{(k\rho c)_s}$, and finally
c) the bulk region where the effusivity goes down again, $\sqrt{(k\rho c)_p} \lesssim \sqrt{(k\rho c)_s} < \sqrt{(k\rho c)_b}$, below the value of the first and the second layer at the surface. Here, the index p refers to the bulk region.

For the graphite quality of the ASDEX limiter, the abrupt change of the slope at very low frequencies is observed (Figure 30), but the pronounced transition peak is not found. Due to a gradual transition in a larger roughness structure, it might be smeared out. This, perhaps, could be clarified by measurements at still lower frequencies. To avoid misinterpretations due to three-dimensional heat propagation in the solid, a larger heated spot area then would be necessary in the experiment.

## 2. Surface Roughness and Porosity of Graphite

The effusivity depth profile found for graphite before exposure to plasmas, which schematically is represented in Figure 31, can be related to the surface roughness and porosity (Bein et al., 1986a), or, in other words, to the influence of cracks, open pores, and volume porosity on the surface heating process. In the limit of high frequencies, when the penetration of the thermal wave is very small in comparison to the average depth $d_r$ of the open pores or surface cracks, the surface value $\sqrt{(k\rho c)_s}$ is measured that is characteristic for the material just at the surface. With decreasing frequencies $f$ and a deeper penetration of the thermal wave $\mu \sim f^{-1/2}$, the effusivity shows a linear increase (See Figure 27)

$$\sqrt{(k\rho c)_b} = \sqrt{(k\rho c)_s}(1 + S_R\mu) > \sqrt{(k\rho c)_s}. \tag{62}$$

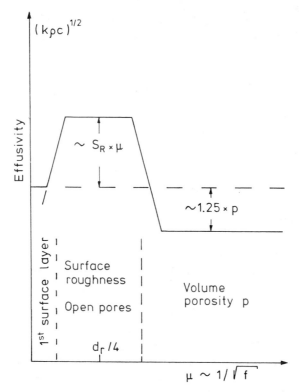

FIGURE 31.  Schematic of the effusivity depth profile of graphite.

This increase can be related to the additional heat capacity of the volumes along the walls of the open pores or cracks (Figure 32) which are shielded from the incident heat flux, but now can be reached by the increasing thermal diffusion length. The quantity $S_R$ has been introduced in Eq. (62) as an empirical constant. In a theoretical description of the influence of roughness and porosity on the effective thermal properties (Bein and Pelzl, 1986; Bein, 1988), $S_R$ can be related to the depth of an ensemble of $N$ open pores or cracks and to the ratio of the nonilluminated surface to the illuminated surface area

$$S_R \sim \frac{1}{A_\perp} \sum_i^N \frac{A_{\|i}}{d_i} \qquad (63)$$

and can be interpreted as "specific internal surface" of the roughness. Here, $A_{\|i}$ is the nonilluminated surface of an open pore $i$, $d_i$ its depth and $A_\perp$ the effective surface perpendicular to the incident heating radiation (Figure 32).

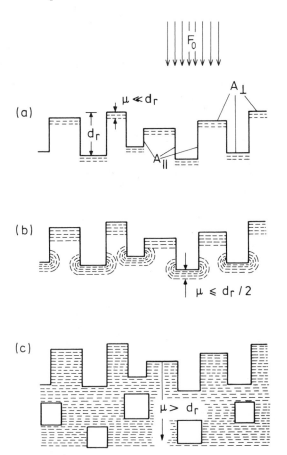

FIGURE 32.   Schematic of the rough and porous solid surface.

At very low frequencies $f$ the increasing thermal wave reaches the bulk of the material, where the heat transfer is affected by the volume porosity $p$: According to theoretical considerations (Brailsford and Major, 1964), empirical formulae, or numerical model calculations (Cunningham and Peddicord, 1981), the thermal conductivity in solids of approximately spherical porosity is reduced to

$$k(p) = k_0(1 - \gamma p), \tag{64}$$

where the empirical constant is limited by $1 < \gamma < 1.5$. The bulk density is given by the definition of the porosity,

$$\rho(p) = \rho_0(1 - p), \tag{65}$$

and the specific heat capacity is not affected (Kelly, 1981),

$$c(p) = c_0. \tag{66}$$

Here $k_0$, $\rho_0$, and $c_0$ are the properties of the respective material without voids. From Eqs. (64)–(66) an effusivity value $\sqrt{(k\rho c)_p}$ can be derived, which, depending on the empirical constant $\gamma$, is limited by

$$\sqrt{(k\rho c)_0}\,(1 - 1.25\,p) \leq \sqrt{(k\rho c)_p}\,(\gamma) \leq \sqrt{(k\rho c)_0}\,(1 - p). \tag{67}$$

Thus, the volume porosity contributes to a reduced effusivity, in contrast to the surface roughness.

Surface roughness and volume porosity also affect the depth profile of the thermal diffusivity. The linear increase of the effusivity with small thermal diffusion lengths, $\mu \approx \sqrt{\alpha/f} < d_r/2$, has here been related to the heat capacity of the material along the unheated shadowed walls of the open pores (Fig. 32b). This additional heat capacity of the open pores also should act on the thermal diffusivity. Starting from the value $\alpha_s = k_s/(\rho c)_s$ of the first surface layer, the thermal diffusivity of the second layer, then, should strongly decrease,

$$\alpha_b = \frac{k_b}{(\rho c)_b} = \frac{k_s}{(\rho c)_s (1 + S_R \mu)^2} = \frac{\alpha_s}{(1 + S_R \mu)^2} < \alpha_s, \tag{68}$$

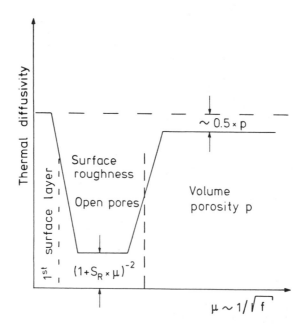

FIGURE 33. Schematic depth profile of the thermal diffusivity related to the open pores.

whereas the thermal diffusivity of the bulk material, calculated from Eqs. (64)–(66),

$$\alpha_{\rm p} = \frac{k_{\rm p}}{(\rho c)_{\rm p}} = \frac{k_0(1 - \gamma p)}{\rho_0(1 - p)c_0} \approx \alpha_0[1 - (\gamma - 1)p], \qquad (69)$$

could be very close to the thermal diffusivity of the respective compact material. Depending on the value of the empirical constant, $1 < \gamma < 1.5$, the bulk value is limited by

$$\alpha_0(1 - 0.5\,p) \le \alpha_{\rm p}(\gamma) \le \alpha_0. \qquad (70)$$

This depth profile for the thermal diffusivity (Figure 33), which initially

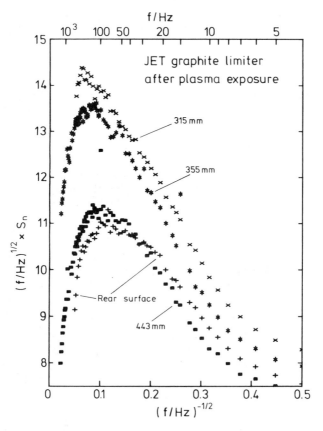

FIGURE 34. Normalized photoacoustic amplitudes from the JET limiter—changes due to plasma surface interactions.

decreases due to the open pores, should act as a thermal barrier at intermediate thermal diffusion lengths. The low value of the thermal diffusivity $\alpha_b$ of the second layer also should explain why rather low modulation frequencies $f < 4\alpha_b/d_r^2$ of heating are needed in the photoacoustic experiment with graphite in order to penetrate the surface roughness of characteristic depth $d_r$, before the bulk region that is characterized by the volume porosity $p$ is reached (Figure 32c).

## 3. Graphite Limiter Plates—Changes Due to Plasma-Surface Interactions

As shown in the previous section, graphite, even before exposure to tokamak plasmas, shows a thermal depth profile due to the surface roughness, cracks, and open pores and volume porosity. After exposure to plasma, this depth profile is changed depending on the position of the measured sample in the tokamak. The graphite limiters analyzed so far were exposed to relatively few discharges, even though the changes due to plasma-surface interaction are perfectly detectable by thermal waves. The tile from the central part of the JET limiter, incorporated in JET from July to December 1983, was exposed only to about 300 discharges, and the graphite plates from the ASDEX toroidal limiter experiment were exposed to approximately 600 discharges.

Figure 34 presents some normalized amplitudes, plotted in the form $\sqrt{f}\,S_n$ versus $f^{-1/2}$, from the JET limiter front surface, in comparison to results from the rear side of the same material without plasma exposure. The location of the sample points on the limiter tile is shown in Figure 35. The normalized amplitudes which for small thermal diffusion length are proportional to the effusivity at the surface, increased in general at those positions that, due to a more intensive plasma contact during incorporation, had regularly been heated to higher surface temperatures.

As Figure 36 shows for higher frequencies, the phases of the plasma-exposed front surface ($* * * *$) of the JET limiter are generally shifted from the value of the material without plasma exposure into the direction of the compact surface-heated solid ($\times \times \times \times$). This means, immediately at the surface, the material behaves after plasma exposure like a more compact solid, with increased density and thermal conductivity. With deeper penetration of the thermal wave, however, for very low frequencies, $f < 20$ Hz, the phase approximates the original value ($+ + + +$) before plasma exposure.

In Figure 37 are shown amplitudes obtained from different positions of the graphite limiter in ASDEX, compared to the amplitude from a plate

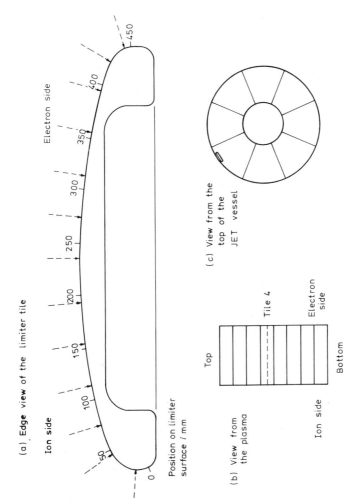

(a) **Edge** view of the limiter tile

Electron side

Ion side

450
400
350
300
250
200
150
100
50
0

Position on limiter
surface / mm

(b) View from
the plasma

Top

Tile 4

Electron
side

Ion side

Bottom

(c) View from the
top of the
JET vessel

FIGURE 35. Position of the sample points on the JET limiter:
(a) Edge view of the graphite limiter tile.
(b) View of the limiter from the plasma.
(c) View from the top of the JET vessel.

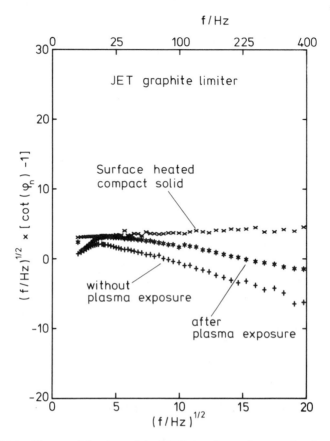

FIGURE 36.   Changes of the phase of the JET limiter due to plasma-surface interactions.

which had not been incorporated in ASDEX-tLE. Figure 38 shows the position of the graphite limiter relative to the position of the plasma. For small thermal diffusion lengths, at higher frequencies, the quantity $\sqrt{f}\,S_n$ $\sim \sqrt{(k\rho c)_s}$ decreased in general. The decrease is larger in regions closer to the main torus axis, farther away from the main plasma-limiter contact region between the distances $R = 160$ cm and 170 cm from the main torus axis.

For the quantitative characterization of the effect of plasma exposure on the effective thermal depth profile, both the amplitudes and phases (Figure 39) at the various measured position have been approximated by theoretical solutions based on the two- and three-layer model. First in the range of the higher modulation frequencies $f$, the thermal diffusion time $l_s^2/\alpha_s$ of the surface layer and the ratio $g_{sb} = \sqrt{(k\rho c)_s}/\sqrt{(k\rho c)_b}$ of the effusivity of

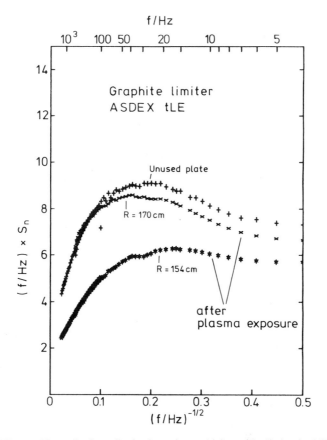

FIGURE 37.   Normalized amplitudes from the toroidal graphite limiter in ASDEX.

the surface layer to that of the second layer below are determined from the two-layer model. Then, in a further step, when also the amplitude and phase values at lower frequencies are considered, the ratio of the effusivities $g_{bp} = \sqrt{(k\rho c)_b} / \sqrt{(k\rho c)_p}$ and the thermal diffusion time of the second layer $l_b^2/\alpha_b$ are determined from the approximation based on the three-layer model. The observed changes due to plasma exposure are then expressed by changed values of the effusivity and thermal diffusion time of the first and second layer, whereas the effusivity $\sqrt{(k\rho c)_p}$ of the bulk region remains unchanged.

Such an attempt of a quantitative description of the changes induced by plasma-surface interactions seems to be rather formal and mathematical, since the various effects on the thermal depth profile, namely, the surface roughness, volume porosity, and the conductivity changes due to plasma

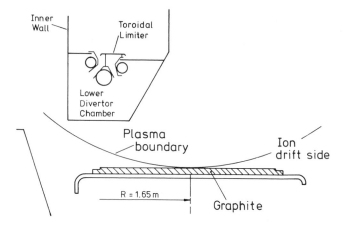

FIGURE 38.   Position of the toroidal limiter relative to the plasma in ASDEX.

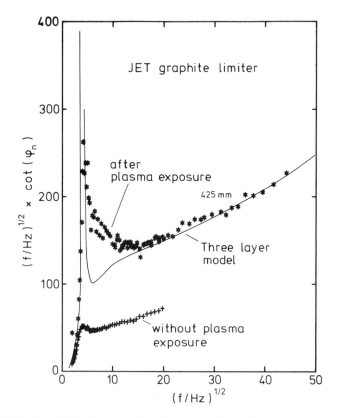

FIGURE 39.   Quantitative interpretation of the normalized phase based on the two-layer and three-layer models.

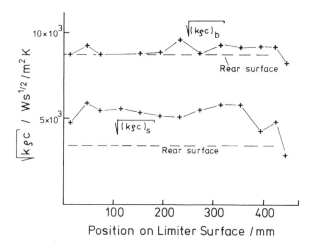

FIGURE 40. Changes of the effusivities of the surface and subsurface layer due to plasma exposure of the JET limiter. Here, index s refers to the surface layer and index b to the subsurface layer.

exposure are mixed up. For practical purposes, however, it is a useful procedure:

a) Based on the parameters $\sqrt{(k\rho c)}$ and $l^2/\alpha$, the thermal response of a limiter surface to a prescribed power deposition easily can be calculated for a two- or three-layer structure (Bein, 1986), and
b) the different effects (plasma exposure, porosity) on the thermal depth profile can be classified within such a limited scheme of a three-layer model.

In principle, the number of layers might be decreased or increased, depending on the length scale of the different effects or depending on the quality of approximation required between measurement and theoretical curve. As can be seen in Figure 39, the experimental phase can well be approximated by the three-layer model, at high and low frequencies, but the gradual transition from the influence of the roughness and plasma exposure effects to the volume properties between 25 and 400 Hz might deserve a more refined theoretical description.

The effusivity of the surface and subsurface layer, plotted in Figure 40 as functions of the position on the JET limiter, are compared to the respective values of the not plasma exposed material. Figure 41 shows the square roots of the diffusion time of the surface and subsurface layer, in comparison to the corresponding values before plasma exposure. A correlation with the rather uniformly distributed maximum surface temperature of the limiter

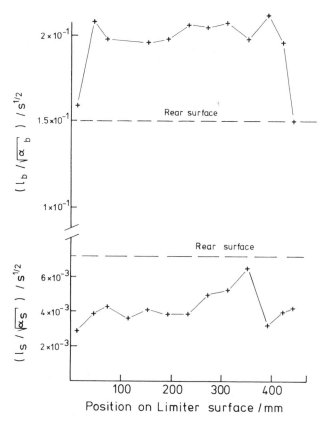

FIGURE 41. Changes of the square root values of the thermal diffusion time of surface (s) and subsurface layer (b) due to plasma exposure of the JET limiter.

(Pick and Summers, 1984) might be possible: In the regions heated during incorporation in JET normally to higher surface temperatures, the effusivity $\sqrt{(k\rho c)}_s$ of the surface layer increased and the square root of the thermal diffusion time $l_s/\sqrt{\alpha_s} = l_s\sqrt{(\rho c)_s/k_s}$ decreased due to the plasma contact. Both these changes may be related to an increased thermal conductivity $k_s$ and densification of the material at the surface due to baking and morphological changes of the micropores, leading to a smaller layer thickness $l_s$.

    The effusivity $\sqrt{(k\rho c)}_b$ of the second layer increased only slightly or remained nearly unchanged, whereas the square root of the thermal diffusion time $l_b\sqrt{(\rho c)_b/k_b}$ increased all over the regularly heated parts of the front surface. The significant increase of $l_b\sqrt{(\rho c)_b/k_b}$ in comparison to the unchanged effusivity is difficult to explain (Bein et al., 1987). Based on the roughness model (Section V.B.2), however, it can be understood that the formation of additional or deeper open macropores, surface cracks due to heating, affects the effusivity only through the square root of the increasing effective heat capacity, whereas the square root of the diffusion time is affected both through the increasing effective heat capacity and the average depth $l_b$ of the open pores. This interpretation is rather speculative here; for a confirmation, the combination of photoacoustics and statistical analysis of the surface morphology based on scanning electron micrographical analysis of heated graphite surfaces would be necessary (Bohdansky et al., 1987).

A direct correlation of the observed changes of the thermal structure with the deposition of metals by the plasma (Ehrenberg and Behrisch, 1984) cannot be seen. The metal content still may be too low.

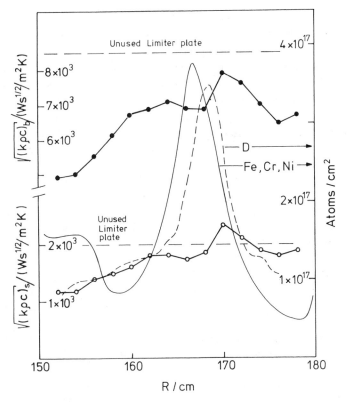

FIGURE 42. Plasma-induced changes of the effusivities of the surface (s) and subsurface layer (b) along the radial direction of the ASDEX limiter (deuterium and metal deposition after Vernickel et al., 1982b).

The effusivity of the surface and subsurface layer, measured on an ASDEX limiter plate, are given in Figure 42. Here, the effusivity of the plasma-exposed limiter is smaller, in general, than that of the unexposed graphite plate. This may be due to a reduced thermal conductivity caused by energetic particle bombardment (ions, charge-exchange neutrals). The decrease of the effusivity could be compensated only near the region of major heat and metal deposition, at a distance $R \approx 170$ cm from the main torus axis (Compare Figure 38). It should be noticed that the temperature of the graphite limiter in ASDEX remained rather low because of the large toroidal limiter area. Thus the baking effect on the thermal properties of this graphite limiter should be small as well.

## C. Thermal Wave Analysis of Metallic Divertor Neutralizer Plates

Here, a first preliminary report is given on photoacoustic measurements on a divertor neutralizer plate which was used in the tokamak device ASDEX, Max-Planck-Institut Garching (Keilhacker et al., 1985). The plate had been incorporated in the lower divertor chamber at the outer side (Figure 43) from the beginning of the experiment (1979) until summer 1985. Thus it accompanied the continuous progress of the experiment during divertor operation to higher plasma temperatures and higher particle and energy fluxes onto the neutralizer plates. During this time, about 20,000 discharges with an average duration of 2.5 seconds were run, initially with ohmic heating alone. About 9000 discharges were run with additional heating, mainly neutral beam injection up to 4 MW during 400 ms, and ultimately with additional wave heating (LH and ICRH). Most of the discharges were run in double-null divertor mode. The intersection between the separatrix and the divertor target plates changed during this time, depending on the plasma current, the current of the divertor field coils, and the varying poloidal $\beta$ due to additional heating (Rapp et al., 1986).

The neutralizer plates consist of a titanium alloy ($TiAl_6V_4$-CONTIMET ALV64). The plates are 2 mm thick and 18 cm wide and between 60 to 80 cm long. The outer neutralizer plates form a toroidal loop of approximately 10 m length. The maximum tolerable heat load for a temperature increase of 800°C was about 400 J/cm². The cooling of the plates between consecutive discharges was only by radiation losses and conduction to the mechanical support structure. The plate analyzed here had to withstand higher temperatures at one of its edges in toroidal direction. Due to thermal stresses the mechanical fixing to the support structure was broken, and the plate was deformed and melted there. The measurements reported here

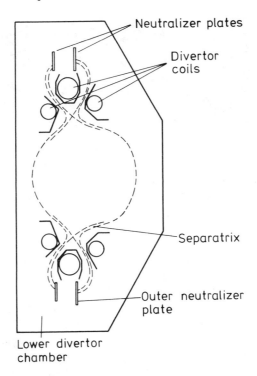

FIGURE 43. Cross-sectional view of the tokamak ASDEX—position of the analyzed divertor neutralizer plate.

were done near the opposite edge, which remained fixed to the support structure and where, thus, the normal conditions of particle and power deposition can be assumed. Optically, the plate shows a bright metallic horizontal trace, about 13 mm large, between the distance of 55 mm and 68 mm from its lower edge (Figure 44), where, probably, the last intensive plasma-solid contact region was before removal from the experiment. Other less bright horizontal traces still are visible in a major distance from the lower edge, former intersection regions between the separatrix and the neutralizer plate. These regions now are covered with a gray layer, probably due to deposition of material by the divertor plasma or due to titanium evaporation. Near the lower edge of the plate, there is a deposition layer of major thickness due to the deposition of carbon. This deposition layer is partially delaminated near the lower edge.

The aim of these measurements was to establish thermal wave analysis in the audioacoustic frequency range as a tool to detect and characterize the changes induced in metallic divertor plates quantitatively (Bein et al., 1989).

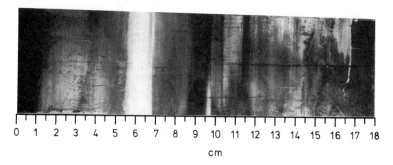

FIGURE 44. ASDEX divertor neutralizer plate—localization of the measurement points (distance from the lower edge of the plate) and visual impression.

For such a test, a neutralizer plate from the ASDEX divertor is the right task due to the rather uniformly distributed particle and power deposition to the large toroidal target plate area in the divertor chambers. The measurements are from a stripe perpendicular to the horizontal intersection between separatrix and neutralizer plate, from the bright main plasma-solid contact region and its neighborhood. The measured points are identified by their distance from the lower plate edge. Data from this region are interesting for two reasons: Here the effect of the plasma on the metal can be measured, free from additional effects of a deposition layer, and, secondly, thermophysical data about this region are interesting for the power deposition calculations based on IR thermography (Section V.D). To compare with material not exposed to the plasma, measurements from the rear surface of the plate and from polished $TiAl_6V_4$ samples are presented where material of 500 $\mu$m thickness has been removed from the surface.

## 1. Experimental Results from Metallic Divertor Plates

In Figure 45, normalized photoacoustic amplitudes are presented from the main intersection region separatrix/divertor neutralizer plate and from a polished $TiAl_6V_4$ sample. For these measurements the modulation frequency $f$ of heating was varied between 40 Hz and 7.8 kHz. If a thermal diffusivity of $\alpha = 2.9 \times 10^{-6}$ m$^2$/s is assumed for this titanium alloy (Mueller et al., 1984), a thermal depth profiling up to a thermal diffusion length of about $\mu = \sqrt{\alpha/f\pi} \approx 150$ $\mu$m below the surface should be possible. For several frequencies, 50 Hz, near 400 Hz, 1.6 kHz, and 5 kHz, the measured signals show small perturbations which arose from network influences. Reference material for the normalization procedure here was a surface-heated

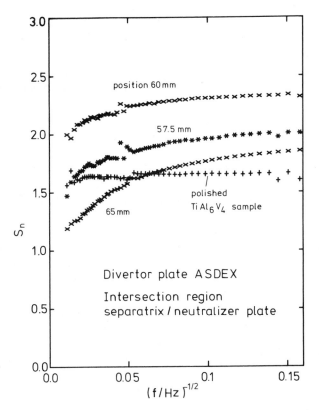

FIGURE 45. Normalized photoacoustic amplitudes from the intersection region between the separatrix and neutralizer plates.

carbon-coated quartz glass sample. The carbon coating of 25 $\mu$g/cm$^2$ is thin enough even for the highest used frequencies to be thermally negligible. The abscissa in Figure 45, $(f/\text{Hz})^{-1/2}$, is proportional to the thermal diffusion length, and thus it is a measure of the penetration of the thermal wave. Assuming a two-layer model for the plasma-exposed metallic surface and a surface-heated compact solid as reference, the normalized amplitude of Eq. (60) can be simplified to give,

$$S_n = \frac{\sqrt{(k\rho c)_s}}{\sqrt{(k\rho c)_r}} \frac{\eta_r}{\eta_s} \left[ \frac{1 - R_{sb}e^{-2\sqrt{\pi f}\,\Lambda_s}\left[2\cos\left(2\sqrt{\pi f}\,\Lambda_s\right) - R_{sb}e^{-2\sqrt{\pi f}\,\Lambda_s}\right]}{1 + R_{sb}e^{-2\sqrt{\pi f}\,\Lambda_s}\left[2\cos\left(2\sqrt{\pi f}\,\Lambda_s\right) + R_{sb}e^{-2\sqrt{\pi f}\,\Lambda_s}\right]} \right]^{1/2},$$

(71)

where the quantity $\sqrt{(k\rho c)_r}$ is the effusivity of the reference material. For high frequencies, in the limit $(f/\text{Hz})^{-1/2} \rightarrow 0$, the amplitude is proportional to the effusivity $\sqrt{(k\rho c)_s}$ just at the surface and inversely proportional to the photothermal conversion factor $\eta_s$. For a deeper penetration of the thermal wave in the low-frequency limit, the normalized amplitude depends on the effusivity value of the bulk material deeper below the surface.

Within the bright reflecting plasma-solid contact region, between 55 mm and 68 mm distance from the lower edge of the plate, the optical reflectivity $R_s$, and consequently the photothermal conversion efficiency, $\eta_s = 1 - R_s$, measured in an additional experiment (Wojczak, 1988) for the used laser light of wave length $\lambda = 488$ nm, does not change very much. Thus, part of the differences of the normalized amplitudes at the positions 57.5, 60, and 65 mm have to be related to differences in the thermal properties. Between 55 mm and 68 mm the relatively highest energy fluxes hit the plate, and the highest temperatures were reached there during incorporation in ASEX. In this region the effusivity $\sqrt{k\rho c}$ increased, in general, at the position 60 mm, about 20% above the original value (Wojczak et al., 1988).

The frequency-dependence of the amplitude found for small thermal diffusion lengths, $\mu \sim (f/\text{Hz})^{-1/2} \leq 0.05$, at the front surface of the target plate, cannot be seen for the polished TiAl$_6$V$_4$ sample or for the measurement from the rear surface. The amplitude of the polished sample $(+ + + +)$ agrees with the special solution

$$S_n = \frac{S_r}{S_s} = \frac{\sqrt{(k\rho c)_s}}{\sqrt{(k\rho c)_r}} \frac{\eta_r}{\eta_s}, \tag{72}$$

following from Eq. (71) under the assumption $g_{sb} = 1$ and $R_{sb} = 0$. The polished sample has constant thermal properties, whereas at the front surface of the target plate a thin layer structure is observed, with an effusivity value slightly decreasing next to the surface. This might be due to lattice damage induced by particle bombardment or due to a thin layer of plasma- or vapor-deposited metallic material. In each case, in this bright reflecting region the thermal depth profile cannot be explained by a thick layer of redeposited material, e.g., carbon.

The phases normalized with neutral-density glass (Schott NG10) plotted in Figure 46 in the forms $\sqrt{f}[\cot(\varphi_r - \varphi_s) \pm 1]$ versus $\sqrt{f}$ are especially well suited to discover qualitative deviations from homogeneous solids of constant thermal properties (Bein et al., 1986a). They confirm the frequency dependence observed for the amplitudes in Figure 45. On the front surface of the target plate between the distance 55 mm and 68 mm from the lower plate edge, slight but systematic deviations from the homogeneous solid of

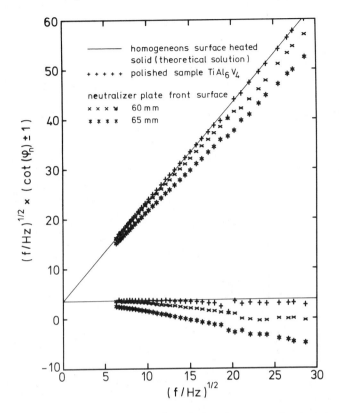

FIGURE 46. Phase test for deviations from the homogeneous surface-heated solid.

constant thermal properties are observed, whereas the polished sample of the titanium alloy $(+ + + \ +)$ coincides very well with the theoretical behavior of the opaque solid of constant thermal properties, derived from Eq. (59),

$$\sqrt{f}\left[\cot(\varphi_r - \varphi_s) + 1\right] = \sqrt{\frac{\alpha_r}{\pi}}\,\beta_r + 2\sqrt{f} \tag{73}$$

or

$$\sqrt{f}\left[\cot(\varphi_r - \varphi_s) - 1\right] = \sqrt{\frac{\alpha_r}{\pi}}\,\beta_r. \tag{73a}$$

In this limit, the theoretical solution only depends on the thermal diffusivity $\alpha_r$ and optical absorption constant $\beta_r$ of the reference body (neutral-density

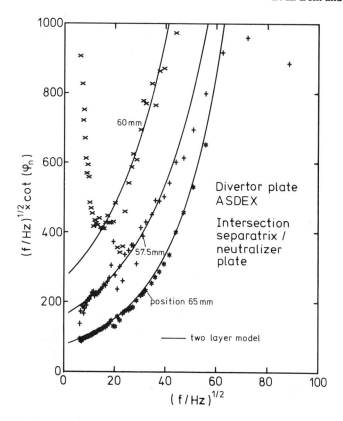

FIGURE 47. Quantitative interpretation of normalized phases based on the two-layer model.

glass) and is independent of any parameter of the opaque homogeneous solid sample.

For a quantitative interpretation, the phases are preferentially normalized with the surface-heated carbon-coated quartz-glass sample and plotted in the form $\sqrt{f} \cot(\varphi_r - \varphi_s)$ versus $\sqrt{f}$. A part of the results in Figure 47 can be interpreted as phases of two-layer bodies, where the theoretical curve is given by Eq. (61). At the distances 57.5 mm and 65 mm from the lower plate edge, the two-layer interpretation is valid for the whole frequency range measured, whereas at the distance 60 mm, the quantitative two-layer interpretation only applies for higher frequencies, $f > 400$ Hz, for a thin layer just at the surface. If deeper frequencies are also considered, corresponding to a larger penetration depth $\mu \approx \sqrt{\alpha/\pi f}$, a transition to a thermal three-layer structure becomes visible (Compare with the theoretical

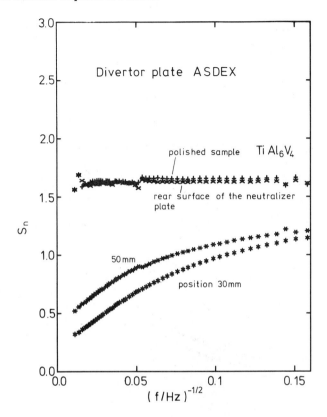

FIGURE 48. Normalized amplitudes from below the main plasma-solid contact region.

three-layer curve in Figure 30). The phase, which in general is more sensitive than the amplitude, shows here a more complicated layer structure, which is not visible in the corresponding amplitudes in Figure 45. Further measurements at lower frequencies could here clarify the situation.

Figure 48 shows normalized amplitudes from below the main plasma-solid contact region, in comparison with the rear surface and the polished sample. The difference between the polished sample $(+ + + \ +)$ and the rear surface $(\times \times \times \ \times)$ is below the error limits of the measurement. The quantitative interpretation of these phases in terms of the two-layer model (Figure 49) gives values for the ratio $g_{sb} = \sqrt{(k\rho c)_s} / \sqrt{(k\rho c)_b}$ of the effusivity of the surface layer to that of the bulk region between 0.25 and 0.40, which are well distinct from the values found for the main plasma-solid contact region between 0.65 and 0.88 (Table 2). At a first glance, the thermal diffusion time $l_s^2/\alpha_s$, which is a measure for the thickness of the

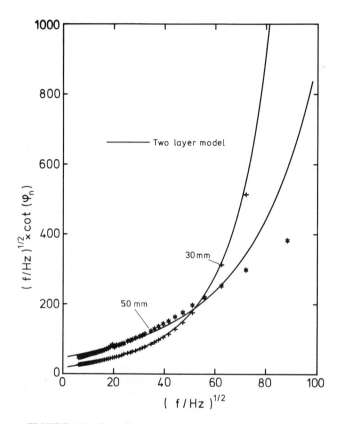

FIGURE 49.  Quantitative interpretation of normalized phases.

surface layer of different thermal properties, does not show such distinct differences between the bright reflecting plasma-solid contact region and the region below with the darker deposited material.

In Figure 50 are presented amplitudes from the region above the main bright plasma-solid contact. The measurements at the distances 67.5 and 82.5 mm from the lower plate edge show a frequency dependence similar to that observed below the main bright plasma-solid contact region, which is typical for a layer of deposited material of lower thermal conductivity on a better heat conductor. The amplitude at 75 mm from the lower plate edge, however, deviates significantly. It initially decreases with the penetration depth of the thermal wave and reaches a minimum near $(f/\text{Hz})^{-1/2} \approx 0.025$; this means after a first surface layer of a certain effusivity value $\sqrt{(k\rho c)_s}$ follows a second layer with a lower effusivity value, and only with

**Table 2.** Photoacoustic Results from a Metallic Neutralizer Plate.

| Distance from lower edge of the neutralizer plate (mm) | Ratio of the effusivities surface layer to bulk region $\sqrt{(k\rho c)_s}/\sqrt{(k\rho c)_b}$ | Square root of the thermal diffusion time of the surface layer $\Lambda_s = l_s/\sqrt{\alpha_s}\ (s^{1/2})$ | Remarks |
|---|---|---|---|
| 30 | 0.24 | 0.0082 | Deposition of |
| 45 | 0.31 | 0.0047 | material, e.g., |
| 50 | 0.39 | 0.0056 | carbon |
| 52.5 | 0.47 | 0.0047 | |
| | | | Main plasma-solid |
| 55 | 0.64 | 0.0049 | contact region |
| 57.5 | 0.785 | 0.0072 | Higher heat load |
| 60 | 0.87 | 0.0076 | Three-layer model |
| 62.5 | | | Three-layer model |
| 65 | 0.655 | 0.0083 | |
| 67.5 | 0.50 | 0.0103 | |
| | | | |
| 70 | | | Thermal resistance |
| 75 | | | below the surface |
| 82.5 | 0.30 | 0.0067 | |

further penetration of the thermal wave at lower frequencies, $\mu \approx \sqrt{\alpha/\pi f}$, the effusivity increases to reach the bulk value.

In the direct plot of normalized phases $\varphi_n = \varphi_r - \varphi_s$ (Figure 51), the position 75 mm shows the change of sign, as expected for delaminated coatings (Section IV.B.3; Patel and Almond, 1985). By comparing with the theoretical solutions for the two-layer model (Figure 21), the phase at this position can be represented in a first approximation by a two-layer model, where the second layer has a very low effusivity value and where the *thermal reflection coefficient* (Eq. 39) has a positive value, $0 < R_{sb} < 1$. Such a behavior has been observed between the distances 70 mm and 75 mm from the lower plate edge. This inversion of the thermal depth profile significantly differs from the other plots of the main plasma-solid contact region (55–67.5 mm) and farther away (position 45 mm and 82.5 mm) from the intersection between separatrix and neutralizer plate.

Such a behavior could be explained by a coating or a layer of plasma-deposited material with reduced thermal contact to the substrate. The thermal resistance may be produced by a thin layer of gas with its lower

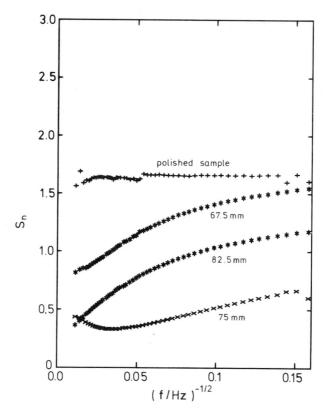

FIGURE 50. Normalized amplitudes from the region above the main plasma-solid contact.

effusivity value $\sqrt{(k\rho c)}$ = 4.71 Ws$^{1/2}$/(m$^2$ K) between the plasma-deposited layer and the metal, a reduction of the thermal conductivity due to hydrogen accumulated below the surface, or cracking parallel to the surface also may be the origin. A correlation with the double peak of heat load observed by Rapp et al. (1986) on some of the plates from the ion side might be possible.

## 2.  Preliminary Interpretation of
Photoacoustic Measurement

The changes of the metallic neutralizer plate produced by the ASDEX divertor plasma and measured here by a thermal wave technique with acoustic detection, can be interpreted in a first approximation in the

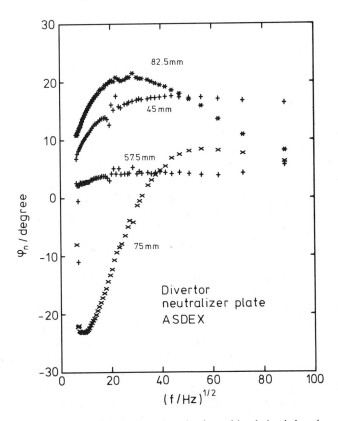

FIGURE 51. Phase test for the detection of a thermal insulation below the surface.

framework of a two-layer model of the thermal properties. The observed modifications can be characterized quantitatively both for

a) the regions farther away from the bright main plasma-solid contact, and also
b) the main plasma-solid contact region, at the intersection between the separatrix and the neutralizer plate.

In the quantitative *thermal characterization*, the ratio $g_{sb}$ of the surface effusivity to that of the volume below shows clear tendencies of variation with the distance from the main plasma-solid contact region and seems to be a practical parameter to describe the changes due to plasma-surface interactions and to distinguish regions of dominant material redeposition from regions of structural changes (Table 2). For the thermal diffusion time $l_s^2/\alpha_s$, such clear tendencies of variation from the dark regions, where mainly the redeposition of material took place, to the bright reflecting main plasma-solid contact region, cannot be recognized immediately. This might be attributable to the fact that this quantity depends both on the layer thickness $l_s$ and on the thermal diffusivity $\alpha_s$, which may vary locally due to structural changes or due to a different chemical composition of the surface layer.

The analysis based on a two-layer model can only be considered as a first approximation. For the main plasma-solid contact region, it might be improved by a three-layer model for the effective thermal properties: a first thin surface layer characterized by both plasma heat treatment and particle bombardment, a second thicker layer where the thermal conductivity and effusivity increased due to the repeated plasma heat treatment, and, only below that, the material with the original bulk properties is found. The frequency dependence of the signal in the high frequency limit, in a very thin first surface layer, indicates an effusivity profile slightly decreasing towards the surface. This effect might be due to lattice damage induced by energetic particle bombardment.

Such thermal depth profiles as found at the front surface after plasma exposure cannot be detected for polished samples of the same material nor for the rear surface of the same plate, far away from the intersection region separatrix/neutralizer plate. The signals from the rear surface and the polished samples are perfectly described by the theory of a homogeneous solid of constant thermal properties. The oxide layer at the rear surface is probably too thin to be thermally detectable, even for the highest modulation frequencies used here, $f = 7.8$ kHz.

The observed depth profile of the thermal properties and the increase of the effusivity up to 20% after plasma exposure might require a reinterpretation of the thermographical measurements at limiter or divertor neutralizer plates (Section V.D). The model of a homogeneous solid of constant thermal properties usually accepted for the power deposition calculations (Bein and Mueller, 1982) seems to be too simple, and the net power deposition onto the neutralizer plates of ASDEX, for example, may be higher by about 20%. As the power deposition to the divertor neutralizer plates is relatively small in a divertor tokamak like ASDEX (Mueller et al., 1983), the global energy balance is only slightly affected by such a correction. In limiter tokamaks, however, the correct interpretation of the energy balance might be more affected by such modified thermal properties of the limiter.

The thermal characterization can be correlated to other phenomena and observations. Thus, in the regions farther away from the main plasma-solid contact, the observed thermal depth structure is due to redeposition of material by the plasma, e.g., carbon, or due to deposition of Ti droplets or vapor. In the bright reflecting main plasma-solid contact region where the thermal structure cannot be caused by material redeposition, it is due to the direct thermal effects and particle bombardment by the plasma. Such changes of the thermal properties might be accompanied by changes of the grain structure and hardness, as observed in the region of main power deposition on the neutralizer plates of the electron side, the inner side, of the divertor (Rapp, 1987).

An improved understanding of the observed thermal depth profiles can be obtained by a direct comparison with other measurements. To this finality, metallurgical observations, electron microscopy, a correlation with the IR thermography at the target plates (Section V.D) and other measurements of the edge and divertor plasma are useful (Bein et al., 1989).

### D. INFRARED THERMOGRAPHY AS A DIAGNOSTIC TOOL IN FUSION DEVICES AND ITS RELATIONSHIP TO THERMAL WAVE TECHNIQUES

Infrared thermography (infrared imaging) is used in various fields as a diagnostic tool to detect temperature differences and to identify heat sources or heat losses due to insulation leakage. Sensing of the thermally induced IR radiation provides an ideal remote technique for in situ studies in hostile, inaccessible environments. In nuclear fusion research, IR thermography is generally used for viewing of limiter or target plates to find areas of maximum surface temperatures and largest energy deposition. These applications, generally, are qualitative and do not explore the quantitative diagnostic possibilities of IR thermography.

Based on solutions of the time-dependent heat diffusion equation, thermographic diagnostics can be developed which serve to determine absorbed net heat flows quantitatively from the temperature distribution and evolution of a solid surface. Such quantitative applications of thermography have first been used in nuclear fusion research to determine the energy deposition on target plates in the bundle divertor of DITE (Goodall, 1977). From thermographical measurements, one can calculate the conduction/convection losses from the plasma and thus get information about the transport in plasmas or the efficiency of additional heating such as neutral beam injection. Inversion solutions of the heat diffusion equation (Bein and Mueller, 1982) in combination with high time- and space-resolved thermography allows one to measure the power losses associated to MHD-instabilities or disruptions and thus can help to enlighten the instability or disruption mechanism. Apart from the emissivity in the infrared, which is necessary to deduce the surface temperature from the measured IR signal, the effective thermal properties of the involved material surfaces have to be known to make such applications really quantative. Thermal wave methods such as frequency-dependent photoacoustics can supply these parameters and thus contribute directly to improve plasma diagnostics in fusion devices. In regard to the classification scheme of thermal wave methods introduced in Section III, the application of IR thermography in nuclear fusion research complies with the pulsed, time-domain IR radiometry where the heat excitation occurs more or less uncontrolled.

### 1. Components of a Thermographic System in Tokamak Devices

In the thermographical measurements, infrared cameras sensitive in the 2 $\mu$m to 5.6 $\mu$m wavelength range monitor the surface temperature along a fixed line or on a limited area of the heated surface with the help of optomechanical scanning devices (one or two rotating prisms). Simultaneous viewing of different limiters is also possible (Bush et al., 1979). The spatial resolution depends on the imaging optics (windows, mirrors, and lens) which are necessary to bridge the distance between the object inside the vacuum vessel and the detector outside the tokamak; a resolution down to a few square millimeters is possible (Mueller et al., 1984a). The time resolution in the line scan mode, e.g., in the range of 1600 to 2500 lines per second, is good enough to account for energy balance events of a duration below 1 ms. The surface temperature can be related to the measured infrared signal by experimental calibration of the whole system, including imaging optics, window, camera, and the emissivity of the monitored solid surface. The amount of data produced for the high time- and space-resolution, e.g., more than 300,000 data per second in the thermographic system in ASDEX (Mueller et al., 1984a), require an automatic real-time data acquisition system and a fully computerized handling of data.

The better resolution limits, the large dynamic range of temperature intervals of over 1000 K (Ulrickson and Pearson, 1982), and the noncontact signal detection, which avoids all problems sensing elements mounted inside the tokamak have with high temperatures and particle fluxes, gamma radiation, and floating electric potential, are the advantages of infrared thermography over thermocouples.

### 2. Interpretation of Time- and Space-Resolved Thermographical Measurements

The net power deposited on a solid by an incident heat or particle flux can be deduced from time- and space-resolved thermographic measurements by solving the heat diffusion problem of the solid. If a homogeneous semi-infinite solid with constant initial temperature distribution in the volume and spatially uniform surface heating is assumed, the surface temperature $T_s(t)$ is related to the absorbed net heat flux $F_s(t)$ by Eq. (53). Goodall (1977) solved this equation for the heat flux by numerical iteration, and most of the quantitative interpretations (Pontau et al., 1982; Chankin et al., 1987) still use such procedures: The time history of the power deposition is approximated by rather schematic curves or expressions, and the resulting calculated surface temperatures are fitted to the measured temperature evolution to adapt the parameters of the initially prescribed power deposition.

A different procedure is possible: The problem of power deposition and heat distribution for a surface-heated solid is governed by the time-dependent heat conduction equation without sources,

$$\rho c \frac{\partial T(\mathbf{x}, t)}{\partial t} = \text{div}[k \,\text{grad}\, T(\mathbf{x}, t)], \tag{74}$$

where the boundary condition at the front surface is given by the temperature evolution $T_s(x = 0, t)$ measured by thermography. If, furthermore, reasonable boundary conditions at the other surfaces and the initial temperature distribution inside the solid can be prescribed, Eq. (74) can directly be solved for the unknown net heat flux $F_x(x \to 0, t)$ absorbed at the front surface. Since it is assumed that there are no volume heat sources or sinks in the solid, inductive heating or phase transitions as melting of the solid are excluded. The energy transfer from the plasma particles to the solid may take place in an infinitely thin plasma-solid interaction layer at the surface $x \to 0$ of the solid (Figure 52) where we have

$$F_i - F_r - F_e = F_s(t) = F_x(x \to 0, t) = -\left[k \frac{\partial T(x, t)}{\partial x}\right]_{x \to 0}. \tag{75}$$

Here, $F_i$, $F_r$, and $F_e$ are the incident, reflected, and emitted heat flux, respectively, and only the net absorbed heat flux $F_s(t)$ can be calculated from the measured surface temperature. This means, that the calculated heat flux $F_s(t)$ is not identical to the conduction/convection losses from the plasma side, the energy of eventually reflected charged or neutral particles have to be considered, as well as cooling due to evaporation of surface material. For the semi-infinite solid, an inversion solution for the heat flux $F_s(t)$ (Bein and Mueller, 1982) is given by

$$F_s(t) = \sqrt{\frac{k\rho c}{4\pi}} \int_0^t \frac{[T(t) - T(t')]}{(t - t')^{3/2}} \, dt' + 2 \frac{[T_s(t) - T_s(t = 0)]}{\sqrt{t}}. \tag{76a}$$

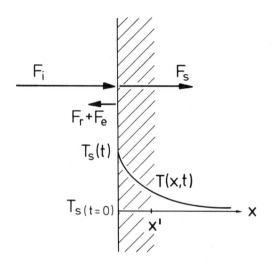

FIGURE 52.    Schematic of surface heating of a solid of finite thickness.

or the equivalent integrals

$$F_s(t) = \sqrt{\frac{k\rho c}{4\pi}} \int_0^\infty \frac{[T_s(t) - T_s(t - t')]}{t'^{3/2}} \, dt' \qquad (76b)$$

and

$$F_s(t) = \sqrt{\frac{k\rho c}{\pi}} \frac{d}{dt} \left\{ \int_0^t \frac{T_s(t')}{\sqrt{t - t'}} \, dt' \right\}. \qquad (76c)$$

Such an inversion solution has been used for the calculation of the heat load on the TFR limiter (TFR group, 1981). For the numerical calculation of the power deposition, these inverse solutions (76) have their inconveniences: Though the integral in Eq. (76a) converges when a linear interpolation between subsequent measured temperature values $T_s(t)$, $T_s(t + \Delta t_m)$ is used, this solution can produce purely numerical peaks in connection with the limited finite time resolution $\Delta t_m$ of the temperature measurement (Figure 53), which cannot be distinguished from real peaks of power deposition. Eq. (76c), on the other hand, requires numerical differentiation, which can also lead to unrealistic purely numerical peaks. If numerical smoothing is then applied, physically real peaks of power deposition can be smoothed out, too.

Numerical peaks of power deposition can be reduced by an "expansion" solution (Bein and Mueller, 1982) based on space-time-dependent heat flux solutions of the diffusion equation by Green's function, given here for the

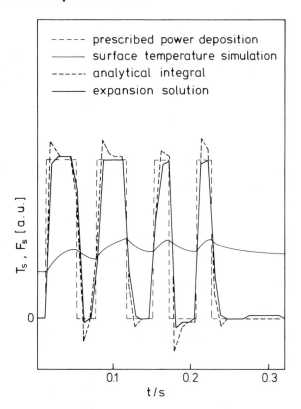

FIGURE 53. Simulation of surface heating and detection of the applied power deposition. Damping of numerical peaks is by the expansion solution (after Bein and Mueller, 1982).

example of the semi-infinite solid by

$$F_x(x, t) = -\sqrt{\frac{k\rho c}{4\pi}} \int_0^t dt' \frac{T_s(t')}{(t - t')^{3/2}} \exp\left[-\frac{x^2}{4\alpha(t - t')}\right]$$

$$\times \left[1 - \frac{2x^2}{4\alpha(t - t')}\right]. \tag{77}$$

By a Taylor expansion for the heat flux $F_x(x, t)$, the value at the surface $F_x(x \to 0, t)$ can be calculated from the value inside the solid, $x' = \sqrt{2\alpha \Delta t_m} > 0$; $\alpha$ is here the thermal diffusivity of the solid material, and $\Delta t_m$ is the time resolution of the thermographical measurement. At the same finite time resolution $\Delta t_m$, the "expansion" solution produces smaller numerical peaks than the integral Eq. (76a), as shown by simulation calculations in Figure 53. Physically, the mathematical expansion solution can be understood to extrapolate the heat flux across the surface from a superposition of thermal waves (Section II), where owing to the finite propagation distance $x' > 0$, the higher frequent perturbations and measuement fluctuations

$$T(x', t; f) = T_s(f) \exp\left[-\sqrt{\frac{\pi f}{\alpha}}\, x'\right] \cos\left(2\pi f t - \sqrt{\frac{\pi f}{\alpha}}\, x'\right) \quad (78)$$

have been damped away; noise of different nature can thus be suppressed.

For the calculation of the power deposition on the divertor neutralizer plates of ASDEX, which due to their thickness $d = 2$ mm and thermal diffusivity $\alpha = 2.9 \; 10^{-6}$ m$^2$/s (titanium alloy Contimet ALV 64) are thermally thin compared to the average discharge duration $t \approx 3$–4 seconds of that tokamak, $d < \sqrt{\alpha t}$, the "expansion" solution has been extended to the model of a solid bounded by two parallel planes (Mueller et al., 1984a). The boundary conditions used in this case are:

a) the measured front surface temperature evolution and
b) the condition of negligible heat losses at the rear surface.

This latter assumption is justified on the time scale of the discharge duration and may substitute the measurement of the rear surface temperature as boundary condition.

A stable analytical solution is achieved by additionally taking the global heat balance of the target plate over the whole discharge duration. In detail, a system of two coupled integrals for the surface heat flux and the rear surface temperature was derived which allows computation of the power deposition $F_s(t)$ from any measured surface temperature evolution $T_s(t)$.

The efficiency of such time- and space-resolved thermography and power deposition calculations is demonstrated in Figures 54 and 55 in connection with other diagnostics in ASDEX. In the discharge presented here, the density is built up (Figure 54c) by pellet injection starting at $t = 1$ second; between $t = 1.82$ and $2.12$ seconds, the plasma is additionally heated by neutral beam injection (Zasche et al., 1986). Figure 54a shows the measured front surface temperature at a fixed position of the outer neutralizer plate of the upper divertor, and Figure 54b the calculated power deposition. This evolution is compared with the electron line density (Figure 54c) and the $D\alpha$ intensity in the upper divertor chamber (Figure 54d).

Both the quality of the space- and time-resolution of the power deposition profile are demonstrated in Figure 55. In this discharge the intersection between separatrix and divertor neutralizer plate changed due to the variation of the divertor multiple currents. Figure 55a shows the position of the separatrix derived from magnetic loop signals as a function of time,

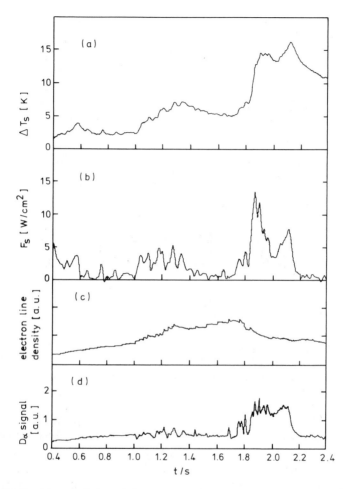

FIGURE 54. Time evolution of the measured front surface temperature (a) and power deposition (b) at a fixed position of the neutralizer plate, compared with electron line density (c) and the $D\alpha$ signal in the upper divertor (after Zasche et al., 1986).

Figure 55b shows the position of the separatrix as derived from the power deposition profile (position of the steepest gradient), and Figure 55c shows the position of the surface temperature maximum. The separatrix position derived from the power deposition profile agrees with the magnetic measurement within a difference of about 0.5 cm, whereas the position of the maximum surface temperature does not directly correlate with the separatrix position due to the inertia of the surface heating or cooling process.

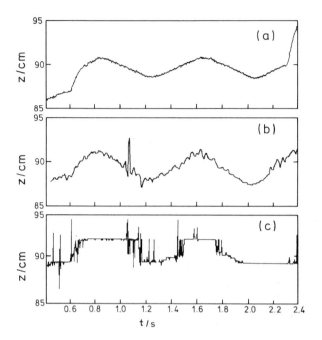

FIGURE 55. Vertical position of the separatrix derived from magnetic loop signals (a) as a function of time, position of the separatrix as derived from the power deposition profile (b) and position of the surface temperature maximum (after Zasche et al., 1986).

For the homogeneous semi-infinite solid with temperature dependent thermal properties and for the semi-infinite solid with a surface layer of different thermal properties, coupled analytical integrals have been derived (Bein, 1986) which allow computation of the power deposition from thermographic data. These inverse solutions may be useful when higher surface temperatures are reached on limiters or neutralizer plates and when, owing to plasma surface interactions, the thermal properties in a surface layer differ from those of the bulk material. The thermophysical properties necessary for the inversion solution of the composite layered solid, the effusivity of the surface layer, $\sqrt{(k\rho c)_s}$ , that of the bulk material below, $\sqrt{(k\rho c)_b}$ , and the thermal diffusion time of the surface layer, $l_s^2/\alpha_s$, govern the signal generation process in frequency-dependent thermal wave analysis (Section II.D) and thus can be gained from photoacoustics or another thermal wave technique.

For actively cooled neutralizer plates of finite thickness it is also possible to derive analytical integrals (Würz et al., 1989), similar to the problem of the plate adiabatically insulated at the rear surface. For more complicated geometries, however, such analytical solutions may no longer be possible or useful and a purely numerical solution of the corresponding differential equation, boundary, and initial conditions by finite-element techniques will become necessary.

## 3. *Quantitative IR Thermography and its Relationship to Photoacoustics*

For the study of the overall or time-resolved energy balance of tokamak discharges, the power losses due to charge exchange neutrals and electromagnetic radiation are measured by bolometry. The losses due to conduction/convection localized in regions where the magnetic field lines intersect with material surfaces should be ascertained by a separate measurement, e.g., from the measured time evolution of the surface temperatures of limiters and divertor neutralizer plates. The advantage of such an indirect measurement of the conduction/convection losses is that the edge plasma is not disturbed by a material probe additionally exposed to the high temperature and heat flux plasma environment. The effort undertaken since 1979 to study the energy balance in tokamak, was only partially successful. Thus in ASDEX, the quantitative thermography with its space and time resolution of 1.3 mm and 0.4 ms, respectively, could contribute to analyze the transition from the $L$ to the $H$-mode of confinement (Mueller et al., 1984a, 1984b), but the energy balance, time-resolved or global, could not be closed: about 30% to 40% of the power is missing, e.g., in neutral beam heated discharges (Mueller et al., 1983). Among the various effects responsible for this apparent energy gap, such as beam halo (Succi and Lister, 1982), toroidal asymmetries, etc., only two will be subsequently discussed. These are related to the power deposition at the limiter or divertor targets and to the surface heating process, namely,

(i) misinterpretations of the thermographic measurement due to incorrect thermophysical data and
(ii) reflection of energy due to extreme grazing incidence of the plasma particles.

The quantitative interpretation of time-resolved thermography, as described in this section, and thermal wave analysis of solid state materials can be considered as related, complementary techniques. In thermal wave analysis of solid-state materials, the solid surface is heated by an intensity-modulated heat source. The thermal response of the solid is measured by photothermal or photoacoustic detection. If the heat source is calibrated by a reference measurement of a solid of known thermal parameters ("normalization" procedure, Section V.B), the unknown thermal properties of the solid limiter or divertor target plate can be determined. In applications in nuclear fusion devices, where the thermal properties of the observed surfaces can change with time due to plasma exposure (Section V.B.3 and V.C.1), the quantitative thermography needs information about the changing actual effusivity value. This information can be obtained from thermal wave methods such as frequency dependent photoacoustics.

The equivalence between the thermal properties measured in the thermal wave experiment and the properties required for the quantitative interpretation of thermography may not be identical *a priori*, as the heat source in the thermal wave experiment and at the divertor neutralizer plates in a Tokamok are different. The energy transfer from the plasma to the solid by electrons or ions hitting the target and absorbed or reflected, eventually as neutral particles, takes place in a thin layer of finite thickness. As long as the thickness of this plasma-solid interaction layer is comparable to the absorption length $\beta^{-1}$ of the radiation in the thermal wave experiments and as long as both these lengths are small in comparison to the thermal diffusion lengths involved in the thermal-wave experiment and in the quantitative thermography, the measured effusivity is appropriate and the model of surface heating is justified for both measurements. The thermal diffusion length for the thermography here is given by $\mu \approx \sqrt{\alpha \Delta t_m}$, where $\alpha$ is the thermal diffusivity of the limiter material and $\Delta t_m$ the time resolution required for the power deposition calculations.

A more complicated situation may arise when rough surfaces are considered, and the thermal properties which are measured in a thermal wave experiment with vertical incidence of heating radiation are applied for grazing incidence of the radiation or particle power deposition. The surface region in the shadow (Section V.B.2; Bein and Pelzl, 1986) then might give rise to differing effective thermal depth profiles. In magnetic confinement-based fusion, the power deposition by charged particles always takes place under oblique incidence. This is due to the gyration motion of the charged particles along the magnetic field lines and partly due to the electric space charge built up between a solid surface and the plasma, nearly independent of the incidence angle of the magnetic field (Daybelge and Bein, 1981). In the case of vertical incidence of the magnetic field, the electrons and ions hit the surface under oblique incidence; for grazing incidence of the magnetic field, which usually is the case on large area limiters or neutralizer plates, the average incidence of the particles is less grazing than the field lines (Chodura, 1982).

Thermal wave measurements with heat generation by incident ion or electron beams, respectively, could help in understanding the power deposition by charged particle. Such experiments also could verify to what extent the thermal properties measured with electromagnetic radiation induced heat sources are identical to the effective thermal properties in surface heating by charged particles and provide information about the amount of energy deposited on solid surfaces by charged particles under oblique or grazing incidence. The gap in the energy balance in thermography monitored discharges of ASDEX is partly explained by such particles, which hit the neutralizer plates under grazing incidence, eventually neutralize, and then escape as charged or neutral particles to the top or the bottom of the divertor chambers, where the energy deposition is not measured (Mueller et al., 1983, 1984a). Available data on light ion reflection (Eckstein and Verbeek, 1979) probably could also be complemented by thermal wave experiments for lower incidence energies and oblique incidence.

## VI. Conclusions

This contribution cannot be a review in the usual sense since the application of thermal wave methods to the analysis of plasma-materials interaction problems is a rather new field of work. Only in one sector, the analysis of plasma-sprayed coatings (Section IV.B), the experience now covers more than two decades. The other fields treated here, application to plasma-exposed semiconductor materials (Section IV.A) and to plasma-surface interactions in controlled fusion (Section V), only started very recently. So, this contribution is more an outlook to the application potential of thermal wave techniques in plasma-materials interaction.

Thermal wave techniques are nondestructive techniques which can be easily handled. Qualitative information about a large variety of phenomena can be rapidly obtained. The signal generation process, however, is rather indirect and involved, and a thorough theoretical analysis is required to deduce quantitative, reliable information. With respect to the quantitative interpretation, the degree of development is different in the various fields. In the applications to plasma sprayed coatings and plasma-surface interactions in controlled fusion, the theoretical understanding of the experimental phenomena is relatively high and better than in the applications to plasma-treated semiconductor materials, where the signal generation process is more complex (Section II.F).

The relations between the physics of plasmas and thermal wave techniques still are not very strong. In the work done up until now, they are more developed in nuclear fusion applications where the infrared diagnostics in fusion devices are the link (Section V.D). The potential of thermal wave methods for in situ control of plasma processing of materials and plasma-solid interactions has been recognized for the three fields (plasma-spraying, semiconductors, and nuclear fusion research), but the practical utilization is far from being explored. The possibilities of using thermal wave methods directly for in situ analysis of plasma-solid interactions are good in nuclear fusion research, where the necessary sensing hardware, e.g., infrared cameras, and recording systems, exist, in part, already. The necessity to develop such noncontact, remote-handling diagnostics sensitive to thermal features, also exists in controlled fusion, since thermal phenomena, high heat fluxes, and temperatures have to be controlled at the material walls surrounding the hot plasma.

## VII. Acknowledgements

Many colleagues, both from the fields of photoacoustics and plasma-surface interactions, helped us, and it is not possible to mention all of them. We at least want to acknowledge the valuable advice of P. Cielo, National Research Council of Canada, Montreal, and D. P. Almond, University of Bath, from the photoacoustic community. We also acknowledge the advice and support of D. Zasche, H. Rapp, G. Haas, and F. Wagner, Max-Planck-Institut für Plasmaphysik Garching, for the work on metallic divertor plates and the support and interest for the work on graphite limiter plates by H. Vernickel, J. Bohdansky, and R. Behrisch, also from Garching.

## References

Aamodt, L. C., Murphy, J. C., and Parker, J. G. (1977). *J. Appl. Phys.* **48**, 927.
Aamodt, L. C. and Murphy, J. C. (1981). *J. Appl. Phys.* **52**, 4903.
Aamodt, L. C. and Murphy, J. C. (1982). *Appl. Opt.* **21**, 111.
Aamodt, L. C. and Murphy, J. C. (1983). *J. Appl. Phys.* **54**, 581.
Adams, M. J., Highfield, J. G., and Kirkbright, G. F. (1980). *Anal. Chem.* **52**, 1260.
Aithal, S., Rousset, G., Bertrand, L., Cielo, P., and Dallaire, S. (1984). *Thin Solid Films* **119**, 153.
Almond, D. F. and Reiter, H. (1980). *Surfacing Journ.* **16**, 4.
Almond, D. P., Patel, P. M., Pickup, I. M., and Reiter, H. (1985). *NDT International* **18**, 17–24.
Amer, N. M. (1983). *J. Phys.* (*Paris*) **44**, C6-185.
Angström, A. J. (1863). *Phil. Mag. a. J. Science*, S.4. **25**, 130.
Auciello, O., Haasz, A. A., and Stangeby, P. C. (1983). *Phys. Rev. Lett.* **50**, 783–786.
Balageas, D., Krapez, J. C., and Cielo, P. (1986). *J. Appl. Phys.* **59**, 348–357.
Balk, L. J. and Kultscher, N. (1983). *Beitr. Elektronenmikr. Direktabb. Oberflächen* **16**, 107.
Balk, L. J. (1986). *Can. Journ. Phys.* **64** (9), 1238.
Bechthold, P. S., Campagna, M., and Chatzipetros, J. (1981a). *Opt. Commun.* **36**, 369.
Bechthold, P. S. and Campagna, M. (1981b). *Opt. Commun.* **36**, 373.
Bechthold, P. S. (1984). In: *Photo-Acoustic Effect* (E. Lüscher, P. Korpiun, H. J. Coufal, and R. Tilgner, eds.), Vieweg, Braunschweig, pp. 375–411.

Behrisch, R. (1972). *Nuclear Fusion* **12**, 695–713.

Behrisch, R. (1980). *J. Nucl. Mater.* **93 & 94**, 498–504.

Bein, B. K. and Mueller, E. R. (1982). *J. Nucl. Mat.* **111 & 112**, 548.

Bein, B. K. and Pelzl, J. (1983). *J. Phys. (Paris)* **44**, C6-27.

Bein, B. K., Krueger, S., and Pelzl, J. (1984). *J. Nucl. Mat.* **128 & 129**, 945–950.

Bein, B. K., Krueger, S., and Pelzl, J. (1985). *Preliminary Report on Photoacoustic Measurements of the JET Graphite Limiter*, JET-TN(85)04, JET Joint Undertaking, Abingdon, UK.

Bein, B. K., Krueger, S., and Pelzl, J. (1986a). *Can. J. Phys.* **64**, 1208–1216.

Bein, B. K. (1986). *Can. J. Phys.* **64**, 1190–1194.

Bein, B. K., Krueger, S., and Pelzl, J. (1986b). *J. Nucl. Mater.* **141–143**, 119–123.

Bein, B. K. and Pelzl, J. (1986). In: *CARBON* 86 (Proc. 4th Int. Carbon Conf. Baden–Baden, 1986), Deutsche Keramische Gesellschaft, Bonn/Bad Honnef, FRG, pp. 268–270.

Bein, B. K., Krueger, S., and Pelzl, J. (1987). *J. Nucl. Mater.* **145–147**, 458.

Bein, B. K. (1988). In: *Photoacoustic and Photothermal Phenomena* (P. Hess and J. Pelzl, eds.), Springer Series on Optical Sciences, Vol. 58, pp. 308–311, Springer Heidelberg/Berlin.

Bein, B. K., Wojczak, M., and Pelzl, J. (1989). *J. Nucl. Mater.*, to be published.

Bell, A. G. (1880). *Am. J. Sci.* **20**, 305.

Bell, A. G. (1881). *Phil. Mag.* **11**, 510.

Bennett, C. A. and Patty, R. R. (1982). *Appl. Opt.* **21**, 111.

Beyer, W., Wagner, H., and Mell, H. (1981). *Sol. St. Comm.* **39**, 375.

Birnbaum, G. and White, G. S. (1984). In: *Research Techniques in Nondestructive Testing*, Vol. 7. Academic Press, London, p. 259.

Boccara, A. C., Fournier, D., and Badoz, J. (1980). *Appl. Phys. Lett.* **36**, 130.

Bohdansky, J. and Roth, J. (1985). *Radiation Effects* **89**, 49–62.

Bohdansky, J., Croessmann, C. D., Linke, J., McDonald, J. M., Morse, D. H., Pontau, A. E., Watson, R. D., Whitley, J. B., Goebel, D. M., Hirooka, Y., Leung, K., Conn, R. W., Roth, J., Ottenberger, W., and Kotzlowski, H. E. (1987). *Nucl. Instr. and Meth. in Phys. Res.* **B23**, 527.

Boulitrop, F., Bullot, J., Gauthier, M., Schmidt, M. P., and Catherine, Y. (1985). *Sol. St. Comm.* **54**, 107.

Brailsford, A. D. and Major, K. G. (1964). *Brit. J. Appl. Phys.* **15**, 313.

Brandis, E. and Rosencwaig, A. (1980). *Appl. Phys. Lett.* **37**, 98.

Brandt, R. (1981). *High Temp.–High Press.* **13**, 79–88.

Bush, C. E., Isler, R. C., Madison, J. M., and Overbey, D. R. (1979). In: *ORNL Fusion Energy Division Annual Progress Report* 1979 (Oak Ridge National Laboratory), ORNL-5645, 56–57.

Busse, G. (1979). *Appl. Phys. Lett.* **35**, 759.

Busse, G., Rief, B., and Eyrer, P. (1986). *Can. J. Phys.* **64**, 1195.

Bustarret, E., Jousse, D., Chaussat, C., and Boulitrop, F. (1985). *J. Non-Crystaline Sol.* **77 / 78**, 295.

Cahen, D. and Halle, S. D. (1985). *Appl. Phys. Lett.* **46**, 446.

Cargill III, G. S. (1980). *Nature (London)* **286**, 691.

Carslaw, H. S. and Jaeger, J. C. (1959). *Conduction of Heat in Solids*. Oxford Univ. Press, London.

Chadwick, P. (1964). In: *Progress in Solid Mechanics*, Vol. 1. (I. N. Sneddon and R. Hill, eds.). North-Holland, Amsterdam.

Chankin, A. V., Chicherov, V. M., Efstigneev, S. A., Grashin, S. A., Grote, H., Grunow, C., Günther, K., Lingertat, J. et al. (1987). *J. Nucl. Mater.* **145–147**, 789.

Cielo, P., Dallaire, S., Lamonde, G., and Johar, S. (1986). *Can. J. Phys.* **64**, 1217.

Cielo, P. (1985). In: *International Advances in Nondestructive Testing*, Gordon and Breach, U.K., Vol. 11, pp. 175–217.

Chodura, R. (1982). *J. Nucl. Mater.* **111 & 112**, 420–423.

# 322
B. K. Bein and J. Pelzl

Coufal, H. (1984). *Appl. Phys. Lett.* **44**, 59.

Coufal, H. (1986). *IEE Transact. Ultras. Ferroel. Frequ. Contr. UFFC*-33, 507.

Cowan, R. D. (1961). *Journ. Appl. Phys.* **32**, 1363.

Cox, R.L., Almond, D. P., and Reiter, H. (1981). *Ultrasonics* **19**, 17.

Cunningham, M. E. and Peddicord, K. L. (1981). *Int. J. Heat a. Mass Transfer* **24**, 1081.

Daybelge, U. and Bein, B. K. (1981). *Phys. Fluids* **24**, 1190–1194.

DeConinck, R. and Snykers, M. (1978). *J. Nucl. Mater.* **76 & 77**, 629–633.

Delle, W., Haag, G., Linke, J., Nickel, H., Sonnenberg, K., and Wallura, E. (1986). In: *CARBON* 86 (Proc. 4th Int. Carbon Conf. Baden–Baden, 1986), Deutsche Keramische Gesellschaft, Bonn/Bad Godesberg, FRG, pp. 804–807.

Dietz, K. J., Bartlett, D., Baeumel, K., and the JET team. (1984). *J. Nucl. Mater.* **128 & 129**, 10–18.

Donnelly, V. M. (1988). In this book series *Plasmas Diagnostics*, Vol. 1 (O. Auciello and D. L. Flamm, eds.), Academic Press, Boston.

Doyle, B. L. and Brice, D. K. (1985). *Radiation Effects* **89**, 21–48.

Eckstein, W. and Verbeek, H. (1979). Data on Light Ion Reflection, Report IPP 9/32. Max-Planck-Institut fuer Plasmaphysik, Garching, FRG.

Ehrenberg, J. and Behrisch, R. (1984). Surface Analysis of a Central Part of the JET Graphite Limiter, IPP-JET Report 23, Max-Planck-Institut fuer Plasmaphysik, Garching, FRG.

Ermert, H., Dacol, F. H., Melcher, R. L., and Baumann, T. (1984). *Appl. Phys. Lett.* **44**, 1136.

Evora, C., Landers, R., and Vargas, H. (1980). *Appl. Phys. Lett.* **36**, 864.

Fauchais, P., Vardelle, M., Couvert, J. F. (1988). In this book series *Plasma Diagnostics*, Vol. 1 (O. Auciello and D. L. Flamm, eds.), Academic Press, Boston.

Florian, R., Pelzl, J., Rosenberg, M., Vargas, H., and Wernhardt, R. (1978). *Phys. Stat. Sol. (a)* **48** K35.

Fournier, D., Saint-Jacques, R. G., Ross, G. G., and Terreault, B. (1987). *J. Nucl. Materials* **145–147**, 379.

Fournier, D. and Boccara, A. C. (1984). In: *Photo-Acoustic Effect* (E. Lüscher, P. Korpiun, H. J. Coufal, and R. Tilgner, eds.). Vieweg Braunschweig, 80–93.

Fournier, D., Boccara, A. C., Skumanich, A., and Amer, N. M. (1986). *J. Appl. Phys.* **59**, 787.

Frye, R. C., Kumler, J. J., and Wong, C. C. (1987). *Appl. Phys. Lett.* **50**, 101.

Geraghty, P. and Smith, W. L. (1986). In: *Symposium on Plasma Processing*. Material Research Society Spring Meeting, Palo Alto, CA, April, 1986.

Goodall, D. H. J. (1977). In: *Plasma Wall Interaction* (Proc. Int. Symp. Juelich, 1976). Pergamon Oxford, 53–58.

Görtz, W. and Perkampus, H.-H. (1982). *Fresenius Z. Anal. Chem.* **310**, 77.

Green, D. R. (1966). *J. Applied Phys.* **37**, 3095.

Grigull, U. and Sandner, H. (1979). *Waermeleitung*. Springer Verlag, Berlin.

Haasz, A. A., Auciello, O., Stangeby, P. C., and Youle, I. S. (1984). *J. Nucl. Mater.* **128 & 129**, 593–596.

Hassanein, A. M., Kulcinski, G. L., and Wolfer, W. G. (1982). *J. Nucl. Mater.* **111 & 112**, 554–559.

Hata, T., Hatsuda, T., Miyabo, T., and Hasegawa, S. (1985). *Jap. Journ. Appl. Phys. (Suppl. 25-1)* **25**, 226.

Heihoff, K. (1986). Ph.D. Thesis, Ruhr-Universität Bochum, FRG.

Heihoff, K. and Braslavsky, S. E. (1986). *Chem. Phys. Lett.* **131**, 183.

Helander, P., Lundstroem, I., and McQueen, D. (1981). *J. Appl. Phys.* **52**, 1146.

Hartunian, R. A. and Varwig, R. L. (1962). *Phys. of Fluids* **5**, 169–174.

Hussla, I., Coufal, J., Traeger, F., and Chuang, T. J. (1986). *Can. J. Phys.* **64**, 1070.

Hutchins, D. A. (1986). *Can. J. Phys.* **64**, 1247.

Inagaki, T., Kagami, K., and Arakawa, E. T. (1981). *Phys. Rev.* **B24**, 3644.

Iqbal, Z., Sarott, F. A., and Veprek, S. (1983). *J. Phys. C.: Sol. State Phys.* **16**, 2005.

Jackson, W. B. and Amer, N. M. (1980). *J. Appl. Phys.* **51**, 3343.

Jackson, W. B., Amer, N. M., Boccara, A. C., and Fournier, D. (1981). *Appl. Optics* **20**, 1333.

Jackson, W. B. and Amer, N. M. (1982). *Phys. Rev.* **B25**, 5559.

Jackson, W. B. and Nemanich, R. J. (1983). *J. Non-Crystalline Sol.* **59 / 60**, 353.

Jousse, D., Bustarret, E., and Boulitrop, F. (1985). *Sol. St. Comm.* **55**, 435.

Jousse, D., Bustarret, E., Deneuville, A., and Stoquert, J. P. (1986). *Phys. Rev.* **B34**, 7031.

Junge, K., Bein, B. K., and Pelzl, J. (1983). *Journ. de Phys.* **44**, C6-55.

Kamitakahara, W. A., Schanks, H. R., McClellan, J. F., Buchenau, U., Gompf, F., and Pinschovius, L. (1984). *Phys. Rev. Lett.* **52**, 644.

Kanstad, S. O. and Nordal, P. E. (1986). *Can. J. Phys.* **64**, 1155.

Keilhacker, M. and the ASDEX team. (1985). *Nuclear Fusion* **25**, 1045.

Kelly, B. T. (1981). *Physics of Graphite* Appl. Sci. Publ., London.

Kordecki, K., Bein, B. K., and Pelzl, J. (1986). *Can. Journ. Phys.* **64**, 1204.

Kordecki, R., Pelzl, J., and Bein, B. K. (1988). In: *Photoacoustic and Photothermal Phenomena* (P. Hess and J. Pelzl, eds.), Springer Series on Optical Sciences, Vol. 58, pp. 490–491. Springer Heidelberg/Berlin.

Korpiun, P., Hermann, W., Kindermann, A., Rothmeyer, M., and Buechner, B. (1986). *Can. Journ. Phys.* **64**, 1042.

Korpiun, P. and Buechner, B. (1983). *J. de Phys.* (*Paris*) **44**, C6-85.

Korpiun, P. and Tilgner, R. (1983). *J. Appl. Phys.* **51**, 6115.

Kultscher, N. and Balk, J. (1986). *J. Scanning Electr. Microscopy* **7**, 33.

Kuo, P. K., Lin, M. J., Reyes, C. B., Favro, L. D., Thomas, R. L., Kim, D. S., Zhang, S.-Y., Inglehart, L. J., Fournier, D., Boccara, A. C., and Jacoubi, N. (1986). *Can. J. Phys.* **64**, 1165.

Kuo, P. K., Sendler, E. D., Favro, L. D., and Thomas, R. L. (1986b). *Can. J. Phys.* **64**, 1168.

Krueger, S., Kordecki, R., Pelzl, J., and Bein, B. K. (1987). *J. Appl. Phys.* **62**, 55–61.

Lackner, K., Chodura, R., Kaufmann, M., Neuhauser, J., Rauh, K. G., and Schneider, W. (1984). In: *Plasma Physics and Controlled Fusion*, Vol. 26. Pergamon, Oxford, pp. 105–115.

Lai, W. P. and Chan, B. L. (1985). *Appl. Spectr. Rev.* **21**, 179–210.

Langley, R. A. (1984). *J. Nucl. Mater.* **128 & 129**, 622–628.

Lepoutre, F., Bein, B. K., and Inglehart, L. J. (1986). *Can. J. Phys.* **64**, 1037.

Leung, W. P. and Tam, A. C. (1984). *J. Appl. Phys.* **56**, 153.

Liebert, C. H. and Miller, R. A. (1984). *Ind. Eng. Chem. Prod. Res. Dev.* **23**, 344–349.

Low, M. J. and Parodi, G. A. (1980). *Applied Spectr.* **34**, 76.

Low, M. J., Lacroix, M., and Morterra, C. (1982). *Applied Spectr.* **36**, 582.

Luukkala, M. V. (1980). In: *Scanned Image Microscopy.* Academic Press, London, p. 273.

Mandelis, A. and Royce, B. S. (1979). *J. Appl. Phys.* **50**, 4330.

Mandelis, A. (1983). *J. Appl. Phys.* **54**, 3404.

Mascarenhas, M., Vargas, H., and Cesar, C. L. (1984). *Med. Phys.* **11**, 73.

Mast, F. and Vernickel, H. (1982). *J. Nucl. Mater.* **111 & 112**, 566–568.

Mattox, D. M. and Davis, M. J. (1982). *J. Nucl. Materials* **111 & 112**, 819–826.

McCracken, G. M. and Stott, P. E. (1979). *Nuclear Fusion* **19**, 889–981.

McDonald, F. A. and Wetsel, G. D. (1978). *J. Appl. Phys.* **49**, 2313.

McDonald, F. A., Wetsel, G. D., and Jamieson, G. E. (1986). *Can. J. Phys.* **64**, 1265.

McPershon, R. (1984). *Thin Solid Films* **112**, 89–95.

Melcher, R. L. and Arbach, G. V. (1982). *Appl. Phys. Lett.* **40**, 910.

Merte, B., Korpiun, P., Lüscher, E., and Tilgner, R. (1983). *J. de Phys.* **44**, C6-463.

Mikoshiba, N. and Tsubouchi, K. (1981). In: 1981 *Ultrasonic Symposium* (B. R. McAvoy, ed.). IEEE, p. 792.

Mostefaoui, R., Chevalier, J., Meichenin, S., and Auzel, F. (1985). *J. Non-Crystalline Sol.* **77 / 78**, 307.

Mueller, E. R., Bein, B. K., Niedermeyer, H., and the ASDEX team (1983). In: *Controlled Fusion and Plasma Physics.* Europhysics Conference Abstracts (S. Methfessel, ed.). Vol. 7D, part I, pp. 47–50.

Mueller, E. R., Bein, B. K., and Steinmetz, K. (1984a). *Time and Space Resolved Energy Flux Measurements in the Divertor of the ASDEX Tokamak by Computerized Infrared Thermography.* Max-Planck-Institut fuer Plasmaphysik, Garching, FRG, Report IPP III/97.

Mueller, E. R., Keilhacker, M., Steinmetz, K., ASDEX team, and NI team. (1984b). *J. Nucl. Mater.* **121**, 138.

Murphy, J. C. and Aamodt, L. C. (1977). *J. Appl. Phys.* **48**, 3502.

Murphy, J. C. and Aamodt, L. C. (1980). *J. Appl. Phys.* **51**, 4580.

Netzelmann, U. and Pelzl, J. (1984). *Appl. Phys. Lett.* **44**, 854.

Netzelmann, U., Krebs, U., and Pelzl, J. (1984). *J. Appl. Phys.* **44**, 1161.

Netzelmann, U., Pelzl, J., Fournier, D., and Boccara, A. C. (1986). *Can. J. Phys.* **64**, 1307.

Netzelmann, U., Pelzl, J. and D. Schmalbein. (1988). In: *Photoacoustic and Photothermal Phenomena.* (P. Hess and J. Pelzl, eds.) Springer Series on Optical Sciences, Vol. 58, pp. 312–315. Springer, Heidelberg/Berlin.

Nordal, P. E. and Kanstad, S. O. (1981). *Appl. Phys. Lett.* **38**, 486.

Nordhaus, O. and Pelzl, J. (1981). *Appl. Phys.* **25**, 221.

Nowacki, W. (1966). *Dynamical Problems of Thermoelasticity.* Noordhoff Int. Publ., Leyden, NL.

Olmstead, M. A., Amer, N. M., Kohn, S., Fournier, D., and Boccara, A. C. (1983). *Appl. Phys.* **A32**, 141.

Opsal, J. and Rosencwaig, A. (1982). *J. Appl. Phys.* **53**, 4240–4248.

Opsal, J. and Rosencwaig, A. (1985). *Appl. Phys. Lett.* **47**, 498.

Pao, Y. H. (1977). *Opto-acoustic Spectroscopy and Detection.* Academic Press, New York.

Parke, S., Warman, G. P., and Mellor, J. H. (1982). *Phys. Chem. Glasses* **23**, 95.

Parker, W. J., Jenkins, R. J., Butler, C. P., and Abbott, G. L. (1961). *J. Appl. Phys.* **32**, 1679–1684.

Parker, J. G. (1973). *Appl. Opt.* **12**, 2974.

Patel, C. K. and Tam, A. C. (1981). *Rev. Mod. Phys.* **53**, 517.

Patel, P. M. and Almond, D. P. (1985). *J. Materials Sci.* **20**, 955–966.

Pawlowski, L., Lombard, D., Mahlia, A., Martin, C., and Fauchais, P. (1984). *High Temperatures–High Pressures* **16**, 347–359.

Pelzl, J., Klein, K., and Nordhaus, O. (1982). *Appl. Optics* **21**, 94.

Pelzl, J. and Bein, B. K. (1983). *Zeitschr. Phys. Chem., Neue Folge* **134**, 17.

Pelzl, J. (1984). In: *Photo-Acoustic Effect* (E. Lüscher, P. Korpiun, H. J. Coufal, and R. Tilgner, eds.). Vieweg, Braunschweig, pp. 52–79.

Pelzl, J., Fournier, D., and Boccara, A. C. (1988). In: *Photoacoustic and Photothermal Phenomena* (P. Hess and J. Pelzl, eds.) Springer Series on Optical Sciences, vol. 58, pp. 241–244. Springer, Heidelberg/Berlin.

Philipps, V., Flaskamp, K., and Vietzke, E. (1982). *J. Nucl. Mater.* **111 & 112**, 781–784.

Pick, M. A. and Summers, D. (1984). *J. Nucl. Mater.* **128 & 129**, 440–444.

Pichon, C., Le Liboux, M., Fournier, D., and Boccara, A. C. (1979). *Appl. Phys. Lett.* **35**, 435.

Pontau, A. E., Gauster, W. B., Mullendore, A. W., Conn, R. W., et al. (1982). *J. Nucl. Mater.* **111 & 112**, 287–293.

Quimby, R. S. and Yen, W. M. (1979). *Appl. Phys. Lett.* **35**, 43.

Rapp, H., Niedermeyer, H., Kornherr, M., and the ASDEX team. (1986). In: *Proceedings 14th Symposium on Fusion Technology*, Sept. 1986, Avignon, France.

Rapp, H. (1987). Private Communication.

Reynolds, W. N. (1986). *Can. J. Phys.* **64**, 1150.

Rich, J. A., Prescott, L. E., and Cobine, J. D. (1971). *J. Appl. Phys.* **42**, 587–601.

Rosencwaig, A. (1973). *Opt. Commun.* **7**, 305.

Rosencwaig, A. and Gersho, A. (1976). *J. Appl. Phys.* **47**, 64.

Rosencwaig, A. (1978). *Adv. Electr. El. Phys.* **46**, 207–311.

Rosencwaig, A. (1980). *Photoacoustics and Photoacoustic Spectroscopy.* John Wiley, New York.

Rosencwaig, A. and Busse, G. (1980). *Appl. Phys. Lett.* **36**, 725.

Rosencwaig, A. and White, R. M. (1981). *Appl. Phys. Lett.* **38**, 165.

Rosencwaig, A. (1982). *Science* **218**, 223.

Rosencwaig, A., Opsal, J., Smith, W. L., and Willenborg, D. L. (1985). *Appl. Phys. Lett.* **46**, 1013.

Rosencwaig, A., Opsal, J., Smith, W. L., and Willenborg, D. L. (1986). *J. Appl. Phys.* **59**, 1392.

Roth, J., Bohdansky, J., and Wilson, K. L. (1982). *J. Nucl. Mater.* **111 & 112**, 775–780.

Roth, E. P. and Smith, M. F. (1986). *Intern. J. of Thermophysics* **7**, 455–466.

Sablikov, V. A. and Sandomirskii, V. B. (1983). *Phys. Stat. Solidi* **B120**, 471.

Satkiewitz, F. G., Murphy, J. C., and Aamodt, L. C. (1985). *Techn. Digest*, 4th Int. Top. Meeting on Photoacoustic, Ecole Polyt. Montreal, TuC8.

Schram, D. C. (1987). *Europhys. News* **18**, 28.

Shanks, H. R., Kamitakahara, W. A., McClelland, J. F., and Carlone, C. (1983). *J. Non-Crystalline Sol.* **59**, 197.

Skumanich, A., Fournier, D., Boccara, A. C., and Amer, N. M. (1985). *Appl. Phys. Lett.* **47**, 402.

Smith, W. L. and Geraghty, P. (1986). In: *Proc. Quantitative NDE*, La Jolla, CA, August, 1986.

Steffens, H. D. and Crostack, H. A. (1980). In: *Proc. 9th Int. Thermal Spraying Conf.*, The Hague, 1980.

Steffens, H. D. and Crostack, H. A. (1981). *Thin Solid Films* **83**, 325.

Street, R. A., Biegelsen, D. K., and Knight, J. C. (1981). *Phys. Rev.* **B24**, 969.

Street, R. A. (1985). *J. Non-Crystalline Sol.* **77 / 78**, 1.

Succi, S. and Lister, G. G. (1982). In: *Heating in Toroidal Plasmas* (Proc. 3rd Joint Varenna-Grenoble Int. Symp., 1982), Vol. 1, Euratom, Brussels, 1982, p. 137.

Swofford, R. L. and Morell, J. A. (1978). *J. Appl. Phys.* **49**, 3667.

Tam, A. C. (1986). *Rev. Mod. Phys.* **58**, 381.

TFR Group (1981). *Report EUR-CEA-FC*-1114.

Timmermanns, C. J. (ed.) (1985). *Proceedings of 7th Intern. Symp. Plasma Chem.*, Vol. 3. IUPAC, Eindhoven, NL, pp. 954–1100.

Träger, F., Coufal, H., and Chuang, T. (1982). *Phys. Rev. Lett.* **49**, 1720.

Uejima, A., Sugitani, Y., and Nagashima, K. (1985). *Analytical Science* **1**, 5.

Ulrickson, M. (1979). *J. Nucl. Mater.* **85 & 86**, 231–235.

Ulrickson, M. and Pearson, G. G. (1982). *J. Nucl. Mater.* **111 & 112**, 91–94.

Utterback, S., Dacol, F. H., Ermert, H., and Melcher, R. L. (1985). *Appl. Phys. Lett.* **46**, 1054.

Vardelle, A., Vardelle, M., Lombard, D., Gitzhofer, F., and Fauchais, P. (1985). In: *Proc. European Mat. Res. Soc. Conf. on Advanced Material Research and Development for Transport*, Strasbourg, 1985, pp. 85–92.

Varlashkin, P. G., Low, M. J., Parodi, G. A., and Morterra, C. (1986). *Applied Spectr.* **40**, 636.

Veprek, S. (1984). In: *Proc. Mat. Res. Soc. Europe, Strasbourg* 1984 (P. Pinard and S.

Kalbitzer, eds.). Les Editions de Physique, Les Ulis, France, p. 425.

Veprek, S. (1985). *Thin Solid Films* **130**, 135.

Vernickel, H., Behringer, K., Campbell, D., and the ASDEX team. (1982a). *J. Nucl. Mater.* **111 & 112**, 317–322.

Vernickel, H. and P. W. W. Gruppe (1982b). In: *JAHRESBERICHT 1982*, Max-Planck-Institue für Plasmaphysik, Garching, FRG.

Vidrine, D. W. (1980). *Applied Spectr.* **34**, 314.

Vietzke, E., Flaskamp, K., and Philipps, V. (1982). *J. Nucl. Mater.* **111 & 112**, 763–768.

Wampler, W. R. (1982). *Appl. Phys. Lett.* **41**, 335–337.

Wagner, J. J. and Veprek, S. (1983). *Plasma Chemistry and Plasma Processing* **3**, 219.

Wagner, I., Stasiewski, H., Abeles, B., and Lanford, W. A. (1983). *Phys. Rev.* **B28**, 7080.

West, G. A., Barret, J. J., Siebert, D. R., and Reddy, K. V. (1983). *Rev. Sci. Instrum.* **54**, 797–817.

White, R. M. (1963). *J. Appl. Phys.* **12**, 3559–3567.

Wienhold, P., Waelbroeck, F., and Winter, J. (1982). *J. Nucl. Mater.* **111 & 112**, 240–242.

Wong, J. H., Thomas, R. L., and Hawkins, G. F. (1978). *Appl. Phys. Lett.* **32**, 538.

Wojczak, M., Pelzl, J., and Bein, B. K. (1988). In: (P. Hess and J. Pelzl, eds.). Springer Series on Optical Sciences, vol. 58, pp. 445–446, Springer Heidelberg/Berlin. *Photoacoustic and Photothermal Phenomena.*

Wojczak, M. (1988). Diploma thesis, Ruhr-Universitaet Bochum, FRG.

Wronski, C. R., Abeles, B., Tiedje, T., and Cody, G. D. (1982). *Sol. St. Comm.* **44**, 1423.

Würz, H., Bein, B. K., Neuhauser, J., and Zasche, D. (1989). In: *Proceedings 15th Symposium on Fusion Technology, 1988, Utrecht*, Netherlands, North-Holland Physics Publ. Amsterdam, Netherlands.

Zasche, D. and the ASDEX team (1986), private communication, Max-Planck Institut fuer Plasmaphysik, Garching, FRG.

Zhao, W. H., Koch, A., Bauder, U. H., and Behrisch, R. (1984). *J. Nucl. Mater.* **128 & 129**, 613–617.

# Index

## A

Absorption coefficient, as function of photon energy of undoped $a$-Si, 246–247
Acoustic emission analysis, 262
AlGaAs, vacuum transfer analysis, plasma etching, 45–46
Aluminum
  analysis in model beam etching systems, 40–43
  etch rate, 42–43
Amorphous silicon
  chlorinated, 249
  defect density, 248–249
  doping studies, 247–248
  far-infrared region, 250–251
  optical studies of band gap states, 245
  passivation effects, 250
  piezoelectric transducer, 250
  photothermal deflection spectroscopy, 246, 250
  plasma-induced deposition, 245–246
  sputtered films, 247
  sub-band-gap states, 246
ASDEX, 277
  metallic divertor neutralizer plates, 296–298
ASDEX limiter, 279
  plasma-induced changes of effusivity, 295–296
  toroidal, 288
    photoacoustic amplitudes, 288, 291
    position relative to plasma, 290, 292
Auger electron, 29
Auger electron spectroscopy, 22–23
  compared with x-ray photoelectron spectroscopy, 26

InP, 41
processes, 28–29
sensitivity limits, 26

## B

Backscattering, 4–7
  cross-talk, 5
  etch rate measurement disruption, 6
  ion beam operation, 7
Beer's formula, 223
Bohm expression, 162
Bohm formula, 172
Boltzmann relation, 166, 177
Breit–Wigner relationship, 148
Bruggeman expression, 89

## C

Carbon resistance probe, 159
Characteristic Auger series, 29
Chemical bonding analysis, 30
Chemical vapor deposition
  plasma-enhanced, 21–22
    silicon dioxide, 100–101
  real-time monitoring, 106–107
Cleanup etch step, 259
Collision cascade, 47–48
Collision kinematics scattering energy, 119–120
Complex refractive index, 73
Compound semiconductor surfaces, hydrogen plasma etching, 98–100

327

Reflectance
  complex, 74
  ratio, complex, 76
Reflection, thermal waves, 223–225
Reflectivity, 223
Refractive index, oscillating, 257
Resonant nuclear reaction analysis, 147–149
  energy level diagram, 147–148
  hydrogen implanted in low-$Z$ materials, 152–154
rf glow discharge system, filter, 4
rf plasma etching, 43–44
RIBE, 45
Rotating calorimeter probes, 159
Rutherford backscattering spectrometry, 24, 63, 110, 119
  channeling effect, 123
  channeling spectra, 124–125
  collision kinematics scattering energy, 119–120
  energy loss, 122–123
  fusion materials research, 125–130
  microbeam profiling analysis, 126
  plasma etching, 123–125
  scattering cross section, 120–122
  scattering geometry, 120
  stopping cross section, 122
  TiC coated graphite standard, 126–127

## S

Scanning Auger spectroscopy, 26
Scanning electron acoustic microscopy, 252–253
  time resolved, 255–256
  wafer subjected to ion-implantation and etching, 253–254
Scanning electron microscopy, 26
Scattering, thermal waves, 223–225
Search probe, 193, 195
Secondary ion mass spectroscopy, 23, 25, 47–50, 62
  $C_4F_8$ discharge, 54–55
  Cl adsorption, 52–54
  depth profiles, 56–57
  in situ, 52–56
  instrumentation, 48

intensity
  coverage dependence, 53
  as function of surface F atom coverage, 55
ion impact event, 47–48
ion mixing, 49
remote, 56–59
  H and D implantation measurement, 57
  surface modification measurement, 58
secondary ion yield, 49
static, 58
Semiconductors
  interband absorption, 227
  plasma exposure, 244–259
    optical properties, 245–251
    plasma etched, 251–259
    process plasma characterization, 244–245
  processing, ion beam analysis, 118
  see also Plasma etched semiconductors
Sheath transmission factors, 167–170
Silane, 246
Silicon
  air-grown natural overlayer removal, 105
  analysis in model beam etching systems, 39–40
  chemisorption of fluorine, 40
  comparison of XTEM and SE analyses, 71
  depth profile, D implanted, 135–136
  dielectric function, 70
  electron spectroscopy processes, 27
  etch rate, 39
    as function of DC bias voltage, 15–16
  film, plasma-assisted etch rate, 8
  H, Ne and Ar implantation damage distributions, 125
  hydrogen analysis, 135–138
  isotope exchange implant of H into D saturated Si, 136–137
  polycrystalline, time-resolved SEAM image, 255–256
  pseudodielectric function, 70, 84–85
  remote analysis, plasma etching, 33–35
  signal variation with take-off angle, 29
  vacuum transfer analysis, plasma etching, 43–45
  x-ray photoelectron spectroscopy, after etching, 34
  see also Amorphous silicon

## W

Wavelength dispersion, 23
Wave solutions, 214–218

## X

X-ray photoelectron spectroscopy, 22–23, 26
  aluminum, 42

chemical bonding analysis, 30
compared with Auger spectroscopy, 26
contaminating radiation, 31
gold, 28
sensitivity limits, 26
silicon, after etching, 34
spectrometers, 31–32
take-off angle, 37
*see also* Electron spectroscopy

# Plasma–Materials Interactions

Dennis M. Manos and Daniel L. Flamm, *Plasma Etching: An Introduction*

Orlando Auciello and Daniel L. Flamm, *Plasma Diagnostics: Volume 1, Discharge Parameters and Chemistry; Volume 2, Surface Analysis and Interactions*